James D. A. Millington and John Wainwright (Eds.)

Agent-Based Modelling and Landscape Change

MDPI

This book is a reprint of the Special Issue that appeared in the online, open access journal, *Land* (ISSN 2073-445X) from 2015–2016, available at:

http://www.mdpi.com/journal/land/special_issues/abm_landscape

Guest Editors
James D. A. Millington
Department of Geography, King's College London
UK

John Wainwright
Department of Geography, Durham University
UK

Editorial Office
MDPI AG
St. Alban-Anlage 66
Basel, Switzerland

Publisher
Shu-Kun Lin

Managing Editor
Elvis Wang

1. Edition 2016

MDPI • Basel • Beijing • Wuhan • Barcelona • Belgrade

ISBN 978-3-03842-280-8 (Hbk)
ISBN 978-3-03842-281-5 (PDF)

Table of Contents

'

List of Contributors

Sampson K. Agodzo Agricultural Engineering, Kwame Nkrumah University of Science and Technology, Kumasi, Ghana.

Rumana Reaz Arifin Department of Civil and Environmental Engineering and Earth Sciences, University of Notre Dame, Notre Dame, IN 46556, USA.

S. M. Niaz Arifin Department of Computer Science and Engineering, University of Notre Dame, Notre Dame, IN 46556, USA.

Biola K. Badmos Civil Engineering, Kwame Nkrumah University of Science and Technology, Kumasi, Ghana; West African Science Service Centre on Climate Change and Adapted Land Use (WASCAL), KNUST, Kumasi, Ghana.

C. Michael Barton School of Human Evolution & Social Change and Center for Social Dynamics & Complexity, PO Box 872402 SHESC, Arizona State University, Tempe, AZ 85287, USA.

David Bennett Department of Geographical and Sustainability Sciences, University of Iowa, 316 Jessup Hall, Iowa City, IA 52242, USA.

Julia Maria Brändle Department of Environmental Systems Science, Natural and Social Science Interface, and Department of Civil, Environmental and Geomatic Engineering, Planning of Landscape and Urban Systems, Swiss Federal Institute of Technology, Stefano-Franscini-Platz 5, 8093 Zurich, Switzerland.

Philip Brown Landcare Research New Zealand, P.O. Box 69040, Lincoln 7640, New Zealand.

Sibyl Hanna Brunner Department of Civil, Environmental and Geomatic Engineering, Planning of Landscape and Urban Systems, Swiss Federal Institute of Technology, Stefano-Franscini-Platz 5, 8093 Zurich, Switzerland.

Julia K. Clark Department of Anthropology, University of Pittsburgh, Pittsburgh, PA 15260, USA.

Frank H. Collins Department of Biological Sciences, and Department of Computer Science and Engineering, University of Notre Dame, Notre Dame, IN 46556, USA.

Stefani A. Crabtree Department of Anthropology, Washington State University, P.O. Box 644910, Pullman, WA 91164, USA; Maison des Sciences de l'Homme et l'Environnement, Université de Franche-Comté, 32 Rue Megevand, 25030 Besancon Cedex, France.

Adam J. Daigneault Landcare Research New Zealand, Private Bag 92170, Auckland Mail Centre, Auckland 1142, New Zealand.

Dilkushi de Alwis Pitts Center for Research Computing, University of Notre Dame, Notre Dame, IN 46556, USA.

Deng Ding Department of Geographical and Sustainability Sciences, University of Iowa, 316 Jessup Hall, Iowa City, IA 52242, USA; Esri, 380 New York Street, Redlands, CA 92373, USA.

Arjun Heimsath School of Earth and Space Exploration (SESE), Arizona State University, PO Box 876004, Tempe, AZ 85287, USA.

Robert Huber Swiss Federal Institute for Forest, Snow and Landscape Research WSL, Zürcherstrasse 111, 8903 Birmensdorf, Switzerland.

Peter George Johnson Centre for Research in Social Simulation, University of Surrey, Guildford GU2 7XH, UK.

Gaby Langendijk Earth System Science Group, Wageningen University and Research Center, Droevendaalsesteeg 4, 6708 PB Wageningen, The Netherlands.

Arika Ligmann-Zielinska Department of Geography and Environmental Science & Policy Program, Michigan State University, Room 121 Geography Building, 673 Auditorium Rd., East Lansing, MI 48824, USA.

Gregory R. Madey Department of Computer Science and Engineering, University of Notre Dame, Notre Dame, IN 46556, USA.

James D. A. Millington Department of Geography, King's College London, Strand Campus, London WC2R 2LS, UK.

Fraser J. Morgan Landcare Research New Zealand, Private Bag 92170, Auckland Mail Centre, Auckland 1142, New Zealand.

Sara Nowreen Institute of Water and Flood Management (IWFM), Bangladesh University of Engineering and Technology (BUET), Dhaka 1000, Bangladesh.

Samuel N. Odai Civil Engineering, Kwame Nkrumah University of Science and Technology, Kumasi, Ghana; West African Science Service Centre on Climate Change and Adapted Land Use (WASCAL), KNUST, Kumasi, Ghana.

Simon Peter Department of Environmental Systems Science, Agricultural Economics, Swiss Federal Institute of Technology, Sonneggstrasse 33, 8092 Zurich, Switzerland.

M. Sohel Rahman Department of Computer Science and Engineering (CSE), Bangladesh University of Engineering and Technology (BUET), Dhaka 1205, Bangladesh.

Laura Schmitt Olabisi Department of Community Sustainability and Environmental Science & Policy Program, Michigan State University, 151 Natural Resources Building, 480 Wilson Road, East Lansing, MI 48824, USA.

Silvia Secchi Department of Geography and Environmental Resources, Southern Illinois University, Carbondale, IL 62901, USA.

Isaac Ullah School of Human Evolution & Social Change and Center for Social Dynamics & Complexity, PO Box 872402 SHESC, Arizona State University, Tempe, AZ 85287, USA.

Grace B. Villamor Center for Development Research, University of Bonn, 53113 Bonn, Germany.

John Wainwright Department of Geography, Durham University, Durham DH1 3LE, UK.

Ryan Qi Wang Department of Civil and Environmental Engineering, Virginia Tech, 113 Patton Hall (0105), Blacksburg, VA 24061, USA.

About the Guest Editors

James D. A. Millington is a Lecturer in the Physical and Quantitative Geography Department of King's College London. He is a geographer and landscape ecologist with expertise in developing bespoke computational and statistical modelling tools to investigate spatial ecological and socio–economic processes and their interaction. His previous research has focused on human decision-making in multifunctional forest and agricultural landscapes in Europe and North America.

John Wainwright is a Professor in the Department of Geography at Durham University. His research interests focus on past, present and future human interactions with the environment, especially in drylands and with an emphasis on the Mediterranean, the US Southwest and sub-Saharan Africa. Computer modelling is central to this research to link data with theory and laboratory experiments.

Preface to "Agent-Based Modelling and Landscape Change"

The use of agent-based models (ABMs) and modelling for understanding landscape change and dynamics continues to grow. One reason for the popularity of ABMs is that they provide a framework to represent multiple, discrete, multi-faceted, heterogeneous actors (human or otherwise) and their relationships and interactions between one another and their environment, through time and across space. By inviting authors from across various disciplines, with this collection we aimed to showcase innovative uses of ABMs for investigating and explaining landscape change and dynamics and to explore and identify how researchers in different disciplines can learn from one another to further innovate. The diverse range of processes and landscapes that ABMs are currently used to examine is clearly demonstrated by the final collection. Contributions address issues ranging from land-use decision-making in agricultural landscapes, soil erosion in semi-arid environments and forest change in mountainous landscapes, to trade in the 1st Century BC in southern France and social adaptations of herders in northern Mongolia. The authors use a range of different levels of agent-based representation, from the implied presence of agents, through comparing heterogeneous vs. aggregated representation of human activity, to alternative means of parameterizing individual agent behavior. We hope this collection will inform all interested in innovative agent-based modelling to further understand landscape change, its causes and consequences for sustainability in the Anthropocene.

James D. A. Millington and John Wainwright
Guest Editors

Comparative Approaches for Innovation in Agent-Based Modelling of Landscape Change

James D. A. Millington and John Wainwright

Reprinted from *Land*. Cite as: Millington, J.D.A.; Wainwright, J. Comparative Approaches for Innovation in Agent-Based Modelling of Landscape Change. *Land* **2016**, *5*, 13.

In this Special Issue on "Agent-Based Modelling and Landscape Change" we aimed to bring together articles that showcase innovative uses of agent-based models (ABMs) for investigating and explaining landscape change and dynamics. The resulting 10 articles demonstrate the diverse range of processes and landscapes that ABMs are currently used to examine, including: land-use decision making in agricultural landscapes; soil erosion in semi-arid environments; forest change in mountainous landscapes; trade in 1st Century BC southern France; social adaptations of herders in northern Mongolia; and malaria epidemiology in Kenya. The articles (Ding *et al.* 2015 [1], Olabasi *et al.* 2015 [2], Morgan *et al.* 2015 [3], Badmos *et al.* 2015 [4], Barton *et al.* 2015 [5], Johnson 2015 [6], Brändle *et al.* 2015 [7], Crabtree 2015 [8], Clark and Crabtree 2015 [9] and Arifan 2015 [10]) draw on a range of modelling approaches, but one common theme among several of the papers is the use of comparative approaches. Here, we discuss how comparative approaches offer opportunities for future innovation in modelling landscape change, particularly for addressing the challenge of understanding the role of human activity in the Anthropocene.

The issue of comparison in ABMs is not new to the studies in this Special Issue and has been advocated and pursued over many years. Axtell *et al.* (1996) [11] were among the earliest to investigate the alignment of computational models, or 'docking' as they suggested it might be abbreviated. Docking entailed comparing an ABM to another model (whether ABM or otherwise) of the same system to see if the models could reproduce similar results, thereby enabling critical experimentation and the determination of whether one model was better than another, or if one was a special case of the other (*i.e.*, could be subsumed). Since then, model-to-model analysis has continued (e.g., Hales *et al.* 2003 [12], Rouchier *et al.* 2008 [13]), although the rate of comparison has not kept pace with number of ABMs being developed. Robust comparison of models, to the point of trying to 'break' them (*i.e.*, identifying at what point modelled mechanisms are no longer useful for explaining observations), is needed to ensure credible and efficient scientific progress in computational modelling (Thiele and Grimm 2015 [14]). Beyond examining how well different models fit

1

the same set of empirical data, model comparison can aim to reproduce others' models from scratch in new computer code (e.g., Janssen 2009 [15]) or extend analysis including by exploring the sensitivity of model parameters in greater detail (e.g., Miodownik *et al.* 2010 [16], Seagren 2015 [17]). In contrast, articles in this Special Issue examine variations in agent-based representation, from an entire absence of agent representation, through comparing heterogeneous *vs.* aggregated representation of human activity, to alternative means of parameterizing individual agent behaviour.

For example, to investigate the effect of agricultural practices on the formation of deeply incised valley formations in semi-arid Mediterranean landscapes, Barton *et al.* (2015) [5] 'turned off' the human land-use component of their hybrid ABM-cellular model. By using the same model with humans represented versus not, this approach aims to understand the influence of human activity on landscape change (e.g., as discussed by Wainwright and Millington 2010 [18]). Through this experimental use of their model, Barton *et al.* showed that the non-ABM component of their model that represents climate and natural vegetation change is able to capture broad-scale (climate-driven) vegetation-change impacts on gulley incision. Including the agent-based representation of human activity shows how finer-scale, localized vegetation change can have similar effects without climate change. Thus, this example shows how drivers of landscape change acting over different scales may need to be represented through fundamentally different modelling approaches.

Brändle *et al.* (2015) [7] compared agent-based versus aggregated models of agricultural change in a contemporary mountain landscape in Switzerland, examining the trade-offs between model types for considering different temporal extents of simulation. They found that their ABM, based on recent behavioural data, was able to simulate landscape change over short and medium durations better than an aggregated model assuming land optimization, while maintaining equivalent sensitivity to broader socio-economic drivers. The trade-offs identified are between the greater demand for more detailed information about (farming) actor behaviour and decision-making by the agent-based model (making transferability of the model to other landscapes difficult) versus the more realistic spatially explicit simulation of land abandonment over the short and medium term due to better representation of diversity in decision-making. However, over longer simulated durations the advantages of an agent-based approach are less obvious and the results remind us that the choice of modelling approach depends on the questions being investigated and relative advantages of the available approaches.

In a third example from the Special Issue, Morgan *et al.* (2015) [3] compared three different approaches for estimating the likelihood of land-use conversion by agricultural agents in New Zealand: (i) no difference between agents in likelihood (*i.e.*, assumes universally rational, profit-maximisation agents); (ii) the social and geographic network of agents influences likelihood (*i.e.*, representing influence of endorsement and imitation alongside economic considerations), and (iii) empirical estimation of likelihood based on an individual agent's attributes (including age, education, land holdings, *etc.*). The different approaches reflect differing perspectives and traditions in how human activity has been investigated by economists compared to geographers. Results showed that at some broader units of aggregation (catchment level) there was little appreciable difference in simulated land uses between the approaches, whereas at finer units differences were evident.

The Brändle *et al.* (2015, [7]) and Morgan *et al.* (2015, [3]) examples are as good as currently exist for demonstrating how assumptions about agent heterogeneity are comparable to existing accepted modelling approaches. Comparisons such as these, and which investigate how and when ABMs are better for improving understanding than other modelling types, will enable demonstrations of how ABM are useful and robust for understanding change into the future. However, they also highlight that differences in modelling approaches are not fully resolved and that the choice of modelling approach will depend on the scientific and policy questions being asked. Currently, the primary influence on modelling approach seems to be the scales and organizational levels at which answers are required. For example, although ABMs may be designed to provide greater representational fidelity (e.g., fine detail at the level of individuals) implementing such models often comes with costs of development (time and data), use (computational resources) and transferability (between landscapes). In some instances the benefits of developing an ABM may ultimately not outweigh the costs, particularly if there is limited heterogeneity in the decision-making context of actors or limited interaction effects between agents (e.g., O'Sullivan *et al.* 2012, [19]).

Taking an alternative perspective, in the Special Issue, Johnson (2015, [6]) explores using an ABM as a mediator or "interested amateur" in the process of policy making. If constructed independently of the policy context (*i.e.*, not co-constructed with stakeholders), using the model and its output in discussions forces a focus on assumptions but in an impersonal way, not directed at any particular person. The comparisons here are between the way in which the model represents the world, how the policy maker understands the world to be structured, and between expected and unexpected outcomes as shown by the model. Johnson argues that for this approach to work there needs to be a degree of transparency about how the model represents processes (e.g., of landscape change) such that it is not a black box, but also that a detailed model is an advantage because it provides more assumptions about which participants can debate and explore the consequences of. More generally then, Johnson sees ABMs as providing greater benefit than rational utility maximization approaches both because the latter are more 'removed from reality' (e.g., not all actors are perfectly rational) but also because their more simplified worldview (with few assumptions) inhibits discussion about structures and relationships in the real world and how they could change. Johnson found his own particular ABM useful for facilitating discussion about policy options for soil and water conservation in Ethiopia, but more general comparison of ABMs against other model types for policy discussion would be welcome.

In future, it seems likely that beyond comparing different types of model (ABM, regression-based models, systems dynamics models), combining ABM with other modelling types to produce innovative representations will become more prominent. For example, Verburg *et al.* (2015, [20]) argue that if modelling is to assist in designing sustainable solutions to the challenges of the Anthropocene, innovative model architectures that can represent human-environment interactions across many scales and levels of organisation will be needed. O'Sullivan *et al.* (2015, [21]) advocated hybrid forms of land-use modelling in which competing and complementary approaches (beyond ABM) are compared and combined in an iterative approach to improve understanding. O'Sullivan *et al.* (2015, [21]) suggest different 'levels' of hybridity, from comparing different modelling approaches to investigate the same substantive domain, through coupling different types of model to examine a single domain, to actually integrating modelling approaches so that there is no discernible point at which one model ends and another begins (e.g., agents that run regressions dynamically and internally as a proxy for individual decision-making).

Developing such innovative modelling hybridity in land-change science is particularly imperative given the recognition that landscape change can be influenced not only by local circumstances (neighbours' decisions, local environmental conditions) but also by decisions and processes that are far remote and operating at different scales and levels of organization (Liu *et al.* 2013, [22]). However, careful thought will need to go into operationalising hybrid model forms for investigating such systems. Although representing all individual actors in a globalized system of land use and food trade, for example, might theoretically be possible, it is not immediately clear that this would be desirable. For example, the heterogeneity of decision-making and/or interaction at one level of organization (e.g., individual farmers) may be so low as to make little difference to what decisions mean for other levels of organization (e.g., food commodity traders). In such cases if the goal is understanding global interactions, but it is at other levels of organization at which most uncertainty, heterogeneity or influence occur, then it may be appropriate to represent local land use decisions in an aggregated manner and focus individual-level representation at non-global levels or scales. Such considerations for how to structure future hybrid models are important if we are to ensure the hybrids do not become 'monster models'—ever more complicated models that are more and more difficult to evaluate. Such a situation is not an inevitable result of hybridization (nor advocated), but as usual important consideration needs to go into developing models that are fit for the desired purpose.

4

Pursuing innovative and hybrid modelling approaches through iterative approaches to scientific inquiry, as advocated by O'Sullivan *et al.* (2015, [21]), might be usefully facilitated through online platforms that encourage greater collaboration between modellers and engagement with policy- and decision-makers. One example might be an online a community-modelling initiative to act as a clearing house for models and best practice. Contemporary online resources such as openABM.org are valuable as a space to present individual models—complete with a peer-review process—but as structured they currently do little to encourage modellers to think about how they can combine or build upon one another's models. A platform that actively encourages and enables modellers to interact, combine and 'mash-up' their conceptualizations to find synergies and produce novel model architectures that overcome trade-offs between representational fidelity and development costs would be particularly valuable going forward. From the perspective of policy-development, an online space such as this might also host models for policy makers to interact with as "interested amateurs". By better enabling modellers to work together to robustly compare and combine their models, and to discuss with users to learn and improve models, advantages of hybridity might be more readily realised. In turn, the models produced should enable more insightful contributions to the comparative issues discussed above and ensure the continuation of innovative modelling for understanding landscape change, its causes and consequences for sustainability in the Anthropocene.

References

1. Ding, D.; Bennett, D.; Secchi, S. Investigating impacts of alternative crop market scenarios on land use change with an agent-based model. *Land* **2015**, *4*, 1110–1137.
2. Olabisi, L.; Wang, R.; Ligmann-Zielinska, A. Why don't more farmers go organic? Using a stakeholder-informed exploratory agent-based model to represent the dynamics of farming practices in the philippines. *Land* **2015**, *4*, 979–1002.
3. Morgan, F.; Brown, P.; Daigneault, A. Simulation *vs.* Definition: Differing approaches to setting probabilities for agent behaviour. *Land* **2015**, *4*, 914–937.
4. Badmos, B.; Agodzo, S.; Villamor, G.; Odai, S. An approach for simulating soil loss from an agro-ecosystem using multi-agent simulation: A case study for semi-arid ghana. *Land* **2015**, *4*, 607–626.
5. Barton, C.; Ullah, I.; Heimsath, A. How to make a barranco: Modeling erosion and land-use in mediterranean landscapes. *Land* **2015**, *4*, 578–606.
6. Johnson, P. Agent-based models as "interested amateurs". *Land* **2015**, *4*, 281–299.
7. Brändle, J.; Langendijk, G.; Peter, S.; Brunner, S.; Huber, R. Sensitivity analysis of a land-use change model with and without agents to assess land abandonment and long-term re-forestation in a swiss mountain region. *Land* **2015**, *4*, 475–512.

8.   Crabtree, S. Simulating littoral trade: Modeling the trade of wine in the bronze to iron age transition in southern france. *Land* **2016**, *5*, 5.

9.   Clark, J.; Crabtree, S. Examining social adaptations in a volatile landscape in northern mongolia via the agent-based model ger grouper. *Land* **2015**, *4*, 157–181.

10.   Arifin, S.; Arifin, R.; Pitts, D.; Rahman, M.; Nowreen, S.; Madey, G.; Collins, F. Landscape epidemiology modeling using an agent-based model and a geographic information system. *Land* **2015**, *4*, 378–412.

11.   Axtell, R.; Axelrod, R.; Epstein, J.M.; Cohen, M.D. Aligning simulation models: A case study and results. *Computational & Mathematical Organization Theory* **1996**, *1*, 123–141.

12.   Hales, D.; Rouchier, J.; Edmonds, B. Model-to-model analysis. *J. Artif. Soc. Soc. Simul.* **2003**, *6*.

13.   Rouchier, J.; Cioffi-Revilla, C.; Polhill, J.G.; Takadama, K. Progress in model-to-model analysis. *J. Artif. Soc. Soc. Simul.* **2008**, *11*.

14.   Thiele, J.C.; Grimm, V. Replicating and breaking models: Good for you and good for ecology. *Oikos* **2015**, *124*, 691–696.

15.   Janssen, M.A. Understanding artificial anasazi. *J. Artif. Soc. Soc. Simul.* **2009**, *12*, A244–A260.

16.   Miodownik, D.; Cartrite, B.; Bhavnani, R. Between replication and docking: "Adaptive agents, political institutions, and civic traditions" revisited. *J. Artif. Soc. Soc. Simul.* **2010**, *13*.

17.   Seagren, C.W. A replication and analysis of tiebout competition using an agent-based computational model. *Soc. Sci. Comput. Rev.* **2015**, *33*, 198–216.

18.   Wainwright, J.; Millington, J.D.A. Mind, the gap in landscape-evolution modelling. *Earth Surf. Proc. Land.* **2010**, *35*, 842–855.

19.   O'Sullivan, D.; Millington, J.; Perry, G.; Wainwright, J. Agent-based models—Because they're worth it? In *Agent-Based Models of Geographical Systems*; Heppenstall, J.A., Crooks, T.A., See, M.L., Batty, M., Eds.; Springer Netherlands: Dordrecht, The Netherlands, 2012; pp. 109–123.

20.   Verburg, P.H.; Dearing, J.A.; Dyke, J.G.; Leeuw, S.V.D.; Seitzinger, S.; Steffen, W.; Syvitski, J. Methods and approaches to modelling the anthropocene. *Glob. Environ. Chang.* **2015**.

21.   O'Sullivan, D.; Evans, T.; Manson, S.; Metcalf, S.; Ligmann-Zielinska, A.; Bone, C. Strategic directions for agent-based modeling: Avoiding the yaawn syndrome. *J. Land Use Sci.* **2016**, *11*, 177–187.

22.   Liu, J.G.; Hull, V.; Batistella, M.; DeFries, R.; Dietz, T.; Fu, F.; Hertel, T.W.; Izaurralde, R.C.; Lambin, E.F.; Li, S.X.; *et al.* Framing sustainability in a telecoupled world. *Ecol. Soc.* **2013**, *18*.

Simulation *vs.* Definition: Differing Approaches to Setting Probabilities for Agent Behaviour

Fraser J. Morgan, Philip Brown and Adam J. Daigneault

Abstract: While geographers and economists regularly work together on the development of land-use and land-cover change models, research on how differences in their modelling approaches affects the results is rare. Answering calls for more coordination between the two disciplines in order to build models that better represent the real world, we (two economists and a geographer) developed an economically grounded, spatially explicit, agent-based model to explore the effects of environmental policy on rural land use in New Zealand. This inter-disciplinary collaboration raised a number of differences in modelling approach. One key difference, and the focus of this paper, is the way in which processes that shape the behaviour of agents are integrated within the model. Using the model and a nationally representative survey, we compare the land-use effects of two disciplinary-aligned approaches to setting a farmer agent's likelihood of land-use conversion. While we anticipated that the approaches would significantly affect model outcomes, at a catchment scale they produced similar trends and results. However, further analysis at a sub-catchment scale suggests the approach to setting the likelihood of land-use conversion does matter. While the results outlined here will not fully resolve the disciplinary differences, they do outline the need to account for heterogeneity in the predicted agent behaviours for both disciplines.

Reprinted from *Land*. Cite as: Morgan, F.J.; Brown, P.; Daigneault, A.J. Simulation *vs.* Definition: Differing Approaches to Setting Probabilities for Agent Behaviour. *Land* **2015**, *4*, 914–937.

1. Introduction

With an increase in demand for strong, evidence-based environmental policy and management, scientists have called for methods to accurately capture the complex nature of socio-ecological systems [1,2]. This call is driven by the need to understand the likely consequences and trade-offs of proposed policies on economic outcomes, land use, and social well-being [3]. A modelling approach is well suited to this task because the social, economic, and geographic factors that determine the choice and impact of land use are in themselves complex [4–7].

Land use and land cover change (LULCC) models represent a well-developed approach to modelling and understanding processes that shape the environment [8–10] and have developed alongside our understanding of wider economic and social systems. As with most modelling approaches, early implementation of LULCC models focused on mathematical programming and rational utility theory, *i.e.*, individuals are assumed to maximise profits [11–15]. These approaches are still common, and while these models capture trends in LULCC, they may fail to reflect accurately the underlying processes driving the change in LULCC [2].

More specifically, more economically focused LULCC models focus on management practices that maximise net returns for a given land use while omitting key spatial, bio-physical, and social details [2,16–21]. Such abstractions ignore the processes, people, and space within the model, thus making the "optimally derived" solution unrealistic [22]. Geographically defined LULCC models, on the other hand, typically account for heterogeneity across space and individuals, but often simplify the level of economic behaviour [23–27]. As such, geographic models are typically structured to include simplified economic approaches and to exclude explicit representations of land and commodity markets [23].

Geographers and economists have rarely collaborated in undertaking these analyses, leading to calls for modellers from these two disciplines to coordinate efforts in order to build models that better represent the real world [23,28].

LULCC is a complex, adaptive process that can also be explained through the use of computational tools such as agent-based models (ABMs) [2,29,30]. ABMs are well suited to analysing decentralised, autonomous decision making such as that underlying LULCC because they represent complex spatial interactions under heterogeneous conditions [30–32]. In addition, the ABM approach accounts for space, distance, and time in decision making.

However, capturing the social and economic behaviour of farmers via ABMs to analyse LULCC is not without its own complexities and limitations [27,33–36]. For example, Burton [37] outlines numerous social processes that should be evaluated when assessing farmer behaviour, including cultural embeddedness [38], social networks, and technology transfer [39,40], and the dichotomy between social and economic approaches to farming [41,42]. Therefore, capturing the heterogeneity of farmer behaviour is essential when modelling rural land-use change. While this notion is widely supported [38,43], there is significant variation in how heterogeneity in farmer decision making is accounted in ABMs. Examples of such heterogeneity include: variation in different production strategies [35,44,45]; dealing with external factors [46,47]; and simulating key parts of the farming process [48–50]. In all cases, this variation depends on the objective of the ABM [51].

To answer these calls, the authors (two economists and a geographer) developed an economically grounded, spatially explicit ABM to explore the effects of environmental policy on rural land use in New Zealand. The Agent-based Rural Land Use New Zealand (ARLUNZ) is capable of analysing the impact of a variety of policies on land use, net revenue, and environmental indicators such as greenhouse gas (GHG) emissions, nutrient loadings, and soil erosion [36].

This inter-disciplinary collaboration required that two differences in approach be resolved. The first is a disciplinary perspective on how individual agents enter into the model. Geographers have traditionally had a strong preference for defining types of agents within a population according to a typological framework [35,38,43,44,46] to limit complexity while still moving agents towards their predefined goals [52–55]. While economists recognise the need to limit computational complexity [34], they have also called for empirical calibration and validation of decision-making hypotheses through surveys, interviews, participatory modelling, and experimental economics [33,56,57]. Because we have access to a large-scale, nationally representative survey that accounts for demographics, social processes, and land use, we side with economists and rely on empirical distributions of farmer and forester characteristics to simulate a population of agents [58].

The second disciplinary disagreement—and a primary focus of this paper—is the way in which processes that shape the behaviour of agents are integrated within the model. Irwin [23] observes that the methods used for modelling land-use change vary significantly: economists tend to focus on econometric analyses, while geographers tend to base their analyses on simulations.

Farmers' information networks are framed around their social interactions and play a role in shaping their decision making processes [39,40,59,60]. Through the nationally representative survey, we could define the observable effects of each farmer's networks into the agent-based model by directly affecting the likelihood of a certain type of behaviour, in this case land-use conversion. Conversely, we could simulate the agent's interactions with their networks and observe how these interactions affect the agent's likelihood of land-use conversion.

Consequently, this paper analyses how each approach affects the resulting land use, net revenue, and environmental outputs at a catchment scale. We hypothesise that the two approaches will produce significant differences for each of these metrics.

We note that these disagreements relate to representation of people and the empirical characterisation of agents within ABMs [1,51,57,58,61,62]. Specifically, the disagreements relate to how empirical data is used to capture and define the bounds of decision making available to the agents. Greater variety of on-farm management options (e.g., reducing stocking rates, fencing streams, and planting riparian buffers) and more information being made available to farmers (e.g., climate, biophysical and soils data) increase the complexity associated with defining farmer agents. Because of the significant empirical data required to inform the use of on-farm management options and to account for additional information through climate and biophysical models, we constrain farmer decision making in this manuscript to complete farm conversion from enterprise to enterprise.

The remainder of this paper is organised as follows: Section 2 describes the methodology used in this research and the approaches used to define an agent's likelihood of land-use conversion; Section 3 summarises the experimental design for the research specifically the region the model has been applied on; Section 4 presents the results from the experiment; and Section 5 concludes.

2. Methods

The ARLUNZ model was designed to analyse complex environmental issues in the rural landscape, to provide information about how farmers will adapt to change, and to inform policy that seeks to address vulnerability to resource scarcity. Specifically, ARLUNZ focuses on variability in decision making among farmers, moving away from a representative decision-making agent to a spatial and behaviourally heterogeneous population of farmers whose decision making reflects the real world.

ARLUNZ is written in Version 5.0.5 of NetLogo [63] using the GIS, String, and Shell extensions. Python 2.7 is used to facilitate a loose coupling [64] between ARLUNZ and a modified version of the New Zealand Forest and Agriculture Regional Model (NZFARM) that provides economic information. NZFARM is a non-linear, partial equilibrium mathematical programming model of New Zealand land use operating at the sub-catchment scale [65]. The version used within ARLUNZ has been refined to produce an economically optimised result for each farm rather than an optimised landscape for a sub-catchment [66–69].

Morgan and Daigneault [36] provides detailed information on the design, structure, outputs, and parameterisation of ARLUNZ and its coupling with NZFARM as well as its use to estimate the impacts of climate change policy on land use in New Zealand. In addition, Table S1 contains an ODD+D description [70] for the ARLUNZ model.

2.1. Survey Research

Some parametrisation in the ARLUNZ model is based on the Survey of Rural Decision Makers (SRDM), a nationally representative survey of land owners and other decision makers [71]. The survey was conducted online between March and July 2013.

The survey gathered up to 192 data points on each respondent, land characteristics and use, current farm practice, demographics, succession plans, professional networks, sources of information regarding best practice, management objectives, income, risk tolerance, and values [71]. The questionnaire was developed in consultation with the Ministry for the Environment, the Ministry of Primary Industries, Dairy New Zealand, Beef + Lamb New Zealand, HortNZ, regional councils, AgResearch, the New Zealand Institute for Economic Research, farmer discussion groups, and other stakeholders.

The sample was drawn from the AssureQuality AgriBase database [72]. Developed in 1993 to track foot and mouth disease, AgriBase records detailed information on privately held rural land across New Zealand. Inclusion in AgriBase is voluntary and entries are updated irregularly. As such, the median address was entered into the database seven years before the survey and some of the individuals contacted for the survey had left farming, making the true response rate difficult to ascertain. However, a total of 1564 responses were collected, yielding a response rate of at least 21%. Participation was incentivised via a donation made to a charity of each respondent's choice and an invitation to view summary results online after the survey had closed. The primary decision-maker for each property was asked to complete the survey, which, on average, took approximately 20 min to complete.

Summary statistics for the variables of interest are shown in Table 1. The mean property comprises 486 hectares, although this high average is driven by a handful of very large Sheep and Beef stations. The average age of respondents is 56.5, consistent with ages reported in New Zealand's Agricultural Census [73]. The average farmer has 25 years of experience. One-third of farmers hold university degrees, while 27.8% have completed diplomas or post-secondary technical training in farming and/or farm management.

The importance of being highly productive was self-evaluated on a scale of 0 to 10, where 0 indicates that being highly productive is "not at all important' and 10 indicates that being highly productive is "extremely important". The mean score for the importance of being highly productive is 6.53. Risk tolerance is measured by the question "Are you a person who is generally prepared to take risks?", where 0 indicates "don't like to take risks" and 10 indicates "fully prepared to take risks". The mean score for risk tolerance was 5.44, indicating a moderate level of risk-taking.

Some 78.5% of respondents report that their business is either profitable or that it breaks even. Having a "large" professional network is defined by visiting more than the median number of farms (i.e., five) or meeting more than the median of farmers to discuss productivity (i.e., four); by definition, half of the survey respondents meet these criteria. The main reported farm enterprises by area include sheep and beef (44%), dairy (21%), horticulture and viticulture (11%), forestry (8%), dairy support (4%), deer and other livestock (3%), and arable (3%). The average number of different enterprises on the farm is 1.68, although a small number of farms have as many as five different enterprises. Respondents in Canterbury (the region on which this paper focuses) account for 17.2% of the total sample.

11

Table 1. Summary Statistics from the Survey of Rural Decision Makers.

Variables	Mean	Std. Dev.	Min.	Max.
intend to intensify over the following 5 years (1–10)	2.678	3.046	0	10
intend to de-intensify over the following 5 years (1–10)	3.569	3.452	0	10
effective land quantity (hectares)	486.440	1932.137	2	34,000
age (years)	56.471	10.098	24	87
experience (years)	25.100	15.812	1	66
high school education (dummy)	0.393	0.488	0	1
diploma/tech training (dummy)	0.278	0.448	0	1
university or higher (dummy)	0.329	0.470	0	1
importance of being highly productive (1–10)	6.535	2.787	0	10
profitable business (dummy)	0.785	0.411	0	1
respondent exceeds median # of farm/farmer visits (dummy)	0.487	0.500	0	1
risk tolerance (1–10)	5.437	2.403	0	10
enterprise = sheep and beef (share)	0.444	0.497	0	1
enterprise = dairy (share)	0.209	0.407	0	1
enterprise = deer and other livestock (share)	0.035	0.183	0	1
enterprise = horticulture and viticulture (share)	0.107	0.309	0	1
enterprise = arable (share)	0.030	0.171	0	1
enterprise = dairy support (share)	0.045	0.207	0	1
enterprise = forestry (share)	0.079	0.270	0	1
enterprise = other enterprise (share)	0.052	0.222	0	1
number of land uses on this operation (#)	1.684	0.884	1	5
region = Auckland (share)	0.031	0.173	0	1
region = Bay of Plenty (share)	0.054	0.226	0	1
region = Canterbury (share)	0.178	0.382	0	1
region = Gisborne (share)	0.024	0.154	0	1
region = Hawke's Bay (share)	0.084	0.277	0	1
region = Marlborough (share)	0.057	0.232	0	1
region = Manuwatu-Whanganui (share)	0.066	0.249	0	1
region = Northland (share)	0.053	0.224	0	1
region = Otago (share)	0.128	0.334	0	1
region = Southland (share)	0.086	0.280	0	1
region = Tasman and Nelson (share)	0.067	0.250	0	1
region = Taranaki (share)	0.043	0.203	0	1
region = Waikato (share)	0.074	0.262	0	1
region = Wellington (share)	0.036	0.186	0	1
region = West Coast (share)	0.020	0.139	0	1

2.2. Defining the Likelihood of Land-Use Conversion

Decision making within the model rests entirely with the farmer agent. The economic component of the model returns the net revenue-maximising land-use for each farm along with the expected net revenue for each enterprise that could be undertaken on each farm. If the economically optimal land use differs from the enterprise currently undertaken, then the farmer agent chooses whether or not to convert through an evaluation against each farmer's "likelihood of land-use conversion".

The likelihood of land-use conversion is thus fundamental to decision making within the model. We adopt several methods for evaluating this value. The first (the homogeneous approach) ignores individual attributes of farmers and assigns an identical likelihood to each farmer. The second (the network approach) allows farmers to interact with peers and to make decisions that are informed by peer performance, either via networks or imitation. The third (the survey approach) uses empirical data from the Survey of Rural Decision Makers to predict this likelihood based on individual characteristics.

2.2.1. Homogeneous Approach

For this approach, which we class as the baseline, we ignore all farmer attributes and define the likelihood of land-use conversion at 0.2 (or 20%) for all farmer agents. We use this approach to represent a common type of economic LULCC model that uses a single rational, profit-maximising agent to make decisions.

2.2.2. Network Approach

This approach uses social and geographic networks to shape the farmer agent's likelihood of land-use conversion. Farmer's information networks are framed around their social [39,40,59] and geographical [74,75] interactions and play a role in decision-making processes [39,50,75]. Two theoretical frameworks inform how networks influence farmer decision making—endorsements and imitation. While endorsements and imitation in social networks are understood, the scale and impact that these processes have on decisions are difficult to quantify in the farming context.

Studies have found that the proximity to the people in one's network is not as important as the stature of those people [59,76]. Therefore we assume that two-way interactions such as endorsements are preferred by farmers and provide a higher level of information acceptance compared with one-way interactions such as imitation. Consequently, we specify that endorsements obtained through social networks have higher weightings than those obtained through imitation.

Specifying these weightings for these network types required additional experimentation as there was no empirical data on the level of acceptance of information obtained through them. Sociological opinion in New Zealand suggests that the relative weighting of the information provided via farmer networks should range between 0.05 and 0.15 [77]. Based on this, we defined the weightings for the likelihood of information uptake as 0.10 for Endorsement and 0.05 for Imitation. To explore the influence that these weightings have on the outputs of the model, we undertook a local sensitivity analysis (Supplementary Material 2). We found that the model outputs are relatively insensitive to small variations in the weightings used for both endorsements and imitation.

Endorsements work on the concept that information about a product, process, or person (*i.e.*, the endorsed) is transferred from one individual (*i.e.*, the endorser) to another individual (*i.e.*, the receiver) through a social process. The information that is transferred by the endorser is subjective and is validated by the receiver based on his or her understanding of the endorser and the product, process, or person. Thus, endorsements capture a *"subjective but socially embedded agent's reasoning process about cognitive trajectories aimed at achieving information and preferential clarity over another, endorsed agent"* ([78]; p. 1).

With endorsement in ARLUNZ, each farmer agent incorporates information on the success of the farming operation for ten farmer agents who are located closest to the decision maker and who undertake the same enterprise as the farmer agent. Each farmer agent learns the profitability/ha of each of the farmer agents within his or her social network; using these values, a mean profitability/ha value is derived for the farmer agent's network and is then compared to farmer agent's profitability/ha value. If the farmer agent's profitability/ha is higher than the mean profitability/ha of the farmer agent's social network, then his or her likelihood of land-use conversion is decreased by 0.10 percentage points to 10%. If the farmer agent's profitability/ha is lower than the mean profitability/ha of the farmer agent's social network, then his or her likelihood of land-use conversion increases by 0.10 percentage points to 30%. The ARLUNZ model assumes that each farmer in the social network has identical stature.

The theory of Social Learning [79,80] describes imitation as a process in which a person observes another person being rewarded for understandable and reproducible behaviour. The original person might then imitate that behaviour to try to achieve the same reward [81]. Imitation transfers knowledge through a one-way network in which information is "absorbed" from the person's surroundings and then used to inform the decisions they make. Farming practices are visible to all, particularly so to farmers in close proximity because of the regular exposure [75,82].

With imitation in ARLUNZ, the farmer agent incorporates information from the farms that are geographically adjacent to his or her own farm, regardless of the enterprise undertaken. If the economic component of the model proposes a change in land use, then each farmer agent in the geographic network that undertakes the proposed land use provides his or her profitability/ha value. Using these values, a mean profitability/ha value is derived for the farmer agent's network, and this figure is compared with the farmer agent's profitability/ha value. If the farmer agent's profitability/ha is higher than that of his or her geographic network, then his or her likelihood of land-use conversion is decreased by 0.05 percentage points to 15%. If the farmer agent's profitability/ha is higher than that of his or her geographic network, then the agent's likelihood of land-use conversion is increased by 0.05 percentage points to 25%.

2.2.3. Survey Approach

This approach is based on empirical data from the Survey of Rural Decision Makers, which accounts for the decision to de-intensify land use as well as the decision to intensify land use based on the predicted net revenue of each enterprise.

The perceived likelihood of changing your current land use to more intensive or less intensive uses over the following five years was evaluated using an 11-point scale, with 0 representing "extremely unlikely" and 10 representing "extremely likely". The average reported likelihood of intensification was 2.68, which we interpret to mean that there is a 26.8% probability of intensifying in the next five years, on average. Similarly, the average reported likelihood of de-intensification was 3.57, which we interpret to mean there is a 37.5% probability of de-intensifying in the next five years, on average.

Importantly, survey participants were asked about intensification and de-intensification, which could mean a change in management on the farm (such as an increase in the number of livestock per hectare) rather than wholesale conversion of a farmer's land use. As these are the best empirical indicators of intentions, however, we ignore this possibility in the analysis that follows.

For the purposes of this research, we ranked the three most common land uses based on the intensiveness of their land use (Figure 1). Dairy farming represents the most intensive land use, followed by sheep and beef farming. Forestry is the least intensive land use.

Figure 1. Land uses within the ARLUNZ model ranked by the intensiveness of their land use.

Using attributes defined by geospatial information (specifically, predominant land use and farm size) and empirical data from the Survey of Rural Decision Makers (specifically, age, experience, education level, importance of productivity, profitability, and network size), we define the likelihood of a farmer intensifying or de-intensifying his or her land use econometrically. Specifically, the probabilities of moving from a low-intensity activity to a high-intensity activity over the subsequent five years and vice versa are estimated using Tobit models in which the dependent variables are censored at 0 and 10. Specifically,

$$y_i^* = \mathbf{X}_i\boldsymbol{\beta} + u_i, u \sim N(0, \sigma^2) \tag{1}$$

where y_i^* is a latent variable equal to the observed variable, y_i, only when the latent variable falls between the values of 0 and 10, $\mathbf{X}$ is a vector of explanatory variables, and the error term, u, is normally distributed. Thus, we have:

$$y_i = \begin{cases} 10 \text{ if } y_i^* \geq 10 \\ y_i^* \text{ if } 0 < y_i^* < 10 \\ 0 \text{ if } y_i^* \leq 0 \end{cases} \tag{2}$$

In contrast to the ordinary least squares with censored data, the tobit estimator is consistent [83].

Table 2 presents the tobit estimates based on the national level data. The β are interpreted as the expected change in the uncensored latent variable, i.e., the uncensored likelihood of intensification or de-intensification associated with a marginal change in an explanatory variable. For example, increasing age of the decision maker by 1% reduces the predicted (uncensored) perceived likelihood of intensification by 0.034 points on the 11-point scale, an effect that is statistically significant at the 0.01 level. Similarly, having a diploma or technical training increases the predicted (uncensored) perceived likelihood of intensification by 0.570 points, also significant at the 0.01 level. Neither of these explanatory variables has a statistically distinguishable effect on the perceived likelihood of de-intensification.

To use the likelihood of land-use conversion values as defined above, we take a random draw from a uniform distribution between 0 and 1. If the value of the random draw is less than the farmer's likelihood of land-use conversion, then the proposed land-use change is accepted by the farmer agent and the farm is immediately converted to the proposed enterprise. In addition to converting between enterprises, the farmer agent also realises the predicted net revenue for that land use as defined by the economic component. If the random draw exceeds the likelihood of land-use conversion, then the incumbent enterprise is retained until the next time step of the model.

Table 2. Predicted intensification and de-intensification for the survey approach (tobit model).

Variables	Intensify	De-Intensify
log of effective land quantity	0.206 *	0.150
	(0.105)	(0.118)
log of age	−3.374 ***	−0.0697
	(0.767)	(0.923)
log of experience	0.380 ***	0.352 **
	(0.145)	(0.164)
diploma/tech training	0.570 *	−0.337
	(0.329)	(0.370)
university or higher	0.295	−0.159
	(0.323)	(0.373)
importance of being highly productive	0.140 **	−0.0216
	(0.0605)	(0.0657)
profitable business	−0.842 **	−0.342
	(0.373)	(0.410)
respondent exceeds median # of farm/farmer visits	1.272 ***	0.787 **
	(0.286)	(0.328)
risk tolerance	0.179 ***	0.0248
	(0.0619)	(0.0705)
Constant	6.874 **	−3.812
	(3.330)	(3.988)
Enterprise dummies	YES	YES
Region dummies	YES	YES
Observations	1,507	1,507
McFadden's adjusted R-squared	0.0449	0.0182

Note: * $p < 0.10$; ** $p < 0.05$; *** $p < 0.01$.

3. Experimental Section

To illustrate the variation in the predicted land use, economic outcomes, and environmental impacts caused by various approaches to assigning farmers' likelihood of land-use conversion, we explore the effects on landowners in the Hurunui-Waiau catchment in the Canterbury region of New Zealand's South Island (Figure 2). These catchments have a large and diverse set of land uses that are expected to see significant changes in the future.

Figure 2. Location of the Hurunui-Waiau catchment. The catchment is located within the Canterbury region of New Zealand's South Island. Planners in the catchment anticipate significant changes in land use over the next several decades.

To provide a sample that encompasses the range of possible outcomes, the model was run using 50 simulations for each of the three approaches being investigated (*i.e.*, the homogeneous approach, the network approach, and the survey approach). The results in the following section are based on the averaged values over all 50 simulations for each of the three approaches.

The model covers a time horizon of 50 years with ten incremental time steps, each of which represent five years. The scenario assumes a real annual increase in farm commodity prices (*i.e.*, milk, meat, and timber) of 2%, which is commensurate with the last 50 years of commodity prices [84]. Climate and available technology are held constant over the entire model simulation. We note that the economic picture is consistent across scenarios and that the strength of milk prices relative to meat and timber prices produces a trend toward dairy conversion.

The land-use map used in the model was captured in June 2010 [85] (Figure 3a), and although the map includes seven different land uses, the model focuses on the three key enterprises that represent 94% of the productive land available within the catchment: dairy, sheep and beef, and plantation forestry. The cadastral land parcel boundaries used are derived from Land Information New Zealand and represent the cadastral structure of the catchments as at August 2012 (Figure 3b), which was the closest database to the 2010 land use map. Farmer agents are only created for farms exceeding 100 ha in order to focus on commercially operated enterprises. Productivity zones are delineated by land use classification [86] and slope and are classified into productivity zones—flats, foothills, or hills (Figure 3c). Any land owned by the Crown (e.g., native forest) is assumed to be non-productive in use and hence no farmer agent is created.

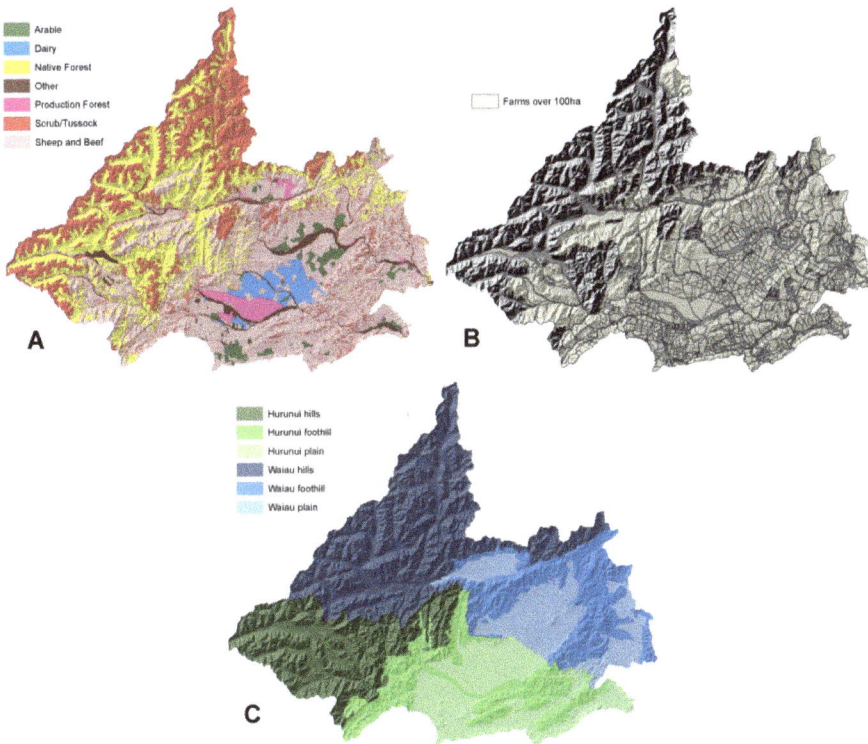

Figure 3. Detailed map of the Hurunui-Waiau catchment by (a) 2010 land use; (b) 2012 cadastral land parcels; and (c) productivity zone. The data layers are used within the model to define the initial land use, farm locations and extents, and the expected productivity for each farm.

The landscape provides a range of geospatial information about the catchment being modelled, such as cadastral boundaries, initial land use, and productivity zones. Using the cadastral boundaries, a farmer agent is generated at the centre of each cadastral parcel and makes decisions for the entire cadastral parcel (*i.e.*, farm). This agent holds a range of social and economic attributes such as age, education, and the size of social networks. These attributes are defined empirically for all farmer agents through the Survey of Rural Decision Makers. For the homogeneous and network approaches, each agent's initial likelihood of land-use conversion is set to 20%; in the case of the survey approach, each agent's likelihood of land-use conversion is defined econometrically as described above.

For the survey approach, there is significant variation in the farmer agents' likelihood of land-use conversion between each of the three enterprises. Table 3 summarises this variation over the 50 randomly generated populations used in the model. Farmers who undertake forestry have the lowest likelihoods of land-use conversion. These values reinforce two characteristics forestry, namely, its low intensity (an average likelihood of de-intensification at 0% highlights a lack of less-intensive options) and that farmers undertaking forestry are less likely to move to more intensive land uses (13% compared with 26% and 31% for sheep and beef and dairy farmers, respectively).

Table 3. Mean values across the 50 simulations for the survey approach using ARLUNZ. Mean values, standard deviation, and confidence intervals for this figure are available in Supplementary Material 2.

	Forestry	Sheep and Beef	Dairy
Intensify	12.99%	26.12%	31.06%
De-Intensify	0.00%	29.10%	26.99%

The simulated likelihoods of land-use conversion to more and less intensive enterprises for sheep and beef farmers are more balanced (26.12% *vs.* 29.10%, respectively), but sheep and beef farmers are more likely to de-intensify their land use over the next five years. The simulated likelihoods of land-use conversion for dairy farmers highlight the production-focused approach commonly associated with the enterprise: alongside a 27% probability that they will de-intensify their land use, there is a 31% chance that they would further intensify their land use over the next five years. However, the model does not include a more intensive land use, so while this intention is accounted for within the model, it is not currently utilised.

4. Results

In this section, we compare results obtained after defining the likelihood of land-use conversion in each of the ways described above. At the catchment level, we project that the area of both dairy and forestry will increase over time (Figure 4). At the production zone level, dairy is estimated to increase in both the plains and foothills, while forestry is estimated to expand in the foothills and hills. The area of sheep and beef farms is estimated to decline in all three productivity zones.

For the homogenous approach, the area allocated to dairy increases from the initial 16,900 ha to 100,450 ha over the 50-year period (with a 95% Confidence Interval, hereafter 95% CI, of 1670 ha). This expansion is split between the highly productive plains region of the catchment and the less productive foothills.

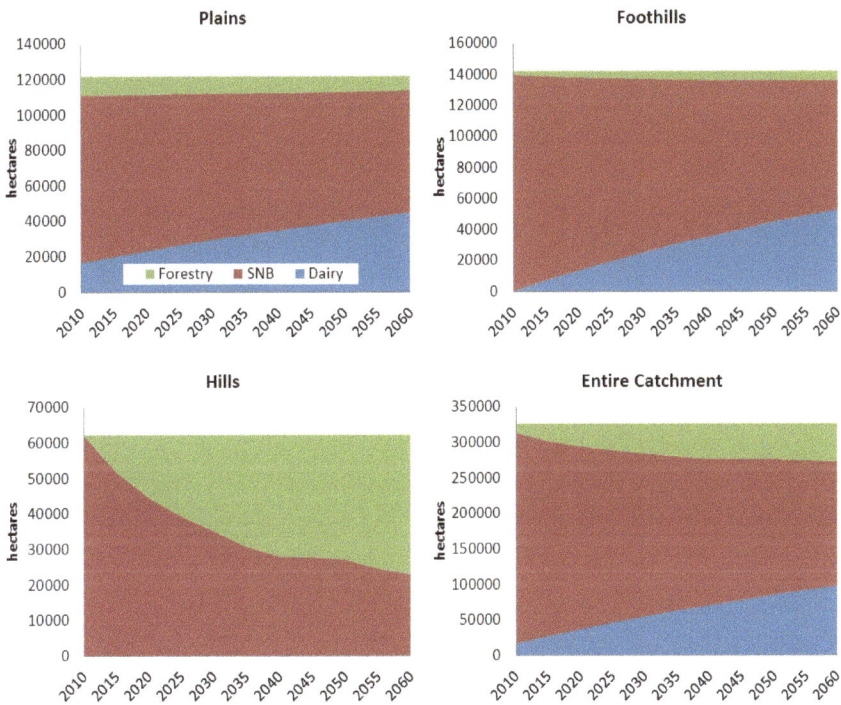

Figure 4. Regional land use area for "Homogenous" baseline projection. Mean values, standard deviation, and confidence intervals for this figure are available in Supplementary Material 3.

Although such an expansion in dairy is large, it is not unrealistic. For example, the area of land allocated to dairy in Canterbury increased by 172% between 1996 and 2008, and it is projected to expand by an additional 51% by 2020 [87]. Moreover, the Hurunui-Waiau catchment have already witnessed conversion to dairy as forests in the highly-productive flat areas of the catchment reached harvest age. Third, there are ongoing discussions of implementing the Hurunui Water Project, which would expand the area of irrigated land by an additional 41,500 ha, bringing the total irrigated area of the Hurunui-Waiau catchment to over 72,000 ha [88].

The modelled change in land use relative to the homogeneous approach is shown in Figure 5. For both dairy and sheep and beef operations, both approaches to defining the likelihood of land-use conversion trend positively and begin to converge by 2060 (Figure 5, C4). Divergence between the network and survey approaches by 2060 can only be found in forestry and then only for the network approach which results in a level of land use similar to the homogeneous baseline.

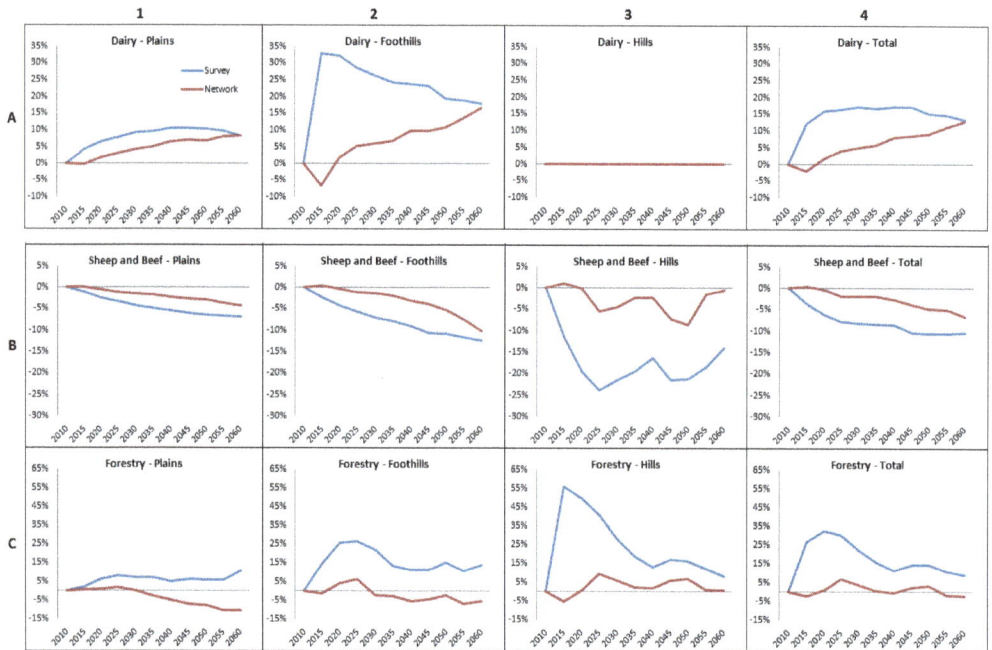

Figure 5. Relative change in regional land use area from "Homogenous" baseline projection. Mean values, standard deviation, and confidence intervals for this figure are available in Supplementary Material 3.

Using a sub-regional scale by productivity zone, the network approach results in impacts that are similar to the homogenous approach. In addition, the survey approach always results in a more rapid change in land use relative to the baseline and to the network approach. Under the survey approach, farmer agents make changes to their farms sooner than under the network approach, which may explain, which explains the rapid growth in dairy in the foothill productivity zone (Figure 5, A2) and forestry in the hills productivity zone (Figure 5, C3).

In the model, forestry at a sub-regional scale is significantly affected by the change in approach to define the farmer's likelihood of land-use conversion. For example, the area of forestry within the plains productivity zone significantly reduces for the network approach when compared with both the survey approach and the homogenous baseline (Figure 5, C1). This outcome is because the more intensive land uses on the plains (such as dairy or sheep and beef) are more profitable, which is translated into the farmer's networks, particularly their geographic network. The increased profitability within their geographic network alters the behaviour of the forestry agents to increase the likelihood of their conversion to more intensive land uses.

4.1. Farm Net Revenue

For reference, all monetary figures from the model are reported in 2012 New Zealand dollars. Farm net revenue is estimated to increase over time under all approaches (Table 4), increasing from $153 million/yr in 2010 to between $745 and $807 million/yr by 2060 (95% CIs of $8 million and $6.8 million respectively). The increase over time stems from the assumption that all commodity prices and yield combinations (*i.e.*, farm revenue per hectare) increase by 2%/yr and the expansion of dairy. Interestingly, while the survey and network approaches vary, the difference in net revenue for the two approaches at a catchment scale over time is negligible ($807 *vs.* $804 million/yr by 2060 respectively, with 95% CI's of $6.8 million/yr and $8.2 million/yr respectively).

Table 4. Total annual farm profits for Hurunui-Waiau catchment (million NZD/yr). Mean values, standard deviation, and confidence intervals for this figure are available in Supplementary Material 3.

Year	Total			Dairy			Sheep and Beef			Forestry		
	Homo	Networks	Survey	Homo	Networks	Survey	Homo	Networks	Survey	Homo	Networks	Survey
2010	$153	0.0%	0.0%	$30	0.0%	0.0%	$115	0.0%	0.0%	$7	0.0%	0.0%
2020	$214	0.9%	6.6%	$83	1.2%	15.3%	$110	1.8%	−6.8%	$21	0.0%	25.0%
2030	$308	3.1%	7.8%	$156	4.3%	15.7%	$118	1.7%	−10.3%	$34	2.9%	19.0%
2040	$421	4.5%	8.7%	$246	7.2%	16.0%	$126	1.6%	−10.5%	$49	−2.1%	9.3%
2050	$568	4.9%	8.4%	$369	7.1%	14.0%	$139	0.0%	−14.9%	$60	1.6%	13.0%
2060	$745	7.3%	7.7%	$514	11.4%	12.7%	$153	−3.4%	−15.9%	$78	−1.3%	9.3%

While net revenue at a catchment scale is similar, the distribution in net revenue between the three enterprises differs. Total dairy profits are estimated to increase from $30 million in 2010 to between $514 and $589 million (95% CIs of $9.1 million and $8.3 million respectively) in 2060 (Table 4). The largest increases are estimated to occur under the survey approach, where annual profits are, on average, 15% larger than under the homogenous approach.

The expansion of forestry in the catchment also yields increased net revenue from an initial $7 million in 2010 to between $77 and $86 million (95% CI of $3.4 million and $3.4 million respectively) in 2060. Again, the survey approach is estimated to yield increased profits (10% or more/yr) relative to the homogenous approach. Profits in sheep and beef farming are estimated to increase from $115 million in 2010 to between $132 and $153 million (95% CI of $2.9 million and $3.3 million respectively) in 2060 even though the area devoted to this land use is expected to contract because of yield and commodity price changes that increases the per hectare revenue of sheep and beef farming.

Interestingly, the distribution in net revenue for sheep and beef under the network approach is estimated to be higher than under the homogenous approach up until 2060 (Table 4). The opposite is found for the survey approach, which sees a reduction in the level of profit for sheep and beef compared with the baseline.

4.2. Environmental Outputs

As land-use change impacts key environmental indicators, the ARLUNZ model captures a range of environmental outputs such as GHG emissions, forest carbon sequestration, nitrogen (N) and phosphorus (P) loss [36]. As seen in Table 5, the expansion of dairy—which often has a higher stocking rate per hectare than sheep and beef—causes an 18% increase in livestock GHG emissions compared with the homogeneous approach. The network and survey approaches show additional livestock GHG emissions of 6% and 3%, respectively.

This growth in livestock GHGs are offset by increases in carbon sequestration through forestry under two of the three approaches. Specifically, the networks approach shows a reduction in forest carbon sequestration of 2% relative to the baseline while the survey approach shows a 9% increase in the amount of carbon sequestered. This result highlights the detrimental effects of the network approach on forestry in more marginal production zones. Even so, the overall expansion of forestry in all three approaches results in a net GHG reduction of between 50% and 68% between 2010 and 2060.

In contrast, the expansion of dairy farming in the plains and foothills, results in a large increase in nutrient loadings that could impact the environmental quality of the local waterways (Table 6). We estimate that N and P will increase by 86% and 43%, respectively, over the next 50 years under the homogenous baseline. The networks approach predicts increases in both N and P by an additional 8% relative to the baseline, while the survey approach predicts additional increases of 8% and 5%, respectively. Consequently, while the expansion of dairy farming in the catchment produces economic benefits, the negative impacts on water quality are non-negligible.

Table 5. GHG outputs for Hurunui-Waiau catchment (tons/yr). Mean values, standard deviation, and confidence intervals for this figure are available in Supplementary Material 3.

Year	Livestock GHGs			Forest Carbon Sequestration			Net GHGs		
	Homogenous	Network	Survey	Homogenous	Network	Survey	Homogenous	Network	Survey
2010	988,619	988,619	988,619	−200,686	−200,686	−200,686	787,933	787,933	787,933
2020	1,009,062	1,027,099	1,007,832	−517,043	−502,681	−677,464	492,020	524,418	330,368
2030	1,052,232	1,080,009	1,060,729	−682,562	−697,085	−833,184	369,670	382,924	227,545
2040	1,080,496	1,125,738	1,113,336	−807,773	−792,856	−896,421	272,723	332,881	216,915
2050	1,134,640	1,179,598	1,164,413	−810,228	−821,304	−930,562	324,412	358,294	233,851
2060	1,166,265	1,238,713	1,201,127	−867,407	−847,212	−947,868	298,859	391,501	253,259

Table 6. Nitrogen and Phosphorus outputs for Hurunui-Waiau catchment (tons/yr). Mean values, standard deviation, and confidence intervals for this figure are available in Supplementary Material 3.

Year	Nitrogen			Phosphorus		
	Homogenous	Network	Survey	Homogenous	Network	Survey
2010	4039	4039	4039	37	37	37
2020	4899	4970	5171	40	41	41
2030	5701	5882	6115	44	45	45
2040	6339	6652	6899	46	49	49
2050	7009	7370	7599	50	53	53
2060	7517	8136	8111	53	57	56

5. Conclusions

We anticipated that the two different approaches to defining the likelihood of land-use change would significantly affect model outcomes. However, the results from this experiment show that the approach has a limited effect at a catchment scale for both the distribution of land use and the resulting total net revenue. Nevertheless, the differing distributions of land use across productivity zones suggest that the approach to estimating the likelihood of land-use change ultimately does matter.

Simulating the social process of information transfer between agents through the network approach resulted in less economically optimal land use for all three enterprises. For dairy, the network approach predicts less conversion vis-à-vis the survey approach. Still, results for both approaches converge by 2060.

The network approach to defining the likelihood of land-use change profoundly impacts model outcomes for sheep and beef farmers in the hill productivity zone. Within this zone, it appears that the level of productivity is very similar at a per hectare level for each sheep and beef farmer. This similarity of profitability for these farmer agents reduces the likelihood of conversion to the more profitable enterprise of forestry within the network approach. For the small amount of forestry on the

25

highly productive plains in the initial state of the model, the network approach enabled the farmer agents to easily recognise significant increases in average net revenue if they converted to different enterprises, primarily dairy. This resulted in a steady decrease in the area of forestry on the plains. In the foothill zone, many sheep and beef farms convert to dairy.

Defining the agents' likelihood of land-use conversion through the survey approach provides greater deviation from the baseline for all sub-regional results. This deviation is particularly evident in the early steps of the model simulation, where variations in area of up to 55% compared with the baseline are observed. This result highlights the design of the survey approach, where land that suits the two ends of the spectrum of land-use intensity—dairy and forestry—quickly sees the influx of new farms, in contrast to the slow changes that occur under the network approach.

The results for the survey approach highlight two interesting observations. First, even with the significant increase in net revenue when converting forestry to other enterprises in the plains productivity zone, the amount of forestry in this zone increases under the survey approach. This result is embedded in the initial land-use conversion values defined for a move to a more intensive enterprise. For example, regardless of the significant increase in net revenue, the survey found that foresters are unlikely to convert to a more intensive enterprise, highlighting the static nature of the survey approach, which—unlike the dynamic network approach—is unable to account for changes in preference over time.

A second interesting result is the fact that substantively different approaches to defining the likelihood of land-use conversion produce similar trends. This observation raises questions that cannot be fully answered in this paper: If the attributes of a population can be reliability and accurately captured through surveys, can the agent-based model that uses these data be simplified by removing the need to simulate social processes? And would this facilitate greater uptake among end users because the process used to define behaviour within an agent-based model is easier to understand?

Malanson and Walsh [29] recently noted the problems of calibration and validation stemming from complex interactions in agent-based models. While agreeing that the challenge for applied agent-based models is in correctly parametrising the agents, we found that the survey approach provided enough detail to generate reliable populations of agents and did not over parametrise the model.

We believe that model design should be informed by model purpose. Where comprehensive surveys are available, we advocate using empirical data to define an agent's likelihood of a type of behaviour. However, if there is limited information about the size and importance of a farmer's social and geographical networks or if a key purpose of the model is to account for the effects of these networks, then developing the structure within the model to simulate, test, and document the effects of changes in the networks is preferred.

For LULCC models—especially those in applied settings—exploring the impacts of different approaches to modelling the likelihood of land-use conversion is critical. Understanding the effects of key model parameters on economic, social, and environmental factors will facilitate the continued acceptance of ABM of LULCC among end users and will improve the results generated by both the economic and geographic ABMs of LULCC. While the results outlined here will not fully resolve the disciplinary differences, they do outline the need to account for heterogeneity in the predicted agent behaviours for both disciplines.

Acknowledgments: The development of ARLUNZ was supported by Landcare Research (CF1112-83-02 and CF1011-83-11) and by the New Zealand Ministry of Business, Innovation, and Employment (C09X1307). The Survey of Rural Decision Makers was supported by the Ministry for the Environment (20443) and the Ministry of Business, Innovation, and Employment (C09X1003). The founding sponsors had no role in the design of the study; in the collection, analyses, or interpretation of data; in the writing of the manuscript, and in the decision to publish the results.

The authors acknowledge the contributions of Oshadhi Samarasinghe and Suzie Greenhalgh, who supported development of the ARLUNZ model, and Florian Eppink for assistance in R. Anne Austin, Suzie Greenhalgh, and William Wright commented on drafts of this paper. Errors and omissions are the responsibility of the authors alone.

Author Contributions: Fraser Morgan had the original idea for the paper, developed and ran the experiments, and led the preparation of the manuscript. Philip Brown led the development and analysis of the survey of rural decision makers and the writing in relation to the survey and statistics derived from the survey. Adam Daigneault analysed the results from the experiments and provided assistance with the economic components of the model. All authors participated in the writing and revision of the manuscript. In addition, all authors approved the final manuscript.

Conflicts of Interest: The authors declare no conflict of interest.

References

1. Smajgl, A.; Brown, D.G.; Valbuena, D.; Huigen, M.G.A. Empirical characterisation of agent behaviours in socio-ecological systems. *Environ. Model. Softw.* **2011**, *26*, 837–844.

2. Parker, D.C.; Manson, S.M.; Janssen, M.A.; Hoffmann, M.J.; Deadman, P. Multi-agent systems for the simulation of land-use and land-cover change: A review. *Ann. Assoc. Am. Geogr.* **2003**, *93*, 314–337.

3. Villamor, G.B.; van Noordwijk, M.; Troitzsch, K.G.; Vlek, P.L. Human decision making for empirical agent-based models: Construction and validation. In Proceedings of the International Environmental Modelling and Software Society (iEMSs)—2012 International Congress on Environmental Modelling and Software, Leipzig, Germany, 1–5 July 2012.

4. Hersperger, A.M.; Gennaio, M.-P.; Verburg, P.H.; Bürgi, M. Linking land change with driving forces and actors: Four conceptual models. *Ecol. Soc.* **2010**, *15*, 1.

5. Lambin, E.F.; Turner, B.L.; Geist, H.J.; Agbola, S.B.; Angelsen, A.; Bruce, J.W.; Coomes, O.T.; Dirzo, R.; Fischer, G.; Folke, C.; *et al.* The causes of land-use and land-cover change: Moving beyond the myths. *Glob. Environ. Chang.* **2001**, *11*, 261–269.

6. Rindfuss, R.R.; Walsh, S.J.; Turner, B.L.; Fox, J.; Mishra, V. Developing a science of land change: Challenges and methodological issues. *Proc. Natl. Acad. Sci. USA* **2004**, *101*, 13976–13981.

7. Brown, D.G.; Verburg, P.H.; Pontius, R.G., Jr.; Lange, M.D. Opportunities to improve impact, integration, and evaluation of land change models. *Curr. Opin. Environ. Sustain.* **2013**, *5*, 452–457.

8. Agarwal, C.; Green, G.M.; Grove, J.M.; Evans, T.P.; Schweik, C.M. *A Review and Assessment of Land-Use Change Models: Dynamics of Space, Time, and Human Choice*; U.S. Department of Agriculture Forest Service, Northeastern Forest Research Station: Burlington, VT, USA, 2002.

9. Baker, W.L. A review of models in landscape change. *Landsc. Ecol.* **1989**, *2*, 111–133.

10. Veldkamp, A.; Lambin, E.F. Predicting land-use change. *Agric. Ecosyst. Environ.* **2001**, *85*, 1–6.

11. Chuvieco, E. Integration of linear programming and GIS for land-use modeling. *Int. J. Geogr. Inf. Syst.* **1993**, *7*, 71–83.

12. Longley, P.; Higgs, G.; Martin, D. The predictive use of GIS to model property valuations. *Int. J. Geogr. Inf. Syst.* **1994**, *8*, 217–235.

13. Sklar, F.H.; Costanza, R. The development of dynamic spatial models for landscape ecology: A review and prognosis. In *Quantitative Methods in Landscape Ecology*; Tuner, M.G., Gardner, R.H., Eds.; Springer-Verlag: New York, NY, USA, 1991; pp. 239–288.

14. Weinberg, M.; Kling, C.L.; Wilen, J.E. Water markets and water quality. *Am. J. Agric. Econ.* **1993**, *75*, 278–291.

15. Leggett, C.G.; Bockstael, N.E. Evidence of the effects of water quality on residential land prices. *J. Environ. Econ. Manag.* **2000**, *39*, 121–144.

16. Parker, D.C.; Filatova, T. A conceptual design for a bilateral agent-based land market with heterogeneous economic agents. *Comput. Environ. Urban Syst.* **2008**, *32*, 454–463.

17. O'Sullivan, D.; Haklay, M. Agent-based models and individualism: Is the world agent-based? *Environ. Plan. A* **2000**, *32*, 1409–1425.

18. Nolan, J.; Parker, D.; van Kooten, G.C.; Berger, T. An overview of computational modeling in agricultural and resource economics. *Can. J. Agric. Econ.* **2009**, *57*, 417–429.

19. Caldas, M.; Walker, R.; Arima, E.; Perz, S.; Aldrich, S.; Simmons, C. Theorizing land cover and land use change: The peasant economy of Amazonian deforestation. *Ann. Assoc. Am. Geogr.* **2007**, *97*, 86–110.

20. Irwin, E.G.; Geoghegan, J. Theory, data, methods: Developing spatially explicit economic models of land use change. *Agric. Ecosyst. Environ.* **2001**, *85*, 7–24.

21. Heckelei, T.; Britz, W.; Zhang, Y. Positive mathematical programming approaches—Recent developments in literature and applied modelling. *Bio-Based Appl. Econ.* **2012**, *1*, 109–124.

22. Evans, T.P.; Sun, W.; Kelley, H. Spatially explicit experiments for the exploration of land-use decision-making dynamics. *Int. J. Geogr. Inf. Sci.* **2006**, *20*, 1013–1037.

23. Irwin, E.G. New directions for urban economic models of land use change: Incorporating spatial dynamics and heterogeneity. *J. Reg. Sci.* **2010**, *50*, 65–91.

24. Janssen, M.; Jager, W. The human actor in ecological-economic models. *Ecol. Econ.* **2000**, *35*, 307–310.

25. Magliocca, N.R.; Brown, D.G.; Ellis, E.C. Cross-Site comparison of land-use decision-making and its consequences across land systems with a generalized agent-based model. *PLoS One* **2014**, *9*, e86179.

26. Valbuena, D.; Verburg, P.H.; Bregt, A.K.; Ligtenberg, A. An agent-based approach to model land-use change at a regional scale. *Landsc. Ecol.* **2010**, *25*, 185–199.

27. Malanson, G.P.; Verdery, A.M.; Walsh, S.J.; Sawangdee, Y.; Heumann, B.W.; McDaniel, P.M.; Frizzelle, B.G.; Williams, N.E.; Yao, X.Z.; Entwisle, B.; *et al.* Changing crops in response to climate: Virtual Nang Rong, Thailand in an agent based simulation. *Appl. Geogr.* **2014**, *53*, 202–212.

28. Farmer, J.D.; Foley, D. The economy needs agent-based modelling. *Nature* **2009**, *460*, 685–686.

29. Malanson, G.P.; Walsh, S.J. Agent-based models: Individuals interacting in space. *Appl. Geogr.* **2015**, *56*, 95–98.

30. Matthews, R.B.; Gilbert, N.G.; Roach, A.; Polhill, J.G.; Gotts, N.M. Agent-based land-use models: A review of applications. *Landsc. Ecol.* **2007**, *22*, 1447–1459.

31. Heard, D.; Dent, G.; Schifeling, T.; Banks, D. Agent-Based models and microsimulation. *Ann. Rev. Stat. Appl.* **2015**, *2*, 259–272.

32. An, L. Modeling human decisions in coupled human and natural systems: Review of agent-based models. *Ecol. Model.* **2012**, *229*, 25–36.

33. Heckbert, S.; Baynes, T.; Reeson, A. Agent-based modeling in ecological economics. *Ann. N. Y. Acad. Sci.* **2010**, *1185*, 39–53.

34. Kaye-Blake, B.; Schilling, C.; Post, E. Validation of an agricultural MAS for southland, New Zealand. *J. Artif. Soc. Soc. Simul.* **2014**, *17*, 5.

35. Valbuena, D.; Verburg, P.H.; Veldkamp, A.; Bregt, A.K.; Ligtenberg, A. Effects of farmers' decisions on the landscape structure of a Dutch rural region: An agent-based approach. *Landsc. Urban Plan.* **2010**, *97*, 98–110.

36. Morgan, F.J.; Daigneault, A.J. Estimating impacts of climate change policy on land use: An agent-based modelling approach. *PLoS One* **2015**, *10*, e0127317.

37. Burton, R.J.F. *Strategic Decision-Making in Agriculture: An International Perspective of Key Social and Structural Influences*; AgResearch: Lincoln, New Zealand, 2009; p. 162.

38. Bakker, M.M.; van Doorn, A.M. Farmer-specific relationships between land use change and landscape factors: Introducing agents in empirical land use modelling. *Land Use Policy* **2009**, *26*, 809–817.

39. Maertens, A.; Barrett, C.B. Measuring social networks' effects on agricultural technology adoption. *Am. J. Agric. Econ.* **2013**, *95*, 353–359.

40. Ramirez, A. The influence of social networks on agricultural technology adoption. *Procedia Soc. Behav. Sci.* **2013**, *79*, 101–116.

41. Smithers, J.; Furman, M. Environmental farm planning in Ontario: Exploring participation and the endurance of change. *Land Use Policy* **2003**, *20*, 343–356.

42. Smithers, J.; Johnson, P. The dynamics of family farming in North Huron County, Ontario. Part I. Development trajectories. *Can. Geogr.* **2004**, *48*, 191–208.

43. Burton, R.J.F. The influence of farmer demographic characteristics on environmental behaviour: A review. *J. Environ. Manag.* **2014**, *135*, 19–26.

44. Millington, J.; Romero-Calcerrada, R.; Wainwright, J.; Perry, G. An agent-based model of mediterranean agricultural land-use/cover change for examining wildfire risk. *J. Artif. Soc. Soc. Simul.* **2008**, *11*, 4.

45. Valbuena, D.; Verburg, P.H.; Bregt, A.K. A method to define a typology for agent-based analysis in regional land-use research. *Agric. Ecosyst. Environ.* **2008**, *128*, 27–36.

46. Acosta-Michlik, L.; Espaldon, V. Assessing vulnerability of selected farming communities in the Philippines based on a behavioural model of agent's adaptation to global environmental change. *Glob. Environ. Chang.* **2008**, *18*, 554–563.

47. Ziervogel, G.; Bithell, M.; Washington, R.; Downing, T. Agent-based social simulation: A method for assessing the impact of seasonal climate forecast applications among smallholder farmers. *Agric. Syst.* **2005**, *83*, 1–26.

48. Berger, T. Agent-based spatial models applied to agriculture: A simulation tool for technology diffusion, resource use changes and policy analysis. *Agric. Econ.* **2001**, *25*, 245–260.

49. Deffuant, G.; Huet, S.; Bousset, J.P.; Henriot, J.; Amon, G.; Weisbuch, G. Agent-based simulation of organic farming conversion in Allier département. In *Complexity and Ecosystem Management: The Theory and Practice of Multi-agent Systems*; Janssen, M., Ed.; Edward Elgar Publishing: Cheltenham, UK, 2002; pp. 158–187.

50. Deffuant, G.; Skerratt, S.; Amblard, F.; Ferrand, N.; Chattoe, E.; Gilbert, N.; Weisbush, G. Agent-based simulation of decision process mixing rational reasoning and influences from socio-informational networks: Case studies of agri-environmental measures adoption by farmers. In Proceedings of the Fifth International Conference on Social Science Methodology, Cologne, Germany, 1 October 2000. [CD-ROM]

51. Robinson, D.T.; Brown, D.G.; Parker, D.C.; Schreinemachers, P.; Janssen, M.A.; Huigen, M.; Wittmer, H.; Gotts, N.; Promburom, P.; Irwin, E.; *et al.* Comparison of empirical methods for building agent-based models in land use science. *J. Land Use Sci.* **2007**, *2*, 31–55.

52. Gigerenzer, G.; Todd, P. *Simple Heuristics That Make Us Smart*; Oxford University Press: Oxford, UK, 1999.

53. Manson, S.M. Bounded rationality in agent-based models: Experiments with evolutionary programs. *Int. J. Geogr. Inf. Sci.* **2006**, *20*, 991–1012.

54. Simon, H.A. Behavioral economics and bounded rationality. In *Models of Bounded Rationality*; Simon, H.A., Ed.; MIT Press: Cambridge, MA, USA, 1997; pp. 267–433.

55. Tversky, A.; Kahneman, D. Rational choice and the framing of decisions. In *The Limits of Rationality*; Cook, K.S., Levi, M., Eds.; University of Chicago Press: Chicago, IL, USA, 1990; pp. 60–89.

56. Dawid, H.; Fagiolo, G. Agent-based models for economic policy design: Introduction to the special issue. *J. Econ. Behav. Organ.* **2008**, *67*, 351–354.

57. Janssen, M.A.; Ostrom, E. Empirically based, agent-based models. *Ecol. Soc.* **2006**, *11*, 37.

58. Berger, T.; Schreinemachers, P. Creating agents and landscapes for multiagent systems from random samples. *Ecol. Soc.* **2006**, *11*, 19.

59. Nelson, K.; Brummel, R.; Jordan, N.; Manson, S. Social networks in complex human and natural systems: The case of rotational grazing, weak ties, and eastern US dairy landscapes. *Agric. Hum. Values* **2014**, *31*, 245–259.

60. Manson, S.M.; Jordan, N.R.; Nelson, K.C.; Brummel, R.F. Modeling the effect of social networks on adoption of multifunctional agriculture. *Environ. Model. Softw.* **2014**.

61. Heckbert, S.; Bishop, I. Empirical calibration of spatially explicit agent-based models. In *Advanced Geosimulation*; Marceau, D., Benenson, I., Eds.; Bentham Books: Oak Park, IL, USA, 2011; pp. 92–110.

62. Kennedy, W.G. Modelling human behaviour in agent-based models. In *Agent-Based Models of Geographical Systems*; Heppenstall, A.J., Crooks, A.T., See, L.M., Batty, M., Eds.; Springer: Dordrecht, The Netherlands, 2012; pp. 167–179.

63. Wilensky, U. *NetLogo*; Center for Connected Learning and Computer-Based Modeling, Northwestern University: Evanston, IL, USA, 1999; Available online http://ccl. northwestern.edu/netlogo (accessed on 24 September 2015).

64. Brandmeyer, J.E.; Karimi, H.A. Coupling methodologies for environmental models. *Environ. Model. Softw.* **2000**, *15*, 479–488.

65. Daigneault, A.; Greenhalgh, S.; Samarasinghe, O. A response to Doole and Marsh (2013) article: Methodological limitations in the evaluation of policies to reduce nitrate leaching from New Zealand agriculture. *Austr. J. Agric. Resour. Econ.* **2014**, *58*, 281–290.

66. Daigneault, A.; Greenhalgh, S.; Samarasinghe, O. Economic impacts of GHG and nutrient reduction policies in New Zealand: A tale of two catchments. In Proceedings of the Australian Agricultural and Resource Economics Society 2012 Conference (56th), Freemantle, WA, Australia, 7 February 2012.

67. Daigneault, A.; Greenhalgh, S.; Samarasinghe, O.; Jhunjhnuwala, K.; Walcroft, J.; de Oca Munguia, O. *Sustainable Land Management and Climate Change—Catchment Analysis of Climate Change: Final Report*; Ministry of Primary Industries: Wellington, New Zealand, 2012.

68. Daigneault, A.; McDonald, H.; Elliott, S.; Howard-Williams, C.; Greenhalgh, S.; Guysev, M.; Kerr, S.; Lennox, J.; Lilburne, L.; Morgenstern, U.; *et al. Evaluation of the Impact of Different Policy Options for Managing to Water Quality Limits*; Ministry of Primary Industries: Wellington, New Zealand, 2012.

69. Daigneault, A.; Samarasinghe, O.; Lilburne, L. *Modelling Economic Impacts of Nutrient Allocation Policies in Canterbury—Hinds Catchment: Final Report*; Landcare Research: Lincoln, New Zealand, 2013.

70. Müller, B.; Bohn, F.; Dreßler, G.; Groeneveld, J.; Klassert, C.; Martin, R.; Schlüter, M.; Schulze, J.; Weise, H.; Schwarz, N. Describing human decisions in agent-based models—ODD+D, an extension of the ODD protocol. *Environ. Model. Softw.* **2013**, *48*, 37–48.

71. Brown, P. Survey of Rural Decision Makers. Available online www.landcareresearch.co.nz/science/portfolios/enhancing-policy-effectiveness/srdm (accessed on 20 March 2015).

72. AsureQuality New Zealand. AgriBase Database. Available online: http://www.asurequality.com/capturing-information-technology-across-the-food-supply-chain/agribase-database-of-new-zealand-rural-properties.cfm (accessed on 24 September 2015).

73. Statistics New Zealand. New Zealand Agriculture Production Surveys and Censuses. Available online: http://datainfoplus.stats.govt.nz/item/nz.govt.stats/6362a469-f374-412e-ac25-d76fd0962003/106/ (accessed on 24 September 2015).

74. Isaac, M.E.; Erickson, B.H.; Quashie-Sam, S.J.; Timmer, V.R. Transfer of knowledge on agroforestry management practices: The structure of farmer advice networks. *Ecol. Soc.* **2007**, *12*, 32.

75. Schmit, C.; Rounsevell, M.D.A. Are agricultural land use patterns influenced by farmer imitation? *Agric. Ecosyst. Environ.* **2006**, *115*, 113–127.

76. Banerjee, A.; Chandrasekhar, A.G.; Duflo, E.; Jackson, M.O. *Gossip: Identifying Central Individuals in a Social Network*. NBER Working Paper No. 20422. Available online: http://arxiv.org/pdf/1406.2293v2.pdf (accessed on 24 September 2015).

77. Small, B. Personal communication, AgResearch: Hamilton, New Zealand, 2013.

78. Alam, S.J.; Geller, A.; Meyer, R.; Werth, B. Modelling contextualized reasoning in complex societies with endorsements. *J. Artif. Soc. Soc. Simul.* **2010**, *13*, 6.

79. Bandura, A. *Social Learning Theory*; Prentice Hall: Englewood Cliffs, NJ, USA, 1977.

80. Bandura, A. *Social Foundations of Thought and Action: A Social Cognitive Theory*; Prentice Hall: Englewood Cliffs, NJ, USA, 1986.

81. Jager, W.; Janssen, M.A.; de Vries, H.J.M.; de Greef, J.; Vlek, C.A.J. Behaviour in commons dilemmas: Homo economicus and Homo psychologicus in an ecological-economic model. *Ecol. Econ.* **2000**, *35*, 357–379.

82. Gotts, N.M.; Polhill, J.G. When and how to imitate your neighbours: Lessons from and for fearlus. *J. Artif. Soc. Soc. Simul.* **2009**, *12*, 2.

83. Amemiya, T. Regression analysis when the dependent variable is truncated normal. *Econometrica* **1973**, *41*, 997–1016.

84. Ministry for Primary Industries. *Situation and Outlook for Primary Industries*; MPI Policy Publication: Wellington, New Zealand, 2013.

85. Hill, Z.; Lilburne, L.; Guest, P.; Elley, R.; Cuff, J. *Preparation of a GIS Based Land Use Map for the Canterbury Region*; Environment Canterbury Report: R10; Environment Canterbury: Christchurch, New Zealand, 2010; Available online: http://ecan.govt.nz/publications/Reports/gis-based-land-map-canterbury.pdf (accessed on 24 September 2014).

86. Lynn, I.H.; Manderson, A.; Page, M.; Harmsworth, G.; Eyles, G.; Douglas, G.; Mackay, A.D.; Newsome, P.J.F. *Land use Capability Survey Handbook: A New Zealand Handbook for the Classification of Land*, 3rd ed.; AgResearch: Hamilton, New Zealand; Landcare Research: Lincoln, New Zealand; Institute of Geological and Nuclear Sciences: Lower Hutt, New Zealand, 2009.

87. Parliamentary Commission on the Environment. Water Quality in New Zealand: Land-Use and Nutrient Pollution. Available online: http://www.pce.parliament.nz/ publications/all-publications/water-quality-in-new-zealand-land-use-and-nutrient-pollution/ (accessed on 29 September 2014).

88. Environment Canterbury. Hurunui Water Project—Waitohi Proposal. Available online: http://ecan.govt.nz/get-involved/consent-projects/past-notifications/hwp/ Pages/waitohi.aspx (accessed on 29 September 2014).

Sensitivity Analysis of a Land-Use Change Model with and without Agents to Assess Land Abandonment and Long-Term Re-Forestation in a Swiss Mountain Region

Julia Maria Brändle, Gaby Langendijk, Simon Peter, Sibyl Hanna Brunner and Robert Huber

Abstract: Land abandonment and the subsequent re-forestation are important drivers behind the loss of ecosystem services in mountain regions. Agent-based models can help to identify global change impacts on farmland abandonment and can test policy and management options to counteract this development. Realigning the representation of human decision making with time scales of ecological processes such as reforestation presents a major challenge in this context. Models either focus on the agent-specific behavior anchored in the current generation of farmers at the expense of representing longer scale environmental processes or they emphasize the simulation of long-term economic and forest developments where representation of human behavior is simplified in time and space. In this context, we compare the representation of individual and aggregated decision-making in the same model structure and by doing so address some implications of choosing short or long term time horizons in land-use modeling. Based on survey data, we integrate dynamic agents into a comparative static economic sector supply model in a Swiss mountain region. The results from an extensive sensitivity analysis show that this agent-based land-use change model can reproduce observed data correctly and that both model versions are sensitive to the same model parameters. In particular, in both models the specification of opportunity costs determines the extent of production activities and land-use changes by restricting the output space. Our results point out that the agent-based model can capture short and medium term developments in land abandonment better than the aggregated version without losing its sensitivity to important socio-economic drivers. For comparative static approaches, extensive sensitivity analysis with respect to opportunity costs, *i.e.*, the measure of benefits forgone due to alternative uses of labor is essential for the assessment of the impact of climate change on land abandonment and re-forestation in mountain regions.

Reprinted from *Land*. Cite as: Brändle, J.M.; Langendijk, G.; Peter, S.; Brunner, S.H.; Huber, R. Sensitivity Analysis of a Land-Use Change Model with and without Agents to Assess Land Abandonment and Long-Term Re-Forestation in a Swiss Mountain Region. *Land* **2015**, *4*, 475–512.

1. Introduction

Land abandonment and the subsequent re-forestation are key developments with respect to the provision of ecosystem services in European mountain regions [1–5]. Land abandonment is driven by the interaction of environmental and socio-economic factors, such as climate, topography, soil conditions, lack of road-infrastructure development, or degree of part-time farming within a region [6–10]. These interactions result in complex social-ecological systems that can only be investigated by a holistic approach and integrated research [11–13].

Traditionally, land abandonment has often been modeled with comparative sector supply models [14–16]. Land management decisions in these long term modeling studies were usually represented by simplified and uniform mechanisms (e.g., income maximization) on an aggregated level. More recently, agent-based models (ABM) in land-use change [17–20] have been introduced as an opportunity to assess future impacts of land-use change in an integrative framework [21,22]. ABM allow interpretation of agent-specific behavior covering individual preferences or motivations beyond income maximization [23–27] which play an important role in mountain farming [28–34].

Realigning the representation of human decision-making with time scales of ecological processes however presents a major challenge when modeling land abandonment, re-forestation and ecosystem services in mountain regions, especially under climate change [35]. Social-economic behavior, which involves other than purely economic decision-making, is usually based on empirical data from surveys, interviews or role playing games, derived from the existing generation of farmers [36,37]. It therefore has only short and medium term validity. In contrast, reforestation processes and climate change impacts on forests and grassland are only visible in the landscape in the long run, *i.e.*, in several decades [38,39].

Coupled socio-ecological models of land abandonment, such as ABM, therefore often adopt either a short term or a long term perspective. The short term perspective focuses on the agent-specific behavior anchored in the current generation of land users, at the expense of adequately representing longer scale ecological processes. The long-term perspective focuses more on simulating ecological succession, *i.e.*, long term forest development under climate change. By doing so the representation of human behavior is simplified in time and space, also due to large uncertainties about the behavior of the next generation of land users.

Existing ABM studies that address farmland abandonment and that consider individual farm decision-making underline the importance of a spatially explicit examination of the linkage between social behavior and environmental factors, and consequently the dynamic heterogeneity of land abandonment and re-forestation [40–45]. None of these studies, however, explicitly discusses the consequences of implementing a particular representation of decision-making and the associated short or long term perspective into their model structure. With this study, we therefore address the following open research questions: (i) to what extent do different aggregation levels of decision-making, *i.e.*, agent-specific *vs.* sectoral optimization, influence modeling results and (ii) what consequences arise for model-based policy assessment in the context of farmland abandonment and ecosystem services in mountain regions?

To address these questions, we present an extensive sensitivity analysis, *i.e.*, an output space analysis, of two different model versions of the land-use model ALUAM (Alpine Land-Use Allocation Model). The sensitivity analysis is performed without the consideration of agents in a comparative static sector supply model approach based on Briner *et al.* [16] and then compared to a dynamic agent-based version of the same model to assess the different impact of each of these key parameters on the model outcome. We test the importance of exogenous parameters using elementary effects proposed by Morris and a subsequent analysis of a combination of important parameters [46,47]. Our ABM is innovative in that we use a comprehensive coupling of typical farm structures with types of farm decision-making in an economic framework, *i.e.*, based on a constraint income maximization approach. The agent characterization is derived from a cluster analysis based on a survey ($n = 111$) and interviews ($n = 15$) with farmers in the case study region and the model is validated against empirical data.

The study does not intend to solve the problem of decision making processes over multiple generations. The sensitivity analysis, however, provides a quantitative assessment of the role of agents in the context of dynamic and medium term ABM programming models, compared to traditional sector supply modeling approaches in agriculture using a comparative static perspective. This comparison allows us to specify the differences and commonalities between two models that address land abandonment with different time horizons by applying different aggregation levels of human decision-making. The results help to assess and interpret existing [48–51] and future model applications as well as to inform model choice in the context of farmland abandonment and re-forestation in mountain regions.

The manuscript is structured as follows. In Section 2, we present the case study region and describe methods and data sources. In Section 3, we present the results of the agent typology and the implementation of this agent-specific behavior in the existing ALUAM framework. Next, we focus on model performance and validation of the adopted agent-based model and provide the results from an extensive sensitivity analysis with and without agents for changes in prices, direct payments schemes, production costs, labor availability and opportunity costs. In Section 4, we discuss our findings in comparison to existing literature on the assessment of land abandonment and re-forestation in mountain regions.

2. Data and Methods

2.1. Case Study Region

Our study region, the "Visp area", is located in the Central Valais of Switzerland and includes the Saas valley (Saas Fee, Stalden), the region around Visp in the main valley and the Baltschieder valley. It has a total of 15,346 inhabitants and covers an area of 443.3 km². Its main economic characteristics are a century-old, strong industrial sector which is one of the main employers for the whole Upper and Central Valais region, and a marked dependence on snow-based winter tourism in the side-valleys [52]. Unproductive land (*i.e.*, rocky, or glaciated terrain) accounts for 62% of the area, while 20% is covered by forest land and about 16% of the land is used by agriculture (1878 ha). Agriculture and forest land-use play an important role as recreation areas and provide habitats for plants and wildlife. Land-use change is a prominent issue in this region. The importance of agriculture has decreased strongly in the area over recent years, resulting in a decline of agricultural land and an increase of forest cover. Overall, forest land-use increased by 252 ha between 1997 and 2009 [53]. The region comprised 161 farmers in 2012. Between 2000 and 2012, the number of farms decreased annually by 2.8%. On average, farmers in the simulated region currently only cultivate 8 ha of agricultural land and house around seven livestock units (LU). Less than 10% of the farmers work full-time on their farm. The main farming activity is the production of livestock based on grassland. Part-time farming based on small livestock has become a widespread activity and regional tradition, with almost 50% of the farmers (79 out of 161) keeping sheep only. Many farmers are members of organized breeding associations that hold exhibitions, breeding competitions and cow fights. These events are very popular among some of the farmers, inhabitants, and tourists, and root farming firmly into local village traditions. Only 7% of the farms cultivate more than 0.5 ha of arable crops [54]. In this dry, continental inner-Alpine mountain valley region, climate change (rise in temperature, further decrease in precipitation, shifts from snow to rain, and increased glacier melt) is expected to have a particularly strong effect both on ecosystems and tourism [52]. This makes it suitable for studying the combined effects of land-use and climate changes.

2.2. ALUAM

2.2.1. Sector Supply Modeling Approach (ALUAM)

The ALUAM modeling approach has been described and validated in detail in Briner *et al.* [16]. ALUAM is a comparative, static income maximization model which simulates the competition between forest and a range of agricultural land-uses to estimate land-use conversions in a spatially explicit manner at high resolution.

Farmers' decision-making is aggregated on a regional level. Using a modular framework, ALUAM was linked with the forest-landscape model LandClim and a crop yield model that simulate the response of forests and crops to changes in climate. LandClim is a spatially explicit process based model that incorporates competition-driven forest dynamics and landscape-level disturbances to simulate forest dynamics on a landscape scale [55]. The model simulates forest growth in 25 m × 25 m cells using simplified versions of tree recruitment, growth and competition processes that are commonly included in forest gap models. Forest development and ecosystem service indicators can be calculated on the basis of different forest management regimes [5].

An iterative data exchange between the models allows for a detailed assessment of the dynamic changes in land-use and the provision of agriculture and forest based ecosystem services. Land-use and livestock activities on the different levels—parcel, farm and regional—are optimized by a maximization of aggregated land rent. Constraints assure that agronomic and socio-economic restrictions on parcel, farm and sectoral level are met:

(i)   At parcel level, location characteristics influence decisions about the choice of the land-use activities (e.g., extensive or intensive grassland or pasture).

(ii)   Grass must be utilized by livestock to generate value. Decisions about animal husbandry are made on the farm level taking into account fodder and nutrient balances between livestock and land-use. Since different parcels can belong to one livestock activity, single parcels must be summed up to calculate these balances.

(iii)   Resources on a regional level—hirable workforce, number of animals available for grazing on summer pastures, milk quota—are only available to a limited extent and are therefore balanced over the whole region.

The aggregated land rent also considers farmers' opportunity costs to measure benefits foregone due to alternative uses of labor. Working hours are assigned a threshold value. If aggregated land rent from the corresponding land unit drops below this value, the parcel will no longer be cultivated. The model has been applied in various case studies assessing the impact of different climate and socio-economic scenarios [48–51]. In these model applications, a comparative static approach was applied because fixed costs of agricultural production are assumed to be independent from existing structures on a longer term. However, due to its high flexibility for changes between different agricultural activities, this comparative static approach is less suitable to represent short and medium term adjustments to market and policy changes. Abrupt activity switches and corner solutions are typical for this approach [56] and make model validation with short term data challenging. To allow for an ex-post model validation, we used flexibility constraints which restrict the solution space for year to year adjustments in animal numbers. This means that upper and lower bounds constrain the increase or decrease in animal numbers based on the number of animals in the previous simulation year. In addition, investments are also restricted based on the income in the previous year. While these restrictions correspond with farm production cycles and empirical observations, the parameterization of such flexibility constraints is rather subjective [57].

2.2.2. Agent-Based Approach (ALUAM-AB)

The agent-based model is described in Appendix B using the ODD protocol [58,59] to allow for an improved model comprehension [60]. The implementation of the agents in the model is analogous to the protocol presented in Huber *et al.* [45] for the applications in the pasture-woodlands of the Swiss Jura. However, instead of individual farms, we treat different groups of farms as one agent and we couple ALUAM-AB to the forest landscape model LandClim [16,51].

The purpose of ALUAM-AB is to simulate future land-use changes, including farmland abandonment and corresponding re-forestation in mountain landscapes, triggered by the combined effects of climate, market and policy changes and giving due consideration to the individual decision-making of the farmers. The model is defined by interconnected human and environmental/agronomic subsystems. Agents represent groups of farms. An agent has (1) its own state (*i.e.*, land endowment, animal housing capacity, *etc.*) which is updated after every simulation period of one year and (2) decision-making mechanisms for managing farm resources (*i.e.*, a constraint income maximization based on mathematical programming techniques). The state of the agent includes variables for household composition and available resources (land, capital and labor) and a specific type of decision-making based on opportunity costs of labor and a threshold for minimum income and other characteristics. These decision-making characteristics represent the model implementation of the actor types detailed under Section 3.1. The environmental/agronomic subsystem is characterized by the agricultural production cycle in the case study area. Agronomic variables include plant nutrient requirements (N, P), manure production and production coefficients such as fodder intake, growth, birth, deaths of animals and labor requirements that are based on national average data and are the same as in the aggregated model presented by Briner *et al.* [16]. In the modeled farm decision process (income optimization), the environmental variables are considered as material (fodder and nutrients) balances that link land-use activities with livestock activities. As a result, land-use intensities can be defined in a spatially explicit manner. Crop rotation requirements and a labor balance are additional constraints that link the human and environmental/agronomic subsystems.

Structural change is modeled using a land market sub-model [45,61]. The model identifies land units that are no longer cultivated under the existing farm structure.

There are three reasons why fields are attributed to the land market in the model (see Figure A1): (i) units generate a land rent below zero, (ii) the corresponding agent does not reach the minimum wage level, therefore the farm is abandoned and all the assigned land enters the land market or (iii) the farmer retires in the simulation year and has no successor. The land market sub-model randomly assigns the land units to one of the other agents. It is then checked to confirm that this agent shows the two following characteristics: The agent receiving the land unit must want to expand his cultivated area (stated willingness to grow) and his shadow price for the land unit must be positive. If these conditions are not met, the land unit is returned to the land market and assigned randomly to another farm. Once again it is checked to verify that this agent fulfils the conditions for the assignment of land. This procedure is repeated until all land units are assigned to a farm or none of the farms is willing to take the land units left on the market. Land units that are not transferred to other farms are defined as abandoned and natural vegetation dynamics get under way on these units (modeled in LandClim). If land-use allocation is optimal, farmland capacities and livestock are updated and the next annual time step is initialized using the parameters (prices, costs) of the following year.

There are two main differences between the model versions presented in Sections 2.2.1 and 2.2.2:

(i)   In the aggregated model, land and labor can shift between farm activities without additional restrictions. Livestock housing capacities are built in every model run. For model validation, however, flexibility constraints are necessary. In ALUAM-AB, changes in land-based activities are only possible through the land market.

(ii) The agents in ALUAM-AB differ with respect to their opportunity costs, availability of workforce, minimum income, the probability of a successor, their stated intention to grow or not, available farm land and livestock housing capacities. In the aggregated model version, all the activities are weighted with the same amount of opportunity costs and hired labor is restricted on the regional level. There is no interaction between different farm types.

Please note that the sensitivity analysis focuses on the land-use part and thus on the effect of different aggregation levels of decision-making in ALUAM. Forest development is modeled in the forest landscape model LandClim. The two models can be linked to assess the development of agriculture and forest ES in mountain regions under land-use and climate change. As we focus here on the effect of different representations of human decision-making on model performance, we leave the assessment of changes in re-forestation and corresponding changes in ecosystem services provision to future research.

2.3. Developing the Agent Typology

Agent typologies for AMBs should be appropriate to the modeling purpose [37], and reflect the main characteristics of the decision types under study in a parsimonious manner. Policy relevance can be increased if the typology is related or embedded in available farm census data and observed land-use choices [62]. Empirical research increasingly highlights the importance of considering multiple objectives [28,63–65] as well as attitudes and preferences towards more nature-friendly farming when representing farmers' decision-making [62,65–67]. In addition, farm diversification and associated constraints on labor availability (and other aspects of part-time farming) are thought to strongly affect farming system development and decisions on land abandonment, particularly in mountain regions, and have been highlighted as important elements in recent farmer typologies [68,69]. Historical accounts of land abandonment in the study region and interviews with farmers and the agricultural extension office confirm the importance of this aspect and of considering socio-economic factors alongside environmental (*i.e.*, parcel) characteristics when assessing land abandonment and reforestation. Therefore we based our agent decision typology on three main aspects: Objectives for farming, attitudes towards extensive land-uses, and attitudes towards taking on off-farm employment.

2.3.1. Farm Household Survey

From October 2011 to January 2012 we conducted both a mail survey and 15 semi-structured interviews to collect data on (i) farming objectives, (ii) farmer attitudes towards off-farm labor, and extensive land-uses, (iii) management intentions, and (iv) farm structural characteristics. The mail survey was sent out in November 2011 to all farmers registered in the livestock census of the municipalities within the modeled region and also the adjacent Matter valley to allow for a larger sample size. Of the 121 questionnaires returned (response rate 25%), 119 contained full decision-making information. Data on farming objectives and attitudes was collected on five-point Likert scales. The survey data was subsequently linked to agricultural farm census data. This enabled a cross-validation of information on livestock types, livestock numbers and farmed area, and also provided additional parcel-level data on land use, land-use intensities and enrollment in agri-environmental compensation schemes. After excluding survey responses where the census data indicated a farming area of 0 (bee keepers, retirees), 111 cases were retained for further analyses.

2.3.2. Actor Typology and Translation into Model Agent Types

Methodologically, the agent typology generation followed four steps: Firstly, we performed a principal component analysis (PCA) with a quartimax rotation on 19 questionnaire items relating to farming objectives and attitudes. The PCA served to condense the information in the data to a lower number of dimensions and to generate uncorrelated components for subsequent cluster analysis. The number of components retained was determined by analyzing the scree-plot, Very Simple Structure statistics (VSS) and the total explained variance. Respondents' scores on each component were computed directly by regression using the principal function of the "psych package" of the R statistical computing environment [70]. Secondly, PCA regression coefficients were clustered by applying k-means clustering. Silhouette statistics were employed to select the number of clusters for further analysis. Thirdly, the typology was refined by describing the resulting clusters with respect to additional survey data on farm structure and management (labor use, household income, age, intentions for future management) and farm census data (land use, livestock types, parcel characteristics, participation in agri-environment schemes). Fourthly, the characteristics of the actor types were translated into model agents, including modeling constraints and guidelines for initial allocation of the decision-making types to model agents.

2.4. Model Validation and Sensitivity

Validation of ABM is a demanding task due to the theoretical as well as empirical challenges involved [71,72]. There are different methods of validating ABM such as comparison to real world data [73], an indirect approach [71], role playing games [74] or extensive sensitivity analysis [22,75,76]. The present study adopts a stepwise sensitivity analysis of model performance. Firstly, we use error decomposition as proposed by Sterman [77] to assess the best-performing model outcome of the agent-based model version and we compare it with modeled values of the aggregated model version as well as observed values of the number of animals and land-use in the case study region. Secondly, we use elementary effects (EE) defined by Morris [47] to determine the most important exogenous factors affecting model outcome and compare these EE's between the two model versions. Thirdly, we test the impact of different policy measures on each model outcome.

2.4.1. Error Decomposition in Single Best-Performance

We perform a behavior reproduction test to assess the model's ability to reproduce the behavior of observed data in our case study region. To achieve this, we describe the error between observed data and simulation output, measured point by point for each simulation run, and provide a decomposition of the error using the Theil inequality statistics [77]. The root mean square percentage error (RMSPE) represents the mean percentage difference between simulation and observed data with n as the number of observations, x_m as the simulation output and x_o as the values of the observed data.

$$RMSPE = \sqrt{\frac{1}{n} \sum \left(\frac{x_m - x_o}{x_o} \right)^2}$$ (1)

The Theil inequality statistics allow this error to be decomposed into three components, so-called bias (U^M), unequal variation (U^S) and unequal covariation (U^C) based on the mean square error (MSE), see Equations (2) and (3), with $\bar{x}$ as the mean value and s as the standard deviation. A bias arises when simulation output and observed data have different means. A large bias refers to a systematic error which should be corrected by adjusting parameters. Unequal variation implies that the variance of the two series differ, *i.e.*, the model and the observed data have different trends (or amplitude fluctuations). Unequal covariation (with r = correlation coefficient) indicates that model and data are imperfectly correlated, *i.e.*, they differ point by point but may have the same mean and trend. The sum of the three components is 1. Thus, the inequality statistics provide an easy interpretation breakdown of the sources of error.

$$MSE = \frac{1}{n} \sum (x_m - x_o)^2$$ (2)

$$U^M = \frac{(\bar{x}_m - \bar{x}_o)}{MSE} \; ; \; U^S = \frac{(s_m^2 - s_o^2)}{MSE} \; ; \; U^C = \frac{2(1-r)s_m s_o}{MSE}$$ (3)

For model calibration, we use census data from the Federal Office of Agriculture containing livestock housing capacities and numbers of farms as well as managed land, farmer age, livestock numbers and land in slope categories for each farm type in the year 2000 [78]. Model validation uses the development in exogenous input parameters, *i.e.*, prices, costs and direct payments between the years 2001 and 2012 to test model behavior (see Table A1). The modeling results with respect to the number of animals (cattle and sheep) and land-use intensities (area of intensive and extensive land-uses) are then compared to the development of these parameters in the census data to assess the single best performance of the model (validation). To compare the different grazing animals, we use livestock units (LU) which represents a nutritional equivalent between sheep (0.17 LU), dairy cows (1 LU), suckler cows (0.8 LU), calves and heifers (0.4 LU). The total area of extensive grassland and total areas of intensive land-uses serve as indicators for land-use intensities. Extensive land-use covers the land-uses entitled to ecological compensation payments in Switzerland, namely extensively managed hay meadows, less intensively managed meadows and extensive pastures. Extensively managed meadows and pastures can only be cut or grazed after the 15th of July. Only two cuts or grazing rotations are permitted and no fertilizers are allowed on meadows.

2.4.2. Elementary Effects

The purpose of the concept of elementary effects is to determine those model factors that have an important impact on a specific output variable and can be understood as the change in an output y induced by a relative change in an input x_i, e.g., the impact of the milk price on land rent or the number of animals in the simulation results.

$$d_i(X) = \left(\frac{y(X_1, \ldots X_{i-1}, X_i + \Delta, X_{i+1}, \ldots X_k) - y(X)}{\Delta} \right) \tag{4}$$

In Equation (4), X is a vector containing k inputs or factors $(x_1, \ldots, x_i, \ldots, x_k)$ in producing the output y. A factor x_i can take a value in an equal interval set. The symbol Δ denotes a predetermined increment of a factor x_i. The number of levels chosen for each factor can be denoted with p. In the set of real numbers, x_{i1} and x_{ip} are the minimum and maximum values of the uncertainty range of factor x_i, respectively. Each element of vector X is assigned a rational number or a natural integer number. The frequency distribution F_i of elementary effects for each factor x_i gives an indication of the degree and nature of the influence of that factor on the specified output. For instance, a combination of a relatively small mean μ_i with a small standard deviation σ_i indicates that input x_i has a negligible effect on the output. A large mean μ_i and a large standard deviation σ_i indicate a strong non-linear effect or strong interaction with other inputs. A large mean μ_i and a small standard deviation σ_i indicate a strong linear and additive effect.

We calculate the EE for the aggregated land rent, i.e., the objective function of ALUAM, and of the agents in ALUAM AB four the exogenous parameters presented in Table 1, i.e., prices, costs, direct payments and agent characteristics. In addition, we also provide the EE for the number of animals since this output is highly correlated to ecosystem services provision in our case study region [49].

2.4.3. Sensitivity to Changes in Direct Payments

Various additional techniques are available [80] to capture the sensitivity of the model. These often involve a specific experimental design or sampling strategy to reduce the number of model evaluations necessary [22]. EE trajectories are only viewed as a good way of screening single factors in sensitivity analysis but do not inform on effects of factor combinations on modeling outcomes. To further test the sensitivity of ALUAM with respect to policy measures that counteract land abandonment, we combine the most important factor identified in the EE trajectories with different levels of direct payments. Direct payments are the most important policy measure in Swiss agricultural policy to support mountain farming. In 2014, Switzerland enacted a new direct payment system [81]. In this context, payments for animals, *i.e.*, a fixed payment for grazing animals (RFB payments) and animals kept under difficult production conditions in upland and mountain areas (TEP payments) were abolished. Direct payments per hectare (area payments) were assigned to specific objectives such as food security, biodiversity or landscape maintenance. Thus, we extended the sensitivity analysis by running both model versions, with and without animal related direct payments, to assess non-linearities and interactions between policy measures and model behavior.

Table 1. Exogenous input parameters for sensitivity analysis.

Parameters (k); $p = 21$	Sub-Categories	Unit	Absolute Change (Δ)	Min. Values (x_{i1})	Max. Values (x_{ip})
Prices					
Milk price	-		0.085	0	1.70
Lamb price	-	CHF/kg	22	0	443
Beef price	-		232	0	4650
Costs					
Variable costs machines	-		0.095	0.1	1.9
Fixed costs machines	-	CHF in %	0.095	0.1	1.9
Price of diesel fuel	-		0.14	0.1	2.7
Direct payments (DP)					
General DP	-		114	1	2280
Ecological compensation areas [1]	Production zone [2]	CHF/ha	43–143	1	855–2850
DP slope	Slope categories [3]		35–48	1	703–970
Animal RFB payments [4]		CHF/per	86	1	1710
Animal TEP payments [5]		head	92	1	1843
Agent characteristics					
Available family labor		% of 2800 h	0.095	0.1	1.9
Opportunity costs		CHF/hour	0.95	0	19

[1] Ecological compensation areas: Extensive meadowland (not more than one cut and no fertilizers), less intensive meadow-land, extensive pastures (only one rotation in autumn); [2] Administrative zone according to the Federal Office of Agriculture [79]: Valley bottom, hillside; mountain regions I–IV depending on climate conditions, road infrastructure and share of steep agricultural land; [3] Administrative category Slope <18%: 0 CHF; 18%–35%: 370 CHF per ha; <35%: 510 CHF per ha; [4] RFB: Payment per roughage livestock unit, *i.e.*, beef cattle 900 CHF per LU; dairy cows and sheep 400 CHF per LU; [5] TEP: Payment per livestock unit in remote areas, *i.e.*, 970 CHF per LU in mountain production zones.

45

3. Results

3.1. Agent Typology

The PCA yielded six components capturing 71% of the variance (see description and Table A1). The cluster analysis of the regression based principal component scores identified five different farming types. Figure A1 shows the relationship between the factor scores (calculated from items with PCA loadings > 0.5) and the five actor types. The five types locate farmers on a gradual scale between more production-oriented full-time and leisure-oriented farming, and varying dependencies on off-farm work and income opportunities. In the following, we briefly describe the actor types including the most important results of the cross-tabulation with farm structure and census data as presented in Table 2.

Type 1: Production-oriented farmers

This type of farmer attaches great importance to generating an adequate income, high yields and innovative products from their farming activities. They tend to be less involved in local traditions, breeding competitions, or providing ecosystem or landscape services. With a few exceptions, farming is their primary source of income and most or all available labor is devoted to farming. Many also have access to hired labor. Opportunity costs are low, as they farm largely independent from work commitments outside of agriculture. Average farm size for this type is significantly higher than for the other types, both with respect to area farmed and livestock kept. The farming systems are mostly specialized, consisting of larger dairy, beef/suckling cattle, mixed or commercial sheep enterprises. Overall, however, the proportion of small livestock is low. Production-oriented farmers regard the financial and ecological benefits of extensive land-uses and the provision of ecosystem services as considerably less attractive than the other farming types. They have the lowest share of extensive land-use which is consistent with their attitudes and production-orientation. On average, they farm significantly higher quality farm land both with respect to slope as well as agricultural production zones.

Table 2. Key characteristics of farmers, farm structure and land use for the five actor types. Numbers in brackets refer to median and standard deviation respectively.

		Full-Time Farmers		Part-Time Farmers		Leisure Farmers
		Production-Oriented Farmers (*n* = 16)	Ecological and Landscape Stewards (*n* = 19)	Part-Time/Leisure-Oriented Breeders (*n* = 30)	Traditionalist Leisure Farmers (*n* = 17)	Leisure-Oriented Farmers (*n* = 29)
Total managed land in cluster	ha	365.8	275.9	274.2	130.2	200
Farmer's age	y	46 (48; 9.53)	45 (47; 7.58)	50 (52.0; 8.36)	46(47; 10.3)	47 (47; 8.19)
Household income	kCHF	60 (55; 27)	82 (85; 30)	66 (55; 22)	68 (55; 23)	76 (85; 16)
Household income from agriculture	%	52 (70; 37)	35 (30; 32)	19 (10; 17)	17 (10; 20)	13 (10; 8)
Labor hours farm manager	h/day	6.5 (7; 2.82)	4.6 (3; 3.26)	4.5 (5; 2.11)	3.7 (3; 1.76)	2.8 (3; 0.89)
Additional available labour (family or hired)	h/day	10.2 (2.9; 19.54)	7.4 (5; 7.91)	5.4 (3.6; 7.38)	2.9 (2.5; 3.11)	4.9 (3; 4.95)
Managed agricultural land	ha	22.9 (18.6; 17.82)	14.5 (8.8; 13.40)	9.1 (7.8; 5.12)	7.7 (6.4; 2.43)	6.9 (6.5; 2.89)
Total livestock	LU	25.7 (14.8; 32.50)	16 (9.8; 15.85)	8.5 (7.2; 5.52)	6.4 (5.2; 3.15)	6.4 (5.9; 3.63)
of which small livestock	LU	5.2 (4.3; 5.10)	3.5 (2.7; 3.92)	4.9 (4.3; 4.19)	4.1 (4.1; 3.45)	4.1 (4.3; 4.14)
of which large livestock	LU	20.5 (3.6; 35.05)	12.5 (4.4; 16.56)	3.5 (0; 5.91)	2.3 (0; 3.24)	2.3 (1.2; 2.26)
of which dairy cows	LU	14.7 (0; 22.26)	2.1 (0; 4.99)	1.4 (0; 2.91)	0.5 (0; 1.5)	0.8 (0; 1.55)
Small livestock	%	20	22	58	64	65
Land in severely disadvantaged production zone 54	%	52	83	72	63	58
Land in production zone 53	%	15	9	26	32	35
Land in best production zone (hill zone 41)	%	29	6	1	0	2
Steep land (> 18°)	%	51	74	86	78	82
Extensive grassland and pastures	%	20	38	28	30	31

Type 2: Ecological and landscape stewards

Farmers in this cluster place a stronger emphasis on the social, ecological and landscape aspects of their farming activities than on the achievement of high yields or profits. They consider extensive land-use and the provision of ecological services to be both an adequate source of income and an effective measure to increase biodiversity. Farmers of this type engage mainly in medium sized suckling cow/beef, mixed, or horse enterprises or small to medium scale sheep and goat farming with an increasing focus of their farming activities towards ecological direct payments. While suckling cow and beef enterprises derive the bulk of their household income from agriculture, the average share of agricultural income in the overall cluster amounts to 35%. On average, farmers of this type devote about 4.6 h per day to farm work, with some variation between farms keeping large or small livestock. Perceived dependence on off-farm labor and income however is low, indicating a certain amount of flexibility in labor use due perhaps to extensification of production. Some of the farmers in this group have access to additional hired labor, and all of them to family labor. On average, this provides them with an additional 7.4 h per day of help on the farm. This cluster exhibits the highest proportion of extensive land uses among the five farming types which is consistent with the stronger ecological orientation of these farmers.

Type 3: Part-time or leisure-oriented breeders

Farmers in this cluster share a strong interest in being recognized as "good" farmers or breeders within their respective (farming) communities and like to share their farming passion by participating in exhibitions, competitions, or cow fights. By engaging in these activities, they also aim to maintain local traditions and contribute to village life. They derive their main income off-farm and devote on average about 4.5 h a day to farming. The stronger off-farm engagement of the farm manager is also reflected in higher perceived opportunity costs compared to the "Ecological and landscape stewards". Most farmers in this group can count on additional family labor of on average 5.4 h per day (median 3.6 h). On average, this farm type houses 8.45 LU and manages an area of 9 ha. In addition to many small to medium scale enterprises focused on breeding small livestock, this cluster also includes farmers who keep low numbers of a specialized cattle breed often used for fighting and small to medium scale suckling/beef enterprises. On average, the proportion of steep land is highest in this cluster.

Type 4: Traditionalist leisure farmers

Farmers of this type undertake small-scale farming as a way to maintain local traditions. Compared to type 3, they do not aim for such a strong involvement in breeding, competitions and local decision-making, and perceive their opportunity costs to be much higher. All of the farmers are employed outside of agriculture and their farming activities depend strongly on off-farm work commitments and income. Therefore, labor invested in agricultural activities is low, as is the share of household income derived from agriculture. Of the 6.4 livestock units housed on average, the overall proportion of small livestock is high (64%). The main farming activities include sheep farming, horses and low numbers of suckling or dairy cows. Farm sizes are among the smallest in the survey.

Type 5: Leisure-oriented farmers

Farmers in this group place a high importance mainly on being involved in local decisions and village life. They are significantly less focused on achieving high income and yields than the other clusters but do not place a strong focus on ecological or competition objectives either. All of the farmers work outside of agriculture and, with an average of 2.8 h, labor invested in agricultural activities is very low. It is however complemented by a few hours of additional family labor (4.9 on average a day, median 3 h). Agriculture only contributes 13% to the total household income. Perceived opportunity costs are midway between the two other leisure-oriented farming types. Farms are small and the majority of the leisure farmers keep sheep only, occasionally mixed with low numbers of suckling cows or beef cattle. On the few farms which keep large livestock, the workload is carried by family members rather than by the farm manager himself.

Table 3 shows how the actor characteristics are implemented into agent types in ALUAM-AB. For the full parameterization of these characteristics we refer to Table A2. Consistent with the actor typology, opportunity-cost levels were introduced as a main proxy to reflect non-economic objectives and attitudes in our income optimization model. The level of opportunity costs represent a measure of benefits forgone due to alternative uses of labor. Each agent type is assigned a specific threshold level as a percentage of a fixed monetary value *i.e.*, the opportunity costs in the aggregated model version. Low opportunity costs imply that farm activities are maintained even though the income generated by these activities is low.

Another important restriction for part-time and leisure-oriented farmers is the available work force. Additional work force, other than the family labor available, can only be hired by production-oriented farmers. In addition, farm growth in the model is only possible for the "Production-oriented farmers" and the "Part-time and leisure breeders" as well as "Ecological and landscape Stewards" since these agents are either more production-oriented or the survey has shown that they are more interested in farm growth. An agent type specific minimum income threshold was introduced as an additional proxy for non-economic farming objectives. In the optimization process, farms exit if they fail to achieve this minimum income threshold. For leisure-oriented farm types however this threshold level is set very low. The succession rate defines the probability that the farm will be taken over when the farmer retires (at the age of 65) and was derived from the farm survey and interviews. Farm extensification describes a maximum level of extensive meadows and pastures on the corresponding farm type. Parameters for farm and livestock housing capacities are derived from census data. Finally, the agents are assigned different production system flexibility, based on their stated preferences for specific farm activities in the survey and the interviews. "Ecological and landscape stewards" and "Traditionalist leisure farmers" can switch between cattle and sheep production.

The other farm types, which are currently specialized, may invest in new fixed assets, *i.e.*, farm buildings but cannot switch their production system. Changes in farm activities are further mediated through the land market module.

Table 3. Translation of empirical farm type characteristics into parameter levels for implementation into the agent-based model.

| | Full-Time Farmers | Part-Time Farmers | | Leisure Farmers | |
	1. Production Oriented Farmers	2. Ecological and Landscape Stewards	3. Part-Time/ Leisure Breeders	4. Traditionalist Leisure Farmers	5. Leisure Farmers
Opportunity costs	Low	Low	Medium	High	Low
Available family labor	High	Medium	Medium	Low	Low
Farm growth possible	Yes	Yes/No	Yes	No	No
Additional hired workforce	Yes	No	No	No	No
Minimum income	High	Medium	Medium	Low	Zero
Succession rate %	High	Medium	Medium	Low	Low
Extensification	Low	High	Low	High	High
Farm size	High	Medium	Low	Low	Low
Livestock housing capacity	High	Medium	Small	Small	Small
Production system flexibility	Specialized	Mixed	Specialized	Mixed	Specialized

3.2. Model Validation: Best-Performing Simulation Output

Table 4 shows the results from the error decomposition to assess the single best output performance of ALUAM-AB with respect to the total number of animals measured in livestock units (LU), the number of sheep and cattle and the aggregated areas of intensive and extensive land use. To summarize, the overall errors of the model performance and the unequal variation error are small, and thus the model captures the mean and trends of the observed data satisfactorily. The mean percentage error of the simulation with respect to these output variables ranges between 1.5% for the number of sheep and 10.9% for the total amount of extensive land use.

Table 4. Error decomposition in the single best-performing output of ALUAM-AB.

	Unit	RMSPE %	Bias (U^M)	Unequal Variation (U^S)	Unequal Covariation (U^C)
Animal production unit	LU	0.035	0.808	0.042	0.150
Sheep	Nr.	0.015	0.003	0.000	0.997
Cattle	Nr.	0.082	0.821	0.059	0.120
Land-use (intensive)	ha	0.057	0.810	0.002	0.188
Land-use (extensive)	ha	0.109	0.092	0.000	0.908

The remaining error in the case of sheep production can be attributed to an unequal covariation, *i.e.*, the simulation shows small lags in the reproduction of observed data (see also Figure 1a). In contrast, the mean errors in total amount of animals (3.5%), cattle (8.5%) and intensive land use (5.7%) are associated with bias. The simulation results for the total amount of cattle and intensive land use are consistently lower than the actual number of dairy cows, sucklers and beef cattle (see Figure 1b) and the total amount of intensive grassland in the case study region (see Figure 1c), *i.e.*, there is a systematic error between simulation results and observed data.

This bias is associated with the aggregation of agents' resources, such as livestock housing capacities and workload, as well as fixed assumptions concerning technical parameters, such as nutrient requirements or mechanization. These assumptions are inevitable and could only be replaced by a data intensive expansion of model parameters to smooth the linear production functions in the model, *i.e.*, by adding more production activities and sub-types of these activities. The unequal variation error for these output categories however, is small and thus no deviation from the trend could be detected.

The largest gap between model and observed data can be found for the aggregated area of extensive land use (see Figure 1d). The error can be attributed to the unequal covariation between simulation results and observed data indicating that the error is unsystematic. The model may not be able to fully capture the changes in the amount of extensive land use. In general, however, there is no systematic deviation from the trend. With respect to land abandonment, a year by year comparison is not possible since observed data on forest regrowth is only available for the whole period (+252 ha of forest). Compared to the initial distribution of parcels, ALUAM-AB abandoned 227 parcels (or ha). Thus, land abandonment is slightly underestimated in our approach.

Figure 1 also illustrates the differences between the sector supply model ALUAM and the agent-based model ALUAM-AB in the single best performing output. Without a specification of the agents, lamb production is not profitable and the number of sheep is continuously decreasing.

The same development can be observed for cattle between the years 2001 and 2008. The increase in prices in 2009 leads to a reversal of this trend. Due to the flexibility constraints, however, the increase is restricted to 10% of the number of cattle in the previous year. For intensive grassland use the aggregated model performs similar to the agent-based model version. In contrast, the amount of extensively used grassland is much lower in ALUAM. More land is abandoned which does not correspond well with the observed data. Overall, the agent-based model shows a better validation to observed data than the sector supply model ALUAM.

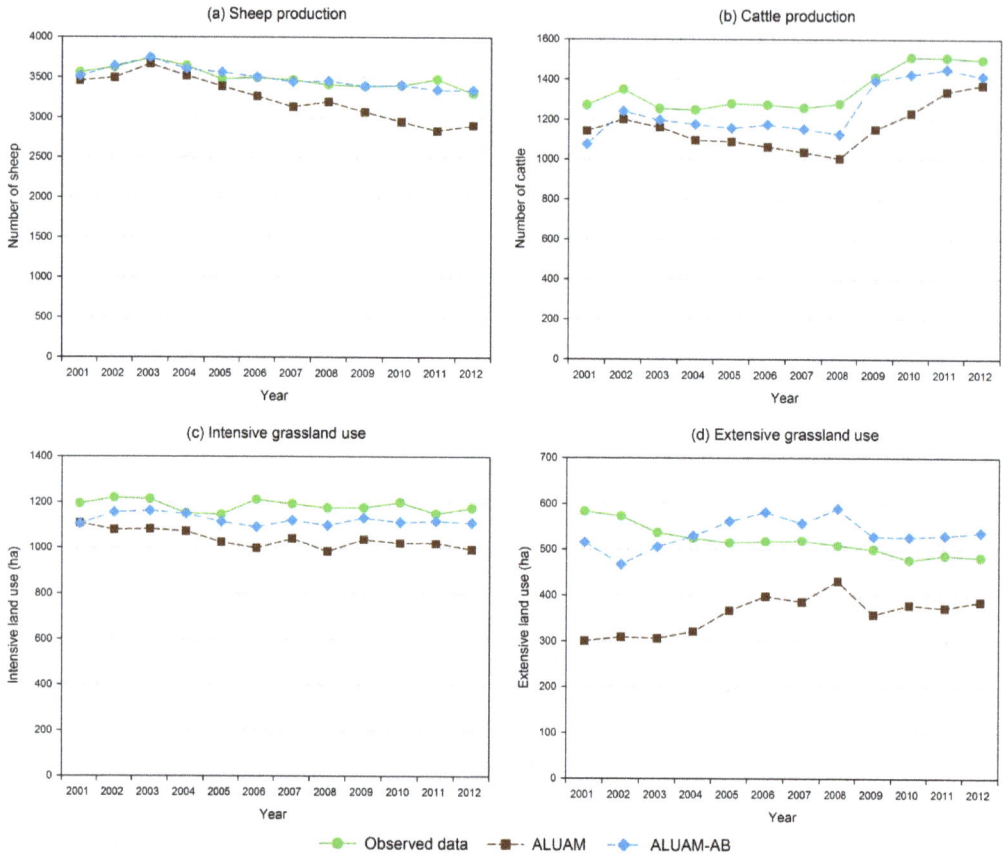

Figure 1. Best-performing model outcome comparing simulation data from ALUAM and ALUAM-AB with observed changes between 2001 and 2012 in animal production (sheep and cattle) and land-use intensities. (**a**) Sheep production; (**b**) cattle production; (**c**) intensive grassland use; (**d**) extensive grassland use.

3.3. Elementary Effects

Figure 2 visualizes the mean and standard deviation of the elementary effects of the 13 exogenous parameters on land rent (the objective function of the model) and the number of animals in both model versions, *i.e.*, with and without agents ($n = 520$ model runs). A detailed overview of EE effects for all parameters is provided in the Appendix (Table A3). Figure 2 shows that the same four parameters emerge as the main exogenous drivers in both model versions: Opportunity costs of labor, milk and lamb prices and the price for energy (fuel).

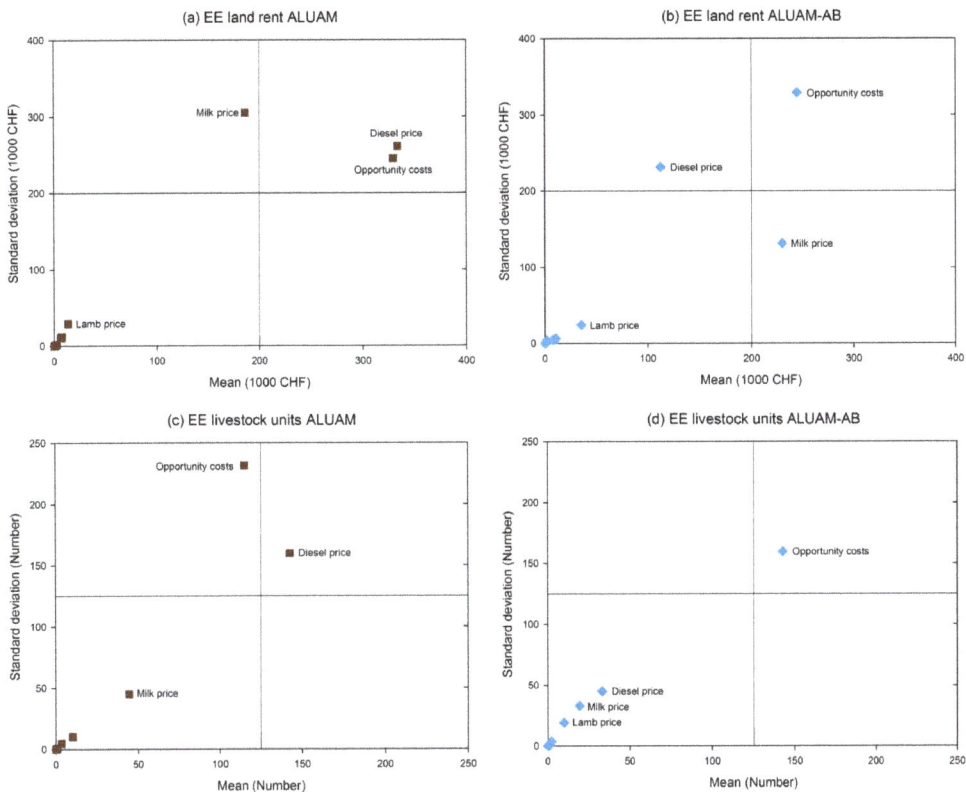

Figure 2. Elementary effects (*i.e.*, mean and standard deviation) of land rent and livestock units in the two model approaches for exogenous input factors. (**a**) EE land rent ALUAM; (**b**) EE land rent ALUAM-AB; (**c**) EE livestock units ALUAM; (**d**) EE livestock units ALUAM-AB.

Compared to these four main drivers, both the mean and standard deviations of other parameters are relatively small, indicating that individual changes in these parameters result in a negligible effect on model outcome (all else being equal). In the sector supply model ALUAM, the mean and standard deviations are large for the impact of diesel price, opportunity costs and milk price on the aggregated land rent (Figure 2a). Simulations imply that the milk price results in the highest variability with respect to the objective function of the model. The impact of the milk price on the number of livestock units (Figure 2c) is much smaller since the model can switch its activities, *i.e.*, from dairy cows to beef and breeding cattle or to sheep production, which overall compensates for the reduction in dairy cows. Such substitution effects are smaller for diesel price and opportunity costs which also have a high impact on the number of livestock. The lamb price has only a small impact on land rent and the number of livestock units in the aggregated model.

In the agent-based model ALUAM-AB, opportunity costs have the highest impact on land rent with respect to mean and standard deviation (Figure 2b). Milk price has a large impact on the mean, but exhibits a much lower variability compared to the sector supply model. The importance of the fuel price decreases in that it has a lower effect on the variability of the outcome compared to ALUAM. With respect to livestock units (Figure 2d), the results show that only opportunity costs have a large impact on mean and standard deviation. The influence of other exogenous inputs is reduced. This exemplifies the reduced flexibility in the agent-based model: Since farm types cannot switch to alternative farm activities, the impact of price and costs on the number of livestock units is small while the effect on the land rent, *i.e.*, their agricultural income, is still high. The extent of this effect depends on the profitability of the corresponding farm activity.

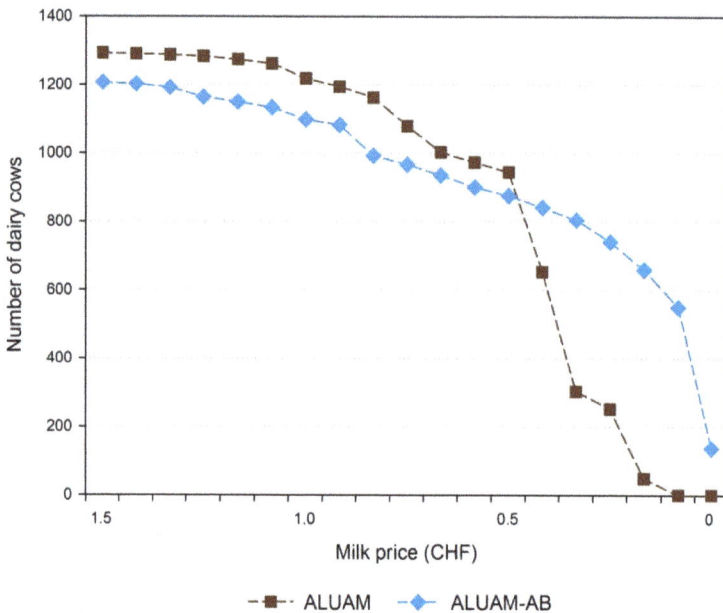

Figure 3. Comparison of changes in the number of dairy cows with one at a time changes in the milk price between 0 and 1.5 CHF in both model approaches.

The higher the profitability, the larger the impact it has on the objective function. Since the productivity of sheep rearing is low, changes in lamb prices have a much lower overall effect on land rent than changes in milk price and diesel costs. Thus, in contrast to the sector supply model, farmers in the agent-based model continue to produce even if prices vary strongly from one year to the next.

This effect is also illustrated in Figure 3 which shows the changes in the number of dairy cows with a one at a time decrease in milk price. In the aggregated model

ALUAM, the number of dairy cows falls drastically if the milk price drops below 0.4 CHF. In contrast, agents in ALUAM-AB continue to produce milk due to structural restrictions (sunk costs in livestock housing capacities and availability of farmland through land market) and farm type characteristics (opportunity costs, intentions to grow and minimum income). This model behavior smoothes the adaptation of farm activities to socio-economic drivers and allows for a more subtle representation of farm structural changes consistent with real world observations (see Figure 1).

3.4. Sensitivity to Changes in Direct Payments

Figure 4 shows the interaction between the three levels of opportunity costs (10, 20 and 30 CHF) and the impact of two different direct payments schemes, *i.e.*, with and without payments per livestock unit. Please note that these levels of opportunity costs are multiplied with the agent-specific levels of opportunity costs (low, medium, high) in the agent-based model (see Tables 3 and A2). The figure directly compares the output from the two model versions with respect to the number of cattle and sheep as well as the amount of intensively and extensively used grassland. The simulation results of the sector supply and the agent-based model are represented with the blue and the brown bars respectively. In general, livestock and intensively used grassland areas in the aggregated model ALUAM are lower compared to the agent-based version ALUAM-AB and the reaction to changes in the direct payments is more pronounced. This is illustrated in the four diagrams:

(1)   Figure 4a shows that the resulting number of cattle is, in general, higher in the agent-based model than in the sector supply model. The only exception is the basic model run with all direct payments and the lowest level of opportunity costs (10 CHF) where the outputs from ALUAM and ALUAM AB show similar numbers. This exception can be explained by the fact that both models were calibrated to this basic combination of input factors. However, with increasing opportunity costs, the number of cattle decreases in the sector supply model irrespectively of direct payments (e.g., by 28% from 758 to 540 livestock units in the simulation runs with direct payments) whereas in the agent-based model opportunity costs have a smaller impact on the number of cattle. Although benefits foregone due to alternative uses of labor increase, cattle numbers remain stable or even increase slightly (e.g., by 3.6% from 611 to 633 livestock units in the simulation runs without area payments).

(2)   With respect to sheep, Figure 4b shows a different simulation behavior of the agent-based model. As in the sector supply model, the number of sheep decreases with increasing opportunity costs (e.g., by 46% from 609 to 328 livestock units in the simulation runs with direct payments). The two models also respond in a similar direction for both direct payment schemes. The abolishment of payments for animals leads to a decrease of sheep in both model versions. In

the sector supply model, the number of sheep even falls to the minimum level, *i.e.*, only production-oriented farmers still produce lamb.

(3) The same pattern can also be observed for the amount of intensively used grassland. An increase in opportunity costs generally leads to a decrease in intensive meadows and pastures in both model versions (e.g., by 40% from 1048 to 632 hectares in the ALUAM simulation runs with direct payments). The discontinuation of payments for animals leads to a decrease in intensively used grassland in both model versions.

(4) The change in the amount of extensively used grassland presented in Figure 4d reflects the opposite pattern of intensively used grassland. In the base simulations with direct payments, the amount of extensively used grassland increases with higher opportunity costs. Without payments for animals, the amount of extensive grassland reaches a threshold level, *i.e.*, a corner solution in both simulation models. The amount of extensively used grassland does not exceed a level of 700 and 1000 ha in ALUAM and ALUAM-AB respectively.

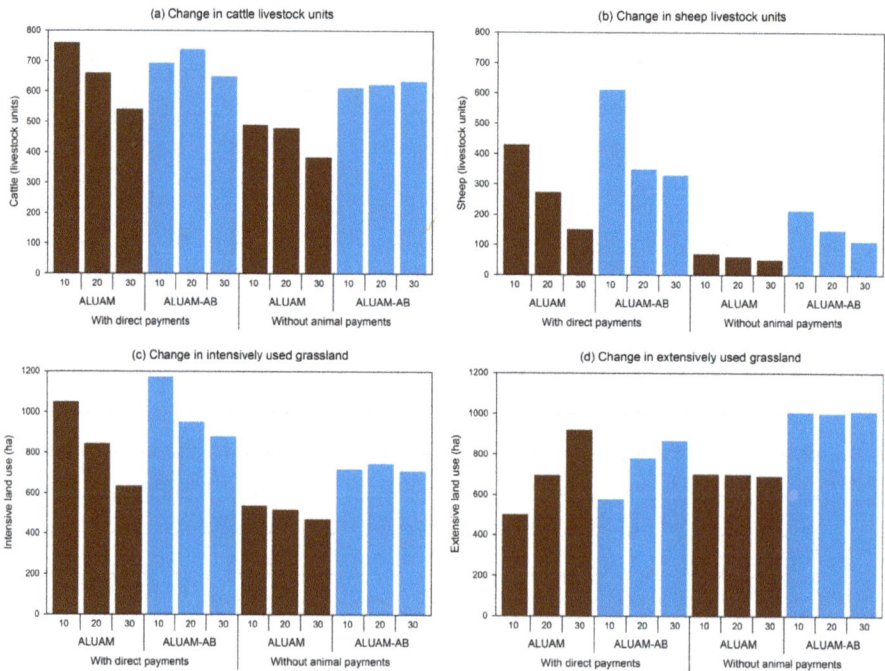

Figure 4. Sensitivity of model outputs ((a) cattle, (b) sheep, and (c,d) land-use intensities) to the abolition of animal based direct payments with three levels of opportunity costs. ALUAM = Sector supply model; ALUAM-AB = Agent-based ALUAM; 10,20,30 = level of opportunity costs in CHF.

The extent of land abandonment can be calculated by adding up areas of intensive and extensive land-use. Without animal based payments, the agricultural surface decreases by 25% in ALUAM whereas simulation results imply land abandonment of 2% in the ALUAM-AB results.

In conclusion, the simulation results presented in Section 3.3 illustrate that although the same exogenous inputs drive the outcome of both models, the interaction of policy measures and opportunity costs strongly influences simulation results. Analogous to the arguments discussed in Section 3.3, these effects can be attributed to the integration of agents' characteristics into the sector supply model. ALUAM-AB is less flexible since areas and fixed assets, *i.e.*, livestock housing capacities do not switch directly to more profitable agricultural activities as in the more aggregated model version. In the agent-based version, land can only be transferred via the land market module and farm type characteristics constrain production flexibility. In ALUAM-AB only "Ecological and landscape stewards" and "Traditionalist leisure farmers" can shift their production from sheep to cattle (or vice versa). However, based on the agent typology, full-time sheep farmers (farm businesses) and leisure-oriented farmers still remain in production as long as they meet their income thresholds. This leads to a more diversified production pattern in the agent-based model version.

4. Discussion

Socio-economic changes will continue to influence land abandonment in mountain regions [3,7,16]. Agent-based models offer the opportunity to include non-economic objectives and attitudes into land-use change models [36,37]. This is of specific importance when addressing farmers' behavior in mountain regions [30,32,33]. The analysis of farmers' decision-making in our case study region in the Valais, Switzerland confirms earlier findings that farmers have multiple values and objectives which translate into different farming strategies whereby profit maximization is only one [62,65–67]. Objectives of part-time and leisure-oriented farmers are particularly diverse and the aspiration to achieve high production and income levels through farming, as assumed by mainstream agricultural policy, is considerably less pronounced. Our analysis also highlights pronounced differences in availability of farm labor and opportunity costs that strongly affect farmers' behavior, in line with findings from other European mountain regions [68]. By relating our analysis of farming objectives and attitudes to farm census data, we were able to develop a farmer typology that could be qualitatively integrated within a simulation framework to assess land-use changes in a mountain region.

To that end, we adapted an existing sector supply model to include specific farm type agents. The existing model uses constraint income maximization based on mathematical programming techniques to simulate an optimal allocation of agricultural production factors while considering a large number of constraints. The farm types identified in the survey are used as an empirical foundation for model restrictions with respect to opportunity costs, farm growth intentions or farm succession. This procedure allows us to take into account both structural characteristics (e.g., fixed assets in land and labor) of existing farms and different types of decision-making separately. Thus, the advantage of this framework is that it allows the consideration of different forms of management, agronomic conditions and locally available production factors restricting the flexibility of farmers to react to socio-economic changes while maintaining the micro-economic footing of the simulations [22]. The constraint agent behavior allows for a good fit of the simulations with observed data (see 4.2). Such behavioral validations are still a challenge in ABM [69,70,72]. In contrast to other ABM studies addressing farmland abandonment [40–42,45], however, we do not model individual farms and remain within the structure of traditional normative farm sector supply modeling approaches [56]. One key challenge in such normative approaches is that corner solutions emerge and these only change if input parameters vary considerably or additional restrictions are introduced into the model structure [56]. Although the integration of agents allows the inclusion of additional constraints, corner solutions may still translate into our framework (see for example the scope of extensive land-use under the sensitivity run without animal based payments). In addition, the integration of empirically grounded data that allows for more flexibility (or more constraints) in the modeling framework requires the acquisition of information on farmers' decision-making. This is very costly and a transfer of the model to other regions demands a new parameterization of the model. This is a disadvantage that our approach shares with other ABM studies. Since our results show, however, that such details are important for model validation, more generic agents [37,82,83] or more flexible model frameworks [84] should still include context specific agents, especially in mountain regions.

A comparison of our ABM (ALUAM-AB) with a sector supply modeling approach [16,48–50], shows that the inclusion of agents allows for a better representation of the short and medium term developments of farm activity changes in mountain regions. At the same time, the findings from the assessment of elementary effects imply that the simulation results are driven by the same exogenous parameters in both model versions. Opportunity cost, *i.e.*, the measure of benefits forgone due to alternative uses of labor, is the most influential factor. The importance of this factor is also supported by other empirical studies which show that farming in Swiss mountain regions would be unprofitable with high labor costs [85,86]. In addition, we find that production prices (milk and lamb) and input prices (fuel price) have a high impact on modeling results. This is in line with other studies that confirm that profitable agricultural activities in mountain and upland regions are very sensitive to these parameters [33,87]. The fact that both simulation approaches are driven by the same exogenous input parameters supports the use of ALUAM-AB to assess short and medium term land-use changes and land abandonment in mountain regions since the economic background of the sector supply model is maintained. On the other hand, it implies that an in-depth sensitivity analysis of opportunity costs is needed when using a comparative static approach to assess forest development in the context of long-term climate change impacts on re-forestation. Such a sensitivity analysis in the aggregated model would allow considering major uncertainties regarding the behavior of the next generation of land-users and the consequences for the provision of forest ecosystem services.

In contrast to Schouten *et al.* [76], we explicitly focus on exogenous parameters which vary over the simulation period and do not present the sensitivity of technical model parameters such as feed required per cow. However, we are aware that the model may be sensitive to these parameters, too. For example, the level of extensive land-use in our model also depends on the percentage of extensive biomass that can be consumed by a cow or sheep without reducing its output *i.e.*, the amount of milk or meat produced. Thus, additional sensitivity analysis may still be required before using our modeling approach to answer more specific research questions.

The sensitivity analysis of the abolition of animal based direct payments presented here reveals that the extent and the form of the direct payment scheme have an essential impact on land abandonment in our simulations. This is in line with other ABM studies addressing land abandonment in marginal areas [41,45,88]. This finding also does not come as a surprise since Switzerland still provides some of the highest support for the agricultural sector worldwide [89] and farm structural change has been slow compared to other European alpine regions [28,90]. More importantly, however, our sensitivity analysis shows that an increase in opportunity costs leads to different simulation outputs for cattle in the two model versions if animal payments are abolished. The assessment of policy measures is thus sensitive to the chosen modeling approach and parameterization. This supports the importance of testing model sensitivity to different levels of opportunity costs. In addition, the extent of land abandonment in the aggregated model ALUAM was always more pronounced compared to the agent-based model version due to higher flexibility in shifts between production activities. This reflects the constraint development within an agent-based model framework which results in more diversified production patterns compared to a purely normative based optimization (see Section 3.3). In our sensitivity analysis, we did not show the spatially explicit consequences of land abandonment as presented in other studies [7,43,44]. For the aggregated model, this has been shown in Briner *et al.* [16,48]. The agent-based model allows for a more realistic spatially explicit representation of land abandonment in the short and medium term, as it better captures the diversity of decision-making in mountain farming. Combined with consistent scenario analysis [91] mountain-specific future developments of land abandonment, re-forestation and ecosystem services can be simulated and compared to other mountain regions such as the Jura mountains [45].

5. Conclusions

Land abandonment, and the subsequent re-forestation are important drivers behind land-use change and losses of ecosystem services in mountain regions. Agent-based models support the development and appraisal of policy and management options to counteract this development. Realigning the representation of human decision-making with time scales of ecological processes such as reforestation presents a major challenge in this context. Our sensitivity analysis comparing a land-use change model with and without agents cannot ultimately answer the question whether to implement agent-specific behavior anchored in the current farming generation or an aggregated optimization model with a focus on long-term ecosystem succession and forest development. Model choice depends on the scientific questions addressed and the corresponding (dis-) advantages of the different approaches. The sensitivity analysis presented here, however, helps to sensitize the model and parameter choice and shows two important directions for the interpretation of model results. Firstly, our agent-based model can capture short and medium term developments in land abandonment better than the aggregated version without losing its sensitivity to important socio-economic drivers. Therefore, also more generic or aggregated modeling approaches should maintain some specific (mountain) characterization of agent types. Secondly, long term and comparative static approaches should assess the sensitivity to opportunity costs or other relevant non-economic drivers in their model framework. This would allow considering some of the variations and uncertainties regarding current and future behavior of mountain farmers also in comparative static approaches and may reveal different reactions to policy changes. Overall, the analysis presented helps to (i) sensitize model and parameter choice (ii) identify important parameters for agent type characterization, and (iii) better interpret existing and future studies when assessing the impact of global change on land abandonment and re-forestation in mountain regions.

Acknowledgments: We would like to thank the farmers for their participation in the survey and the interviews, and the regional agricultural extension office for helpful comments and insights into the local farming systems. We would also like to thank Simon Briner, who developed the first agent-based implementation of ALUAM, the Competence Center of Environment and Sustainability (CCES) of ETH Zurich which funded this study through the projects MOUNTLAND and MOUNTLAND II and Jennifer Bays for English corrections. We are also grateful for the very helpful and constructive comments of two anonymous reviewers.

Author Contributions: R.H. designed research. J.M.B. and R.H. designed the survey, J.M.B. performed survey, farmer interviews and developed farmer typology, R.H., S.P. and S.H.B adapted model code; G.L. and R.H. performed simulation analyses; J.M.B. and R.H. wrote the manuscript; S.H.B. reviewed and commented on various versions of the manuscript. All authors read and approved the final manuscript.

Appendix A.

Table A1. Results of the principal component analysis of farming objectives and attitudes.

	REC	PROD	LOC_INF	COMP_TRAD	EVAL_EXT	OPP_COST	Communalities
Farming Objectives							
With my farming activities, how important is it for me to . . .							
...achieve high financial profit	0.18	0.74	0.09	0.01	0.00	−0.10	0.60
...earn enough for a good living	0.03	0.67	−0.11	0.16	0.17	−0.41	0.68
...realize innovative products, projects, and ideas	−0.15	0.68	0.02	0.15	0.03	0.07	0.51
...achieve high yield and production	0.11	0.76	0.14	−0.06	−0.19	0.11	0.67
...have the best/most beautiful animals, fields	**0.88**	0.21	0.08	−0.07	−0.07	0.10	0.84
...present my achievements (e.g., in breeding animals) and compete with others in exhibitions or cow fights	**0.88**	−0.12	0.05	0.15	0.03	0.09	0.83
...maintain the traditions of the region and the family	0.29	0.07	0.48	**0.56**	−0.03	0.18	0.66
...comply with rules and regulations of society	0.06	0.07	0.24	**0.83**	−0.10	0.06	0.77
...fulfil the demands of the public (e.g., with respect to providing additional services)	0.08	0.01	0.10	**0.89**	0.05	−0.01	0.81
...maintain decision power for important issues in the village	0.01	0.11	**0.88**	0.15	0.03	−0.04	0.82
...contribute actively to economic/social activities in the village	0.19	−0.02	**0.89**	0.18	0.00	0.02	0.85
...earn recognition of other farmers	**0.68**	0.10	0.25	0.31	−0.02	0.00	0.63

	REC	PROD	LOC_INF	COMP_TRAD	EVAL_EXT	OPP_COST	Communalities
Attitudes towards part-time farming							
The time I invest in farming depends on the level of income I can earn outside of agriculture	0.19	0.07	-0.17	0.04	−0.02	**0.82**	0.75
Without employment outside of agriculture which helps support my farming activities I would give up farming	0.06	−0.17	0.17	0.08	−0.08	**0.81**	0.73
Attitudes towards extensive land use							
With extensive use of grassland and pastures I can achieve an adequate (financial) yield	0.05	−0.06	0.14	−0.14	**0.72**	−0.16	0.58
With extensive grass and pasture I can considerably improve biodiversity and landscape quality	0.08	−0.25	0.04	0.03	**0.78**	0.01	0.67
Remuneration for the provision of ecosystem and landscape services represents a good alternative to producing agricultural goods	−0.18	0.20	−0.10	0.02	**0.81**	−0.02	0.74
Proportion of explained variance	*0.13*	*0.13*	*0.12*	*0.12*	*0.11*	*0.09*	
Cronbach Alpha	*0.8*	*0.71*	*0.84*	*0.8*	*0.67*	*0.64*	

[1] The overall KMO value amounted to 0.65 and was considered acceptable for exploratory analysis, as were the KMO values of individual items. The model with 6 extracted components showed a fit based on the diagonal of 0.93, and explained 71% of the variance.

The PCA allowed the variance in the data to be summarized into 6 components: The first component, labeled "Recognition" (REC) reflects the aspiration of farmers to earn recognition within their own farming community, specifically by showing their livestock and skills at competitions and exhibitions. The second component, "Profit and Yield" (PROD), describes the degree to which farmers aim to achieve an adequate income, high profits and high yields from their farming activities. The third component, "Local influence" (LOC_INF), relates to maintaining an influence on, and contributing to, local village life through farming. The fourth component, labeled "Compliance and Tradition" (COMP_TRAD), summarizes farming motivations related to maintaining family traditions and to fulfilling societal expectations, e.g., with respect to providing additional ecological or landscape services. The fifth component, "Evaluation of extensive land-uses" (EVAL_EXT) describes how farmers perceive financial and non-financial benefits of extensive land-uses. Finally, the sixth component, "Opportunity costs" (OPP_COSTS), reflects the dependence of the farming engagement on extra-agricultural work commitments and income sources.

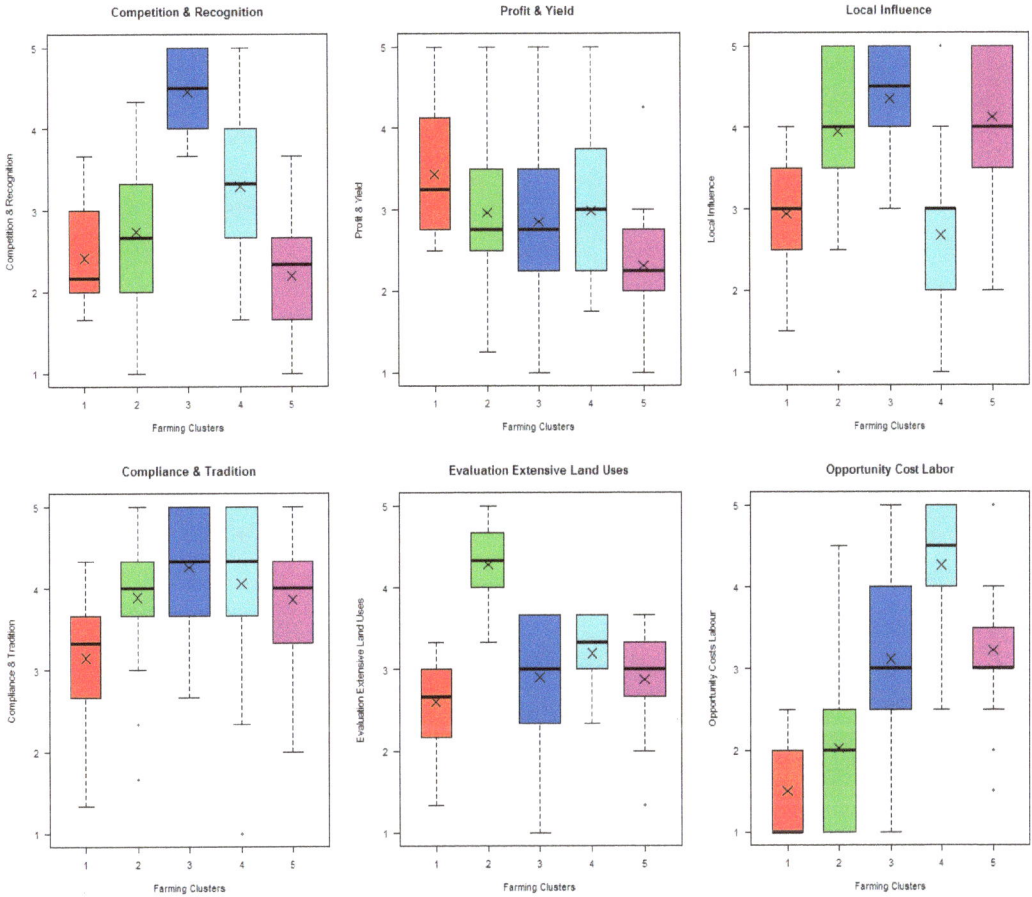

Figure A1. Boxplots and mean values of farming objectives and attitudes for the five farming clusters

Appendix B.

ODD Protocol for ALUAM-AB

B1. *Purpose*

The purpose of ALUAM-AB is to simulate future land-use changes, including farmland abandonment and corresponding re-forestation in mountain landscapes, triggered by the combined effects of climate, market and policy changes giving due considering to the individual preferences of the farmers. Thus, the consequences of changes in prices and policy measures relating to agricultural land-use activities can be simulated and feedback from climate change impacts on grassland and forestry can be considered. Spatially explicit information on agricultural land-use activities allows for a viable linkage with the forest landscape model LandClim.

B2. *State Variables and Scale*

Agents represent groups of farms. A farm agent has (1) its own state (*i.e.*, land endowment, animal housing capacity, *etc.*) which is updated after every simulation period of one year and (2) decision-making mechanisms for managing farm resources (*i.e.*, a constraint optimization based on mathematical programming techniques). The objective function and the set of constraints which define the solution space formally written as:

$$Z = \sum_j (p_j - c_j) \cdot x_j$$

$$\sum_j a_{ij} \cdot x_j \leq b_i \quad \forall i = 1, \cdots, I \tag{B1}$$

$$x_j \geq 0 \quad \forall j = 1, \cdots, I$$

With Z = income per farmer; x_j = agricultural farm activity (j = 1 to I); p_j = returns on activity j; c_j = cost per activity j; a_{ij} = technical coefficients required to produce x_j (of constraint i and activity j); b_{ij} = available resource. The state of the farm agent includes variables for household composition and available resources (land, capital and labor) and a specific type of decision-making based on opportunity costs of labor and a threshold for minimum income (leisure-oriented, part-time, full-time farmer, see Figure 1). Information on decision-making types was derived from surveys and interviews and combined with agricultural census data (see Section 3.1). The smallest landscape unit in ALUAM-AB is an area of 100 m × 100 m as it is used by the individual agent-groups. Natural conditions of the different land-use units and potential fodder production are based on the results presented in Briner *et al.* [16]. Agronomic variables include yield losses, plant nutrient requirements (N, P), manure production and production coefficients such as fodder intake, growth, birth, deaths of animals, labor requirements *etc.* that are based on Swiss average data. Production related variables, e.g., the number of livestock or the amount of hay sold, are aggregated over farm groups and represent aggregated values over one year. In the optimization process, these variables are optimized under the consideration of different fodder and nutrient balances that link land-use activities with livestock activities. As a result, land-use intensities are, as in the sectoral supply approach, defined in a spatially explicit manner.

B3. Process Overview and Scheduling

ALUAM-AB proceeds in annual time steps. The agents allocate their available resources to maximize their income (aggregated land rent). Thereby they consider natural, farm level and individual constraints as well as incentives and regulations from the market and policy instruments. Investments in production capacity made in previous years are considered as sunk costs representing path dependencies of the individual farm groups. Structural change is modeled using a land market sub-model [45,61]. The model identifies land units that are no longer cultivated under the existing farm structure. There are three reasons why fields are attributed to the land market in the model: (i) units generate a land rent below zero, (ii) the corresponding agent does not reach the minimum wage level, therefore the farm is abandoned and all the assigned land enters the land market or (iii) the farmer retires in the simulation year and has no successor. The land market sub-model randomly assigns the land units to one of the other agents. It is then checked to confirm that this agent shows the two following characteristics: The agent receiving the land unit must want to expand his cultivated area (stated willingness to grow) and his shadow price for the land unit must be positive. If these conditions are not met, the land unit is returned to the land market and assigned randomly to another farm (Figure A1). Once again it is checked to verify that this agent fulfils the conditions for the assignment of land.

This procedure is repeated until all land units are assigned to a farm or none of the farms is willing to take the land units left on the market. Land units that are not transferred to other farms are defined as abandoned and natural vegetation dynamics get under way on these units (modeled in LandClim). If land-use allocation is optimal, farm capacities and livestock are updated and the next annual time step is initialized using the parameters (prices, costs) of the following year.

The environmental feedback is based on a "lightweight" coupling between ALUAM-AB and LandClim [26] and is modeled in the following sequence: While each model is driven by a (synchronized) time series of climate or agronomic constraints, land-use change is passed from ALUAM-AB to LandClim. In response, forest development is transferred from LandClim to ALUAM-AB. This data exchange occurs for time steps of 30 years, starting in the year 2010.

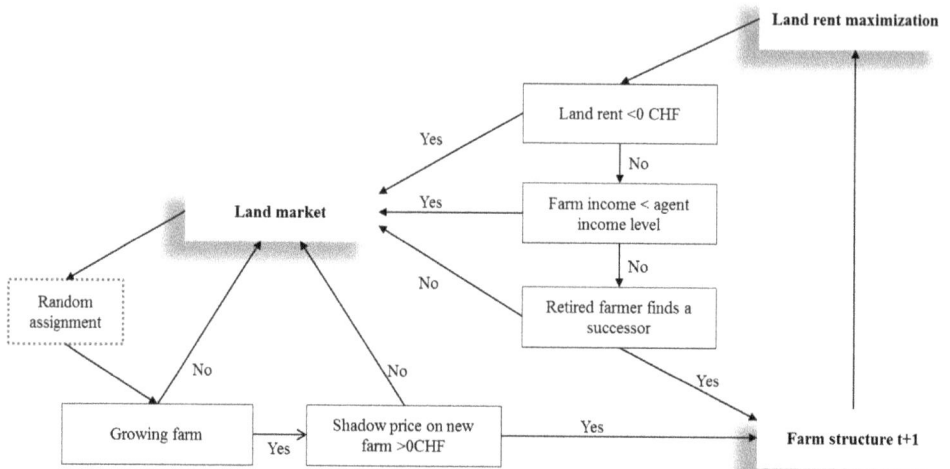

Figure A1. Process overview of land market module in ALUAM-AB. Source: Adapted from [45]

B4. Design Concepts

Emergence: Changes in farm activities emerge from an endogenous development that is determined by prices, policies, and decision-making type which are given exogenously. In addition, land-use patterns (intensity levels of land-use) emerge from the main outcome of the structural changes on agent level. Climate induced changes are also taken into account.

Adaptation: Agents respond to climatic, socio-economic and policy changes by adjusting their production activities, applying new production technologies, increasing (or reducing) land size and adjusting land-use intensities. In addition, agents also exit the sector if their income falls below a minimum threshold.

Prediction: The agent's objectives are characterized by an overall farm income optimization approach. This dictates the allocation of an agent's available resources to production giving due consideration to natural, farm-level and individual constraints as well as incentives and regulations from the market (yearly price and cost parameters) and policy scenarios. Thus, the fundamental concept behind our approach is rational economic behavior (land rent maximization) and no learning patterns exist. However, the consideration of individual constraints, such as opportunity costs, minimum income wage and limited time resources, leads to the inclusion of non-economic goals in the decision-making process.

Interaction: The interaction between the agents is based on the land market described in the process overview. Interaction between agents and the environment is based on the model linkage of LandClim and ALUAM-AB. Detailed information on spatially explicit natural conditions is provided by the LandClim model. Although the LandClim model can provide stochastic output, only mean changes in yields are considered in ALUAM-AB which does not include stochastic variables. The corresponding maps are used as an input for ALUAM-AB. The spatially explicit information following the optimization procedure is then re-entered into the vegetation model. These maps can be used to illustrate the changes in land-use dynamics.

B5. Initialization

Initial attributes for households were defined using information from the survey and interviews along with farm census data of the FOAG (see Section 2.3). Based on the distribution of the farm characteristics in the census data, we assigned the observed age structure to each farm type. Thus, the retirement of farmers within each farm type corresponds with the existing age structure in the case study region. This age structure is updated after every simulation period. The initial allocation of land-units to agents is based on a random assignment of parcels in which the share of parcels according to slope corresponds to the real world distribution [72]. The accumulative share of land cultivated by different agent types reported in the census data was determined for three slope strata (<18°, 18°–35°, >35°). Within these strata, the land-units were then allocated to the agents according to their relative land tenure with the help of a random number. Sensitivity tests with repeated random assignments showed marginal impact on simulation outcomes. Model versions initialized with allocation of land-units based on alternative or multiple stratification criteria (e.g., agricultural zone, municipality, elevation) performed badly compared to the observed data.

B6. Input

Information with respect to natural conditions is derived from the LandClim model and the crop model described in Briner *et al.* [16]. Price and cost developments in the applications of ALUAM were derived from scenarios for the European agricultural sector [92]. Policy and climate changes are based on an interdisciplinary development of scenarios for our case study region [91]. For the validation period, prices and costs were adopted from federal statistics (see Table A1). Table A2 shows the parameterization of the agents' characteristics in ALUAM based on the results from Section 3.1.

Table A1. Observed data: Price and cost assumptions for the period 2000–2012.

Parameters (k)Pric	Unit	2000	2001	2002	2003	2004	2005	2006	2007	2008	2009	2010	2011	2012
Prices														
Milk price	CHF/kg	89.4	89.9	88.4	85.5	84.6	82.4	81.8	80.0	87.7	74.8	71.8	72.7	70.5
Lamb price		232	232	233	233	217	192	194	194	197	213	193	183	202
Beef price		2446	2446	2423	2491	2677	2787	2968	3051	3197	3197	3197	3197	3197
Costs														
Variable costs	%	1	1	1	1	1	1.04	1.05	1.07	1.09	1.11	1.13	1.15	1.17
Fixed costs		1	1	1	1	1	1	1	1	1.01	1.02	1.03	1.04	1.06
Price of diesel	CHF/l	1.44	1.4	1.34	1.36	1.45	1.64	1.74	1.77	2.03	1.6	1.72	1.86	1.93
Direct payments														
General DP	CHF/ha	1200	1200	1200	1200	1200	1200	1200	1200	1080	1040	1040	1040	1020
ECA		700	700	700	700	700	700	700	700	700	700	700	700	700
DP slope		510	510	510	510	510	510	510	510	510	510	620	620	620
RFB	CHF	900	900	900	900	900	900	900	900	860	690	690	690	690
TEP		970	970	970	970	970	970	970	970	970	970	1010	1010	1010

B7. Sub-models

LandClim: Forest dynamics and forest derived ES, such as potential timber harvest are simulated using the forest landscape model LandClim [93]. LandClim is a spatially explicit process based model that incorporates competition-driven forest dynamics and landscape-level disturbances to simulate forest dynamics on a landscape scale. LandClim was designed to examine the impact of climate change and forest management on forest development and structure [94]. The model has been tested in the Central Alps, North American Rocky Mountains, and Mediterranean forests, and has been used to simulate current, paleo-ecological [95–97] and future forest dynamics [55,94]. LandClim simulates forest growth in 25 m × 25 m cells using simplified versions of tree recruitment, growth and competition processes that are commonly included in forest gap models [98]. Forest growth is determined by climatic parameters (monthly temperature and precipitation), soil properties and topography, land-use and forest management and large-scale disturbances. Individual cells are linked together by the spatially explicit processes of seed dispersal, landscape disturbances and forest management. Forest succession processes within each cell are simulated in a yearly time step, while landscape-level processes are simulated in a decadal time step. Forest dynamics within each cell are simulated by following tree age cohorts, where cohorts are characterized by the mean biomass of an individual tree and the number of trees in the cohort. We implemented a forest management regime to evaluate potential timber production within each landscape cell. Forest stands are evaluated every 20 years to determine if they should be entered and timber removed. If the average height of the dominant trees within a stand (largest 100 trees·ha^{-1}) is greater than 15 m, the stand is entered and all trees with a DBH (diameter at breast height) greater than 20 cm are harvested. This yields harvested trees that have an average DBH between 25 and 30 cm. This management routine is used to obtain a timber production value for each cell on the landscape. This can then be returned to ALUAM and used to inform land-use conversion. For this study, the data on forest production and forest ecosystem services was taken from an earlier analysis [16,48–50].

B8. Crop Model

Projected future yields of relevant crops are calculated using FAO (Food and Agriculture Organization of the UN) data on optimal and absolute crop growing conditions. The minimum and maximum temperature and precipitation values that support optimal crop development and the values that define the crops' temperature and precipitation extremes, are extracted from the FAO crop data base EcoCrop (FAO, online http://ecocrop.fao.org). These four values formed the basis for a relative crop yield curve for temperature and precipitation values using an incomplete beta distribution. These species specific crop yield curves are then used to calculate the relative yield for six crops based on monthly precipitation and temperature values for each landscape cell (100 m × 100 m) in the case study landscape. The projected realized yield is taken as the minimum yield value from the temperature and precipitation responses. If land is irrigated, yield is only deemed to be limited by temperature responses. The absolute yield of crops is calculated by standardizing the values against observed yield of crops in 2000.

Table A2. Parameterization of agent characteristics in ALUAM-AB.

Agent Name	Farm Type	Opportunity Costs % of × CHF	Available Work % of 2800	Minimum Income CHF	Number of Farms	Average Farm Size ha	Thereof Slope >18% ha	Land Per Agent ha	Farm Growth	Succession Rate in %	Sheep	Dairy Cows	Beef Cattle	Suckler Cows
											Number in the year 2000			
MILAS	1	0.2	1	25,000	7	42.1	5.4	295	Yes	0.75		237	215	
MASA	1	0.2	0.6	25,000	11	11.7	4.8	129	Yes	0.75	376			
MUK	1	0.5	0.6	25,000	3	24.9	12.5	75	Yes	0.75			86	43
MIAA	2	0.2	0.5	10,000	44	5.2	2.8	227		0.55		156	123	
MILA	2	0.2	0.5	10,000	10	13.1	6.1	131		0.55		93	93	
MIAS	2	0.2	0.8	10,000	14	6.8	2.7	95		0.45	44	41	208	
SCH	2	0.2	0.5	100,000	23	7.1	4.0	164		0.45	870			
MIAAS	2	0.2	0.8	10,000	6	15.6	8.3	93		0.45	26	27	146	
AUR	3	0.5	0.5	0	19	2.8	1.3	52		0.45	208			
LEG	3	0.5	0.5	0	18	6.6	2.5	119		0.45	222			
MISCH	3	1.25	0.3	0	26	6.4	3.0	165	Yes	0.55	558			
MILS	4	1	0.5	10,000	4	26.1	11.0	104	Yes	0.55		38	27	
AK	4	1	0.3	10,000	26	6.5	1.8	170		0.55				
MIL	5	0.2	0.3	0	40	4.0	2.4	162		0.45	932			
Total					**251**	**7.9**	**4.9**	**1981**			**3236**	**592**	**898**	**43**

Table A3. Elementary effects in two model versions.

	ALUAM-AB		ALUAM		ALUAM-AB		ALUAM		ALUAM-AB		ALUAM	
	Land Rent (CHF)				Animal Total				Grassland Intensive			
	Mean	Stdev	Mean	Stdev	Mean	Stdev	Mean	Stdev	Mean	Stdev	Mean	Stdev
Prices												
Milk price	229,915	131,355	305,040	185,823	33	19	45	44	6	5	8	6
Lamb	35,221	24,010	29,203	13,407	19	10	10	10	9	7	7	7
Price beef	822	1365	1542	2269	0.4	0.6	0.6	0.8	0.3	0.4	0.4	0.6
Costs												
Variable costs machines	8	5	14	8	0.0	0.0	0.0	0.0	0.0	0.0	0.0	0.0
Fixed costs machines	66	38	88	58	0.0	0.0	0.1	0.0	0.0	0.0	0.0	0.0
Price of diesel fuel	112,646	231,293	260,699	333,490	45	33	231	115	61	48	194	99
Direct Payments												
General DP	8860	4911	10,533	5910	0.1	0.1	1.0	0.5	0.0	0.0	1.1	0.5
ECA	1315	983	1040	1873	1.0	0.6	1.0	0.5	1.5	0.8	1.2	0.6
DP slope	10,574	5920	11,915	7213	0.0	0.2	0.5	0.4	0.0	0.1	0.8	0.5
RFB Payments	472	514	794	907	0.1	0.0	0.2	0.2	0.1	0.0	0.0	0.1
TEP payments	7371	4407	10,152	6370	3.6	2.0	4.7	3.3	2.8	1.2	2.3	1.2
Agent Characteristics												
Workload	1744	3893			0.6	0.9			1583	2644		
Opportunity costs	244,625	329,172	244,625	329,172	160	143	160	143	170	115	170	115

Abbreviations: DP: Direct Payments; ECA: Environmental compensation area; RFB: Payment per roughage livestock unit, TEP: Payment per livestock unit in remote areas.

Conflicts of Interest: The authors declare no conflict of interest.

References

1. MacDonald, D.; Crabtree, J.R.; Wiesinger, G.; Dax, T.; Stamou, N.; Fleury, P.; Gutierrez Lazpita, J.; Gibon, A. Agricultural abandonment in mountain areas of Europe: Environmental consequences and policy response. *J. Environ. Manag.* **2000**, *59*, 47–69.

2. Tasser, E.; Walde, J.; Tappeiner, U.; Teutsch, A.; Noggler, W. Land-use changes and natural reforestation in the Eastern Central Alps. *Agric. Ecosyst. Environ.* **2007**, *118*, 115–129.

3. Sitzia, T.; Semenzato, P.; Trentanovi, G. Natural reforestation is changing spatial patterns of rural mountain and hill landscapes: A global overview. *For. Ecol. Manag.* **2010**, *259*, 1354–1362.

4. Keenleyside, C.; Tucker, G.M. *Farmland Abandonment in the EU: An Assessment of Trends and Prospects*; Report Prepared for WWF; Institute for European Environmental Policy: London, UK, 2010.

5. Huber, R.; Rigling, A.; Bebi, P.; Brand, F.S.; Briner, S.; Buttler, A.; Elkin, C.; Gillet, F.; Grêt-Regamey, A.; Hirschi, C.; *et al.* Sustainable land use in mountain regions under global change: Synthesis across scales and disciplines. *Ecol. Soc.* **2013**, *18*, 36.

6. Gehrig-Fasel, J.; Guisan, A.; Zimmermann, N.E. Tree line shifts in the Swiss Alps: Climate change or land abandonment? *J. Veg. Sci.* **2007**, *18*, 571–582.

7. Gellrich, M.; Baur, P.; Koch, B.; Zimmermann, N.E. Agricultural land abandonment and natural forest re-growth in the Swiss mountains: A spatially explicit economic analysis. *Agric. Ecosyst. Environ.* **2007**, *118*, 93–108.

8. Gellrich, M.; Baur, P.; Robinson, B.H.; Bebi, P. Combining classification tree analyses with interviews to study why sub-alpine grasslands sometimes revert to forest: A case study from the Swiss Alps. *Agric. Syst.* **2008**, *96*, 124–138.

9. Gellrich, M.; Zimmermann, N.E. Investigating the regional-scale pattern of agricultural land abandonment in the Swiss mountains: A spatial statistical modelling approach. *Landsc. Urban Plan.* **2007**, *79*, 65–76.

10. Pointereau, P.; Coulon, F.; Girard, P.; Lambotte, M.; Stuczynski, T.; Sanchez Ortega, V.; del Rio, A. *Analysis of Farmland Abandonment and the Extent and Location of Agricultural Areas that are Actually Abandoned or are in Risk to be Abandoned*; European Commission-JRC-Institute for Environment and Sustainability: Ispra, Italy, 2008.

11. Huber, R.; Bugmann, H.; Buttler, A.; Rigling, A. Sustainable land-use practices in European mountain regions under global change: An integrated research approach. *Ecol. Soc.* **2013**, *18*, 37.

12. Figueiredo, J.; Pereira, H. Regime shifts in a socio-ecological model of farmland abandonment. *Landsc. Ecol.* **2011**, *26*, 737–749.

13. Claessens, L.; Schoorl, J.; Verburg, P.; Geraedts, L.; Veldkamp, A. Modelling interactions and feedback mechanisms between land use change and landscape processes. *Agric. Ecosyst. Environ.* **2009**, *129*, 157–170.

14. Flury, C.; Gotsch, N.; Rieder, P. Site-specific and regionally optimal direct payments for mountain agriculture. *Land Use Policy* **2004**, *22*, 207–214.

15. Renwick, A.; Jansson, T.; Verburg, P.H.; Revoredo-Giha, C.; Britz, W.; Gocht, A.; McCracken, D. Policy reform and agricultural land abandonment in the EU. *Land Use Policy* **2013**, *30*, 446–457.

16. Briner, S.; Huber, R.; Elkin, C.; Grêt-Regamey, A. Assessing the impacts of economic and climate changes on land-use in mountain regions: A spatial dynamic modeling approach. *Agric. Ecosyst. Environ.* **2012**, *149*, 50–63.

17. Parker, D.C.; Manson, S.M.; Janssen, M.A.; Hoffmann, M.J.; Deadman, P. Multi-agent systems for the simulation of land-use and land-cover change: A review. *Ann. Assoc. Am. Geogr.* **2003**, *93*, 314–337.

18. Heckbert, S.; Baynes, T.; Reeson, A. Agent-based modeling in ecological economics. *Ann. N. Y. Acad. Sci.* **2010**, *1185*, 39–53.

19. An, L. Modeling human decisions in coupled human and natural systems: Review of agent-based models. *Ecol. Model.* **2012**, *229*, 25–36.

20. Le, Q.B.; Park, S.J.; Vlek, P.L.G.; Cremers, A.B. Land-Use Dynamic Simulator (LUDAS): A multi-agent system model for simulating spatio-temporal dynamics of coupled human-landscape system. I. Structure and theoretical specification. *Ecol. Inform.* **2008**, *3*, 135–153.

21. Kelly, R.A.; Jakeman, A.J.; Barreteau, O.; Borsuk, M.E.; ElSawah, S.; Hamilton, S.H.; Henriksen, H.J.; Kuikka, S.; Maier, H.R.; Rizzoli, A.E.; *et al.* Selecting among five common modelling approaches for integrated environmental assessment and management. *Environ. Model. Softw.* **2013**, *47*, 159–181.

22. Berger, T.; Troost, C. Agent-based modelling of climate adaptation and mitigation options in agriculture. *J. Agric. Econ.* **2014**, *65*, 323–348.

23. Valbuena, D.; Verburg, P.; Bregt, A.; Ligtenberg, A. An agent-based approach to model land-use change at a regional scale. *Landsc. Ecol.* **2010**, *25*, 185–199.

24. Schouten, M.; Opdam, P.; Polman, N.; Westerhof, E. Resilience-based governance in rural landscapes: Experiments with agri-environment schemes using a spatially explicit agent-based model. *Land Use Policy* **2013**, *30*, 934–943.

25. Nainggolan, D.; Termansen, M.; Fleskens, L.; Hubacek, K.; Reed, M.S.; de Vente, J.; Boix-Fayos, C. What does the future hold for semi-arid Mediterranean agro-ecosystems? Exploring cellular automata and agent-based trajectories of future land-use change. *Appl. Geogr.* **2012**, *35*, 474–490.

26. Schreinemachers, P.; Berger, T. An agent-based simulation model of human-environment interactions in agricultural systems. *Environ. Model. Softw.* **2011**, *26*, 845–859.

27. Rounsevell, M.D.A.; Arneth, A. Representing human behaviour and decisional processes in land system models as an integral component of the Earth system. *Glob. Environ. Change* **2011**, *21*, 840–843.

28. Flury, C.; Huber, R.; Tasser, E. Future of mountain agriculture in the Alps. In *The Future of Mountain Agriculture*; Mann, S., Ed.; Springer: Berlin, Germany, 2013; pp. 105–126.

29. Schirpke, U.; Leitinger, G.; Tasser, E.; Schermer, M.; Steinbacher, M.; Tappeiner, U. Multiple ecosystem services of a changing Alpine landscape: Past, present and future. *Int. J. Biodivers. Sci. Ecosyst. Serv. Manag.* **2012**.

30. Marini, L.; Klimek, S.; Battisti, A. Mitigating the impacts of the decline of traditional farming on mountain landscapes and biodiversity: A case study in the European Alps. *Environ. Sci. Policy* **2011**, *14*, 258–267.

31. Celio, E.; Flint, C.G.; Schoch, P.; Grêt-Regamey, A. Farmers' perception of their decision-making in relation to policy schemes: A comparison of case studies from Switzerland and the United States. *Land Use Policy* **2014**, *41*, 163–171.

32. Pinter, M.; Kirner, L. Strategies of disadvantaged mountain dairy farmers as indicators of agricultural structural change: A case study of Murau, Austria. *Land Use Policy* **2014**, *38*, 441–453.

33. Garcia-Martinez, A.; Bernués, A.; Olaizola, A.M. Simulation of mountain cattle farming system changes under diverse agricultural policies and off-farm labour scenarios. *Livest. Sci.* **2012**, *137*, 73–86.

34. Huber, R.; Flury, C.; Finger, R. Factors affecting farm growth intentions of family farms in mountain regions: Empirical evidence for central Switzerland. *Land Use Policy* **2015**, *47*, 188–197.

35. Agarwal, C.; Green, G.M.; Grove, J.M.; Evans, T.P.; Schweik, C.M. *A Review and Assessment of Land-Use Change Models. Dynamics of Space, Time, and Human Choice*; General Technical Report NE-29; U.S. Department of Agriculture, Forest Service, Northeastern Research Station: Newtown Square, PA, USA, 2002.

36. Smajgl, A.; Brown, D.G.; Valbuena, D.; Huigen, M.G.A. Empirical characterisation of agent behaviours in socio-ecological systems. *Environ. Model. Softw.* **2011**, *26*, 837–844.

37. Rounsevell, M.D.A.; Robinson, D.T.; Murray-Rust, D. From actors to agents in socio-ecological systems models. *Philos. Trans. R. Soc. B Biol. Sci.* **2011**, *367*, 259–269.

38. Lindner, M.; Maroschek, M.; Netherer, S.; Kremer, A.; Barbati, A.; Garcia-Gonzalo, J.; Seidl, R.; Delzon, S.; Corona, P.; Kolström, M.; *et al.* Climate change impacts, adaptive capacity, and vulnerability of European forest ecosystems. *For. Ecol. Manag.* **2010**, *259*, 698–709.

39. Foster, D.; Swanson, F.; Aber, J.; Burke, I.; Brokaw, N.; Tilman, D.; Knapp, A. The importance of land-use legacies to ecology and conservation. *Bioscience* **2003**, *53*, 77–88.

40. Gibon, A.; Sheeren, D.; Monteil, C.; Ladet, S.; Balent, G. Modelling and simulating change in reforesting mountain landscapes using a social-ecological framework. *Landsc. Ecol.* **2010**, *25*, 267–285.

41. Piorr, A.; Ungaro, F.; Ciancaglini, A.; Happe, K.; Sahrbacher, A.; Sattler, C.; Uthes, S.; Zander, P. Integrated assessment of future CAP policies: Land use changes, spatial patterns and targeting. *Environ. Sci. Policy* **2009**, *12*, 1122–1136.

42. Brady, M.; Sahrbacher, C.; Kellermann, K.; Happe, K. An agent-based approach to modeling impacts of agricultural policy on land use, biodiversity and ecosystem services. *Landsc. Ecol.* **2012**, *27*, 1363–1381.

43. Bakker, M.M.; van Doorn, A.M. Farmer-specific relationships between land use change and landscape factors: Introducing agents in empirical land use modelling. *Land Use Policy* **2009**, *26*, 809–817.

44. Millington, J.; Romero-Calcerrada, R.; Wainwright, J.; Perry, G. An agent-based model of mediterranean agricultural land-use/cover change for examining wildfire risk. *J. Artif. Soc. Soc. Simul.* **2008**, *11*, 4.

45. Huber, R.; Briner, S.; Peringer, A.; Lauber, S.; Seidl, R.; Widmer, A.; Gillet, F.; Buttler, A.; Le, Q.B.; Hirschi, C. Modeling social-ecological feedback effects in the implementation of payments for environmental services in pasture-woodlands. *Ecol. Soc.* **2013**, *18*, 41.

46. Nguyen, T.G.; de Kok, J.L. Systematic testing of an integrated systems model for coastal zone management using sensitivity and uncertainty analyses. *Environ. Model. Softw.* **2007**, *22*, 1572–1587.

47. Morris, M.D. Factorial sampling plans for preliminary computational experiments. *Technometrics* **1991**, *33*, 161–174.

48. Briner, S.; Elkin, C.; Huber, R. Evaluating the relative impact of climate and economic changes on forest and agricultural ecosystem services in mountain regions. *J. Environ. Manag.* **2013**, *129*, 414–422.

49. Briner, S.; Huber, R.; Bebi, P.; Elkin, C.; Schmatz, D.R.; Grêt-Regamey, A. Trade-offs between ecosystem services in a mountain region. *Ecol. Soc.* **2013**, *18*, 35.

50. Huber, R.; Briner, S.; Bugmann, H.; Elkin, C.; Hirschi, C.; Seidl, R.; Snell, R.; Rigling, A. Inter- and transdisciplinary perspective on the integration of ecological processes into ecosystem services analysis in a mountain region. *Ecol. Processes* **2014**, *3*, 9.

51. Hirschi, C.; Widmer, A.; Briner, S.; Huber, R. Combining policy network and model-based scenario analyses: An assessment of future ecosystem goods and services in swiss mountain regions. *Ecol. Soc.* **2013**, *18*, 42.

52. Brand, F.S.; Seidl, R.; Le, Q.B.; Brändle, J.M.; Scholz, R.W. Constructing consistent multiscale scenarios by transdisciplinary processes: The case of mountain regions facing global change. *Ecol. Soc.* **2013**, *18*, 43.

53. Bundesamt für Statistik (Swiss Federal Statistical Office, BFS). *Arealstatistik*; BFS: Neuchâtel, Switzerland, 2014.

54. Bundesamt für Landwirtschaft (Federal Office for Agriculture, BLW). *AGIS*; BLW: Bern, Switzerland, 2013.

55. Elkin, C.; Gutiérrez, A.G.; Leuzinger, S.; Manusch, C.; Temperli, C.; Rasche, L.; Bugmann, H. A 2 °C warmer world is not safe for ecosystem services in the European Alps. *Glob. Change Biol.* **2013**, *19*, 1827–1840.

56. Buysse, J.; van Huylenbroeck, G.; Lauwers, L. Normative, positive and econometric mathematical programming as tools for incorporation of multifunctionality in agricultural policy modelling. *Agric. Ecosyst. Environ.* **2007**, *120*, 70–81.

57. Chen, X.; Önal, H. Modeling agricultural supply response using mathematical programming and crop mixes. *Am. J. Agric. Econ.* **2012**, *94*, 674–686.

58. Grimm, V.; Berger, U.; Bastiansen, F.; Eliassen, S.; Ginot, V.; Giske, J.; Goss-Custard, J.; Grand, T.; Heinz, S.K.; Huse, G.; *et al.* A standard protocol for describing individual-based and agent-based models. *Ecol. Model.* **2006**, *198*, 115–126.

59. Grimm, V.; Berger, U.; DeAngelis, D.L.; Polhill, J.G.; Giske, J.; Railsback, S.F. The ODD protocol: A review and first update. *Ecol. Model.* **2010**, *221*, 2760–2768.

60. Müller, B.; Balbi, S.; Buchmann, C.M.; de Sousa, L.; Dressler, G.; Groeneveld, J.; Klassert, C.J.; Le, Q.B.; Millington, J.D.A.; Nolzen, H.; *et al.* Standardised and transparent model descriptions for agent-based models: Current status and prospects. *Environ. Model. Softw.* **2014**, *55*, 156–163.

61. Lauber, S. *Agrarstrukturwandel im Berggebiet*; Agroscope Forschungsanstalt Reckenholz-Tänikon (ART): Zürich, Switzerland, 2006; p. 217.

62. Guillem, E.E.; Barnes, A.P.; Rounsevell, M.D.A.; Renwick, A. Refining perception-based farmer typologies with the analysis of past census data. *J. Environ. Manag.* **2012**, *110*, 226–235.

63. Gasson, R. Goals and values of farmers. *J. Agric. Econ.* **1973**, *24*, 521–542.

64. Solano, C.; Leon, H.; Perez, E.; Herrero, M. Characterising objective profiles of Costa Rican dairy farmers. *Agric. Syst.* **2001**, *67*, 153–179.

65. Karali, E.; Brunner, B.; Doherty, R.; Hersperger, A.; Rounsevell, M.A. The effect of farmer attitudes and objectives on the heterogeneity of farm attributes and management in Switzerland. *Hum. Ecol.* **2013**, *41*, 915–926.

66. Valbuena, D.; Verburg, P.H.; Bregt, A.K. A method to define a typology for agent-based analysis in regional land-use research. *Agric. Ecosyst. Environ.* **2008**, *128*, 27–36.

67. Daloğlu, I.; Nassauer, J.I.; Riolo, R.L.; Scavia, D. Development of a farmer typology of agricultural conservation behavior in the American Corn Belt. *Agric. Syst.* **2014**, *129*, 93–102.

68. López-i-Gelats, F.; Milán, M.J.; Bartolomé, J. Is farming enough in mountain areas? Farm diversification in the Pyrenees. *Land Use Policy* **2011**, *28*, 783–791.

69. O'Rourke, E.; Kramm, N.; Chisholm, N. The influence of farming styles on the management of the Iveragh uplands, Southwest Ireland. *Land Use Policy* **2012**, *29*, 805–816.

70. Revelle, W. *psych: Procedures for Psychological, Psychometric, and Personality Research*; Version 1.4.8.11; Northwestern University: Evanston, IL, USA, 2014; Available online: http://CRAN.R-project.org/package=psych (accessed on 1 September 2014).

71. Windrum, P.; Fagiolo, G.; Moneta, A. Empirical validation of agent-based models: Alternatives and prospects. *J. Artif. Soc. Soc. Simul.* **2007**, *10*, 8.

72. Filatova, T.; Verburg, P.H.; Parker, D.C.; Stannard, C.A. Spatial agent-based models for socio-ecological systems: Challenges and prospects. *Environ. Model. Softw.* **2013**, *45*, 1–7.

73. Bert, F.E.; Rovere, S.L.; Macal, C.M.; North, M.J.; Podestá, G.P. Lessons from a comprehensive validation of an agent based-model: The experience of the Pampas Model of Argentinean agricultural systems. *Ecol. Model.* **2014**, *273*, 284–298.

74. Villamor, G.B.; Le, Q.B.; Djanibekov, U.; van Noordwijk, M.; Vlek, P.L.G. Biodiversity in rubber agroforests, carbon emissions, and rural livelihoods: An agent-based model of land-use dynamics in lowland Sumatra. *Environ. Model. Softw.* **2014**, *61*, 151–165.

75. Daloğlu, I.; Nassauer, J.I.; Riolo, R.; Scavia, D. An integrated social and ecological modeling framework—Impacts of agricultural conservation practices on water quality. *Ecol. Soc.* **2014**, *19*, 12.

76. Schouten, M.; Verwaart, T.; Heijman, W. Comparing two sensitivity analysis approaches for two scenarios with a spatially explicit rural agent-based model. *Environ. Model. Softw.* **2014**, *54*, 196–210.

77. Sterman, J.D. *Business Dynamics: Systems Thinking and Modeling for a Complex World*; Irwin/McGraw-Hill: Boston, MA, USA, 2000; Volume 19.

78. Bundesamt für Landwirtschaft (Federal Office for Agriculture, BLW). *Agrarpolititsches Informationssystem (AGIS)*; BLW: Bern, Switzerland, 2012.

79. Federal Office for Agriculture (FOAG). *Zonenkarte*; FOAG: Bern, Switzerland, 2014; Available online: http://www.blw.admin.ch/themen/00015/00182/ (accessed on 1 January 2014).

80. Saltelli, A.; Annoni, P. How to avoid a perfunctory sensitivity analysis. *Environ. Model. Softw.* **2010**, *25*, 1508–1517.

81. Mann, S.; Lanz, S. Happy Tinbergen: Switzerland's new direct payment system. *EuroChoices* **2013**, *12*, 24–28.

82. Murray-Rust, D.; Brown, C.; van Vliet, J.; Alam, S.J.; Robinson, D.T.; Verburg, P.H.; Rounsevell, M. Combining agent functional types, capitals and services to model land use dynamics. *Environ. Model. Softw.* **2014**, *59*, 187–201.

83. Arneth, A.; Brown, C.; Rounsevell, M.D.A. Global models of human decision-making for land-based mitigation and adaptation assessment. *Nat. Clim. Change* **2014**, *4*, 550–557.

84. Murray-Rust, D.; Robinson, D.T.; Guillem, E.; Karali, E.; Rounsevell, M. An open framework for agent based modelling of agricultural land use change. *Environ. Model. Softw.* **2014**, *61*, 19–38.

85. Lips, M. Calculating full costs for Swiss dairy farms in the mountain region using a maximum entropy approach for joint-cost allocation. *Int. J. Agric. Manag.* **2014**, *3*, 145–153.

86. Jan, P.; Lips, M.; Dumondel, M. Synergies and trade-offs in the promotion of the economic and environmental performance of Swiss dairy farms located in the mountain area. *Yearb. Socioecono. Agric.* **2011**, *4*, 135–161.

87. Hanley, N.; Acs, S.; Dallimer, M.; Gaston, K.J.; Graves, A.; Morris, J.; Armsworth, P.R. Farm-scale ecological and economic impacts of agricultural change in the uplands. *Land Use Policy* **2012**, *29*, 587–597.

88. Brady, M.; Kellermann, K.; Sahrbacher, C.; Jelinek, L. Impacts of decoupled agricultural support on farm structure, biodiversity and landscape mosaic: Some EU results. *J. Agric. Econ.* **2009**, *60*, 563–585.

89. Huber, R.; Häberli, C. A "beyond WTO" scenario for Swiss agriculture: Consequences for income generation and the provision of public goods. *Yearb. Socioecon. Agric.* **2010**, *2010*, 361–400.

90. Streifeneder, T.; Tappeiner, U.; Ruffini, F.; Tappeiner, G.; Hoffmann, C. Selected aspects of agro-structural change within the Alps. *J. Alp. Res.* **2007**, *95*, 41–52.

91. Walz, A.; Braendle, J.M.; Lang, D.J.; Brand, F.; Briner, S.; Elkin, C.; Hirschi, C.; Huber, R.; Lischke, H.; Schmatz, D.R. Experience from downscaling IPCC-SRES scenarios to specific national-level focus scenarios for ecosystem service management. *Technol. Forecast. Soc. Change* **2013**, *86*, 21–32.

92. Abildtrup, J.; Audsley, E.; Fekete-Farkas, M.; Giupponi, C.; Gylling, M.; Rosato, P.; Rounsevell, M. Socio-economic scenario development for the assessment of climate change impacts on agricultural land use: A pairwise comparison approach. *Environ. Sci. Policy* **2006**, *9*, 101–115.

93. Schumacher, S.; Bugmann, H.; Mladenoff, D.J. Improving the formulation of tree growth and succession in a spatially explicit landscape model. *Ecol. Model.* **2004**, *180*, 175–194.

94. Schumacher, S.; Bugmann, H. The relative importance of climatic effects, wildfires and management for future forest landscape dynamics in the Swiss Alps. *Glob. Change Biol.* **2006**, *12*, 1435–1450.

95. Colombaroli, D.; Henne, P.D.; Kaltenrieder, P.; Gobet, E.; Tinner, W. Species responses to fire, climate and human impact at tree line in the Alps as evidenced by palaeo-environmental records and a dynamic simulation model. *J. Ecol.* **2010**, *98*, 1346–1357.

96. Henne, P.D.; Elkin, C.M.; Reineking, B.; Bugmann, H.; Tinner, W. Did soil development limit spruce (*Picea abies*) expansion in the central Alps during the Holocene? Testing a palaeobotanical hypothesis with a dynamic landscape model. *J. Biogeogr.* **2011**, *38*, 933–949.

97. Elkin, C.; Giuggiola, A.; Rigling, A.; Bugmann, H. Short- and long-term efficacy of forest thinning to mitigate drought impacts in mountain forests in the European Alps. *Ecol. Appl.* **2015**, *25*, 1083–1098.

98. Bugmann, H. A review of forest gap models. *Clim. Change* **2001**, *51*, 259–305.

Investigating Impacts of Alternative Crop Market Scenarios on Land Use Change with an Agent-Based Model

Deng Ding, David Bennett and Silvia Secchi

Abstract: We developed an agent-based model (ABM) to simulate farmers' decisions on crop type and fertilizer application in response to commodity and biofuel crop prices. Farm profit maximization constrained by farmers' profit expectations for land committed to biofuel crop production was used as the decision rule. Empirical parameters characterizing farmers' profit expectations were derived from an agricultural landowners and operators survey and integrated in the ABM. The integration of crop production cost models and the survey information in the ABM is critical to producing simulations that can provide realistic insights into agricultural land use planning and policy making. Model simulations were run with historical market prices and alternative market scenarios for corn price, soybean to corn price ratio, switchgrass price, and switchgrass to corn stover ratio. The results of the comparison between simulated cropland percentage and crop rotations with satellite-based land cover data suggest that farmers may be underestimating the effects that continuous corn production has on yields. The simulation results for alternative market scenarios based on a survey of agricultural land owners and operators in the Clear Creek Watershed in eastern Iowa show that farmers see cellulosic biofuel feedstock production in the form of perennial grasses or corn stover as a more risky enterprise than their current crop production systems, likely because of market and production risks and lock in effects. As a result farmers do not follow a simple farm-profit maximization rule.

Reprinted from *Land*. Cite as: Ding, D.; Bennett, D.; Secchi, S. Investigating Impacts of Alternative Crop Market Scenarios on Land Use Change with an Agent-Based Model. *Land* **2015**, *4*, 1110–1137.

1. Introduction

Midwestern landscapes are dominated by commodity crops including corn, soybean, and wheat. Commodity prices and the policies affecting them are the key drivers of farmers' decisions about agricultural practices, such as crop rotations and fertilizer rates. Federal and state policies on renewable energy production interact with commodity markets to affect decisions about changes in crop type and land management. These decisions, in turn, have significant environmental impacts in terms of water quantity and quality, soil erosion, and carbon sequestration in the

Midwest [1–4]. Farmers' involvement in the production of biofuel crops (e.g., corn stover, miscanthus, and switchgrass) could result in new land use patterns and, thus, altered environmental outcomes. Due to substantial risk associated with the adoption of novel agricultural practices, however, farmers may not follow the same maximum profit rules when considering biofuel crop production as they do when considering traditional crops. To achieve insight into the potential impact of biofuel crop markets on agriculture land use, empirical information about farmers' attitudes towards novel practices must be integrated into land use models.

Secchi *et al.* [5] investigated the potential water quality changes associated with market scenarios of decreasing soybean to corn price ratio in the Upper Mississippi river basin by integrating an economic-driven land use model and the Soil and Water Assessment Tool (SWAT), a surface water quality model. Their results showed that an increase in corn acreage by 14.4% could result in increase in N loadings to the watershed by 5.4% and P loadings by 4.1%. Another similar study was conducted to investigate the impacts of different corn price scenarios on crop rotation patterns and environmental consequences in Iowa, USA, by integrating an economic model and the edge of field environmental impact model EPIC (Environmental Policy Integrated Climate) [6]. The authors found that sustained high corn prices might result in continuous corn in crop rotation patterns on both current cropland and CRP (Conservation Reserve Program) land. This change in land management is associated with increased sediment and nitrogen loss from fields to surface water. Leaving crop residues such as corn stover in field can help recycle nutrients, control surface runoff and prevent water and wind erosion. Potential biofuel markets for corn stover feedstock could force further shifts from corn-soybean rotation to continuous corn and a significant reduction of crop residue, thus, creating further losses of soil, nitrogen (N) and phosphorus (P) [7]. As an alternative, switchgrass, a native perennial grass in the Midwest, has been extensively investigated as a potential feedstock for cellulosic biofuel production. Switchgrass is currently used in riparian buffer strips to reduce sediment, N and P in surface runoff [8], however, its production as a biofuel crop requires the use of fertilizers that can make their way to surface and groundwater [9].

Various approaches, such as statistical techniques, expert models (e.g., Bayesian probability), cellular models (cellular automata and Markov models), and hybrid models that combine multiple techniques have been applied in land use modeling [10]. These approaches, however, focus mainly on spatiotemporal patterns of land use change rather than the decision-making process of individual land managers, despite the fact that these managers are the essential driving force in the complex land use dynamics [11].

Agent-based model (ABM) is a computer simulation approach that can be used for land use modeling [10,12]. Agent-based land use models represent system complexity using a bottom-up approach that characterizes the decision making processes of heterogeneous agents as well as feedback processes among agents and between agents and the biophysical environment. Simulated agents, such as land owners, ranchers, farmers, and policy makers, act and interact following decision rules. They are characterized by sets of parameters that can be derived from theory or empirical data obtained by land use surveys, participatory observations, and field and laboratory experiments [13,14]. Due to the mechanistic nature of this approach and its unique perspective concerning agent-agent and agent-environment interactions, ABMs are often integrated with environment models to explore the impacts of socio-economic driving forces or alternative policies on land use and consequent environment outcomes [15–17]. ABM can be used to represent multiple types or levels of agents (e.g., institutions and individuals) and the co-evolution of and interaction between biophysical and human decision making processes in a coupled human-environment system [10,18–22].

In Midwestern agricultural landscapes, ABM has been applied in studies on how farmer decision making is impacted by agricultural policies, such as the CRP [23], potential market scenarios of biofuel crops [24], and price incentives for nitrogen and carbon abatement [25,26]. In the first two studies [23,24], the research is focused on how land use changes emerge from farmers' decisions under the assumption that the natural environment is static through time. In the latter two studies [25,26], the focus is on the impacts of farmers' management decisions on water quality. The authors integrated ABM with SWAT, and analyzed two-way interactions between farmer agents and the natural environment. The natural environment included two types of processes: biological and hydrological, both of which interacted with agents but at different levels. Agents' decisions on cropping practices not only influenced but also responded to crop yields. Agents' decisions affected watershed water quality, but there was no feedback from water quality into the agents' decisions. The study was considered as "semi-hypothetical" since no empirical information was used to parameterize agents' risk premiums towards novel agricultural practices.

The research objectives of this study are to: (1) develop an ABM to simulate farmers' decisions about land use practices given a risk-averse profit maximization rule; and (2) investigate the impacts of alternative market scenarios on land use change by integrating empirical information about farmers' attitudes into an ABM. Through this study, we address the following research questions: (1) is the current land cover pattern economically optimal given declines in crop yield resulting from continuous corn/soybean rotation? and (2) would biofuel crops be underutilized given farmers current attitudes towards biofuel crop production?

2. ABM of Agricultural Land Use

2.1. Mathematical Programming for Modeling Decision Making in ABM

Farm household modeling often relies on mathematical programming (MP) techniques to simulate land management decision of farmers and their responses to policy. This approach has several drawbacks [12,23]: (1) it generally assumes the maximization of a single objective (e.g., profit), which is often not appropriate when modeling the adoption of new technologies; (2) it ignores the social aspects of farm households such as communication and interaction among farmers in the same community; and (3) it does not properly capture the heterogeneity of the social behaviors and responses of farmers. In ABMs of agricultural land use decision making, mathematical programming is generally applied at the farm level and combined with heuristic approaches and Bayesian inference or Bayesian probability networks [23,25–27]. Heuristic approaches, such as decision trees or rule-based models, assume limited human cognition, while optimization approaches, such as MP, assume that inefficiency in human decisions comes from external factors, such as the failure of institutions, imperfect markets, and lack of infrastructure or limited information [27]. Though the two approaches appear to be different in theory, Schreinemachers and Berger [27] argue that in practice they can be converted into each other. For example, decision rules about production and consumption in heuristic approaches can be incorporated into MP models as constraints. An optimization approach is even more appropriate for policy analysis and planning since it can quantitatively characterize the outcomes of alternative policies.

In the Midwestern U.S., Sengupta *et al.* [23] employed a hybrid approach to model land enrollment in an agricultural land set aside conservation program, CRP. Different types of farmers with distinct decision making rules were modeled in the Cache River watershed of southern Illinois. In another example of ABM for agricultural land use, Ng [25] and Ng *et al.* [26] combined a farm-level stochastic programming model with a Bayesian updating procedure to represent the optimization and adaption processes of farmers' decision in the Salt Creek watershed in Central Illinois. In Ng's ABM [25,26], the farmers are assumed to be economically rational with bounded information. Farmer agents learn new information through time with Bayesian updating. In this ABM, farmers interact with their geographic neighbors to exchange information about crop yields and costs.

2.2. Empirical Information for ABM

Empirically-based ABMs for land use simulation construct and parameterize decision making models from empirical information that characterizes macro-level patterns and micro-level processes. Information about micro-level processes could be directly utilized for developing and testing the structure of decision making models in ABM, while datasets about macro phenomena are usually applied in calibrating and validating the model [13]. In this study, we focus mainly on empirical information about micro-level processes and its use for the parameterization of the decision making model.

To characterize heterogeneous agents and their behavioral responses, empirical information needs to be analyzed using statistical methods (e.g., factor or cluster analysis, regression) and GIS techniques. Based on the statistical descriptions of agents characteristics derived from observation datasets, different populations of unique artificial agents can then be generated with Monte Carlo techniques [28].

Smajgl *et al.* [14] developed a framework for generating and calibrating the parameters that describe agent attributes and behavioral functions. In their framework, the parameterization process is composed of five steps: (1) create an agent typology based on behavioral differences; (2) specify the attribute values for each agent type; (3) specify the behaviors of each agent type by obtaining the parameter values for their behavioral functions; (4) develop agent types from agent attributes or behavior responses; and (5) generate a population(s) of agents based on the agent typology, and relevant attributes and parameters.

Empirical information can be collected using various methods [10,13,14]. These methods include role-playing games [29,30], sample surveys [31], participatory observations [32], field or laboratory experiments [33], and GIS and RS data collection [34]. Of these five methods, sample survey is the most quantitative approach and can be carried out at a relatively large spatial scale [13]. Though qualitative information about agents' interactions and feedbacks could be obtained using different methods, it is still challenging to empirically identify and quantify social, spatial and cross-scale interactions among agents and feedbacks between agents' decisions and the biophysical processes [35].

3. Study Area

The study area is the Clear Creek watershed located in Iowa and Johnson Counties in Eastern-Central Iowa, U.S. (Figure 1). It is a typical Midwestern agricultural watershed dominated by row crops (corn, soybean), with an area of about 267 km^2. The average annual temperature from 2000 to 2011 was 10 °C and average annual precipitation was 886 mm based on the weather records at the Iowa City Municipal Airport station. According to the United States Department of Agriculture National Agricultural Statistical Services (NASS) remote sensing Cropland Data Layer (CDL) images in 2008, the percentages of the predominant cover types were respectively 29.77% corn, 20.11% soybean, and 28.17% grassland/pasture. Additionally, 14.56% of the total watershed was in urban and developed uses and 6.9% was forest including deciduous, evergreen and mixed forest. Urban and forest areas are located mainly in the eastern part of the watershed. According to the Iowa

Soil Properties and Interpretations Database (ISPAID), about 54% of the watershed is covered by soil with slope range of 2%–14%; while soils are distributed among several soil drainage type ranging from well drained soils (26%), moderate-well drained soils (23.5%), to poorly drained soils (11.9%). The dominant soil textures of the surface horizon in the region are silty clay loam (35.8%) and silt loam (35.1%); about 45% of the total land is highly erodible.

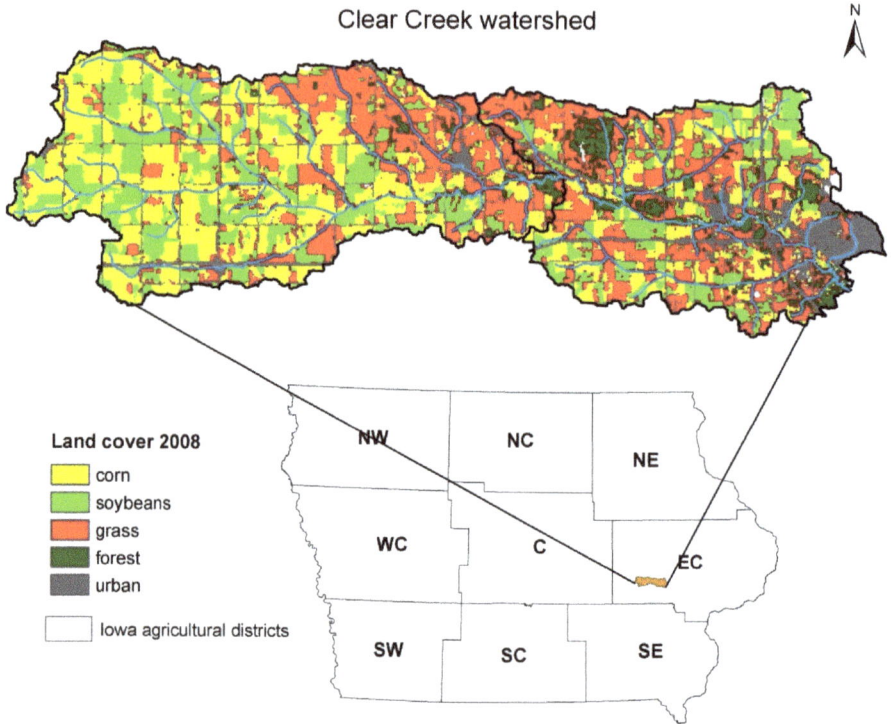

Figure 1. The Clear Creek Watershed in East-Central Iowa.

According to an agricultural land use survey conducted in 2010 [36,37], the dominant crop rotation type is corn-soybean and corn-corn-soybean with conventional tillage used for both corn and soybean. Some farmers adopt conventional tillage practices for corn but no-till or minimum-till practices for soybean. Specific information about actual application rates of inorganic N fertilizers in corn fields is not available for this watershed. According to Iowa State University Extension [38], the highest N rate is applied for corn after corn (168–224 kg/ha), followed by corn after soybean if no manure is applied (112–168 kg/ha), and corn on recently manured soils (0–101 kg/ha), given that all N is applied before corn planting or crop emergence.

4. Model Description

In this section, the model is described following the Overview, Design concept and Details (ODD) protocol [39].

4.1. Overview

4.1.1. Purpose

This model is used to simulate farmers' decisions on agricultural land use including crop type, tillage type, and fertilizer rates. We aim to run simulations to study farmers' decisions on biofuel crop productions in response to biofuel crop prices.

4.1.2. Entities, State Variables, and Scales

Agents in the model are farmers. Each farmer agent corresponds to a farm composed of one or more land parcels that are represented by command land units (CLUs) in this study. Farmer agents have attributes including farm size, and parameters about their attitudes towards biofuel crop production are summarized in Table 1.

We use eight parameters to characterize farmers' responses to the questions in Table A1 and thereby to describe their attitudes towards biofuel crop production and the associated risk premium. Descriptions of the eight parameters are included in Table 1. Additionally, there is a parameter *Area* representing the land acreage of the farm size of each farmer agent.

Each land parcel has attributes including area (in hectares), Corn Suitability Rate (CSR) (dimensionless) which is an index of land suitability for crop production, and potential yield of switchgrass (metric ton/ha).

Based on each land unit, farmer agents make CLU-specific decisions on crop type, tillage type, and fertilizer rates by changing the value of the CLU-specific (i as index) and year-specific (t as index) state variables: $X_{i,j,t}$, and j represents all available options for crop type, tillage type, and fertilizer rate levels. $X_{i,j,t}$ is a binary variable about crop type, tillage type, and N application rate level. Crop types include corn (with or without stover harvest), soybean, and switchgrass. Tillage types include conventional tillage, mulch tillage, and no tillage. Fertilizers include N, P, and K (potassium). For example, $X_{i,j=1,t} = 1$ means corn is planted in land parcel i at year t, $X_{i,j=2,t} = 1$ means soybean is planted in land parcel i at year t, and $X_{i,j=3,t} = 1$ means the land parcel i is fallow at year t. Similarly, $\sum_{j=4}^{6} X_{i,j,t} = 1$ and $j = 4, 5, 6$, respectively, mean conventional tillage, mulch tillage, and no tillage. When $j = 7, 8, \ldots, 16$, X_{ij} is the decision variable of the N application level from the lowest to the highest. In the biofuel crop market scenarios, crop type decision variables also have two extra options, switchgrass, and corn with stover harvested.

85

Table 1. Parameters characterizing farmers' attitudes towards biofuel crop production.

Parameter	Value	Agent Characteristics
Type	0	Not interested in either marketing corn stover or planting switchgrass
	1	Only interested in marketing corn stover
	2	Only interested in planting switchgrass
	3	Interested in both
Profit SWG	Numeric	Profit rate ($/ha) required by the farmer to plant, harvest and market switchgrass *
Percent1 SWG	Numeric	Percent of farm acreage on which the farmer would plant switchgrass if the Profit SWG can be achieved
Percent2 SWG	Numeric	Percent of farm acreage on which the farmer would plant switchgrass if 1.5 times the Profit SWG can be achieved
Profit Stvr	Numeric	Profit rate ($/ha) required by the farmer to harvest and market corn stover *
Percent1 Stvr	Numeric	Percent of farm acreage on which the farmer would consider harvesting and marketing corn stover if the *Profit Stvr* can be achieved
Percent2 Stvr	Numeric	Percent of farm acreage on which the farmer would harvest, and market corn stover if 1.5 times the *Profit Stvr* can be achieved
Portion Stvr	Numeric	Portion of corn stover that the farmer would harvest

* Given that in 2010 when the survey was conducted, the average price for corn was $196.85/metric ton and for soybean was $404.14/metric ton).

One simulation step represents one year. Farmer agents make decisions on each land parcel (CLU) of their farms every year. CLUs are the smallest land units with common crop choice and management, a continuous boundary and the same owner and operator. Thus, in the ABM, land parcels represented as CLUs are the smallest simulation units for which both the physical conditions (soil fertility, slope, *etc.*) and farmer's land use management practices are assumed to be homogeneous. Generally, a farm is composed of multiple CLUs and each farm corresponds to a single land use decision maker represented as a farmer agent in the ABM. Data for CLU boundaries in Iowa are available to the public through the Natural Resources Geographic Information Systems (NRGIS) Library. However, the information about farm boundaries is confidential and thus cannot be publically displayed. In this study, we generate pseudo farm boundaries by utilizing the statistical distribution of farm sizes, and the spatial locations of farm centroids based on real information of farm boundaries.

4.1.3. Process Overview and Scheduling

Each year, farmer agents estimate commodity prices and fuel prices, calculate crop production costs and returns, and make decisions on crop type, tillage type, and fertilizer rates. To make these decisions, farmer agents try to maximize farm profit given the constraints of crop yield drag effect, and their risk aversions towards biofuel crop production.

More specifically, at each time step, for each farmer agent, the land use decision making process is implemented in two parts in the model. In Part I, linear programming is used to look for land use practices that maximize farm profit with planning horizon of 1 year given the crop choices of corn without stover, soybean and fallow. The 1-year planning horizon is used in biofuel crop market scenarios. In model verification, 1-year, 2-year, and 3-year planning horizons are used. In Part II, a decision-tree based algorithm is designed for comparing the profit rates determined from Part I with the profit rate from corn with stover at the simulation step, and with the average yearly profit rate from corn with stover and switchgrass calculated within a 10-year planning horizon (discount rate = 0.05). If the risk premium for biofuel crops is met and expected profits exceed those of traditional crops, the profit maximizing biofuel crop will be selected and planted. Total biofuel production is limited by farm specific acreage constraints. For next time step, the same two processes are implemented except that the land parcel that was previously enrolled in switchgrass is be excluded from the farm profit maximization until it is out of the 10-year enrollment limit.

4.2. Design Concepts

4.2.1. Principles

We assume that farmers are risk averse profit maximizers, and that there is little risk associated with the choice between corn and soybeans, since they are both well-established annual crops. Therefore, in the case of these crops the decision rule collapses to profit maximization. However, there is substantial risk associated with new markets and crops. Further, if these crops are perennials, they lock in farmers for substantial periods of time, thereby increasing both market and production risk.

In all cases, farm profit depends on yield potential for corn, soybean, and switchgrass specific to each land parcel. Corn and soybean yields are lower if the crops are not rotated, and corn yield responds to different levels of N rates. Because of the additional risk associated with cellulosic ethanol markets, farmer agents might require higher profits from biofuel crops than they do from traditional crops and limit the total acreage allocated to biofuel crops.

4.2.2. Emergence

Watershed-scale agricultural land use acreages and fertilizer application rates are modeled as an emergent property of farmer agent decisions. For model results, we expect to observe insensitivity of cropland switched to switchgrass and corn stover harvest in response to the price change due to farmers' risk aversions towards biofuel crop production.

4.3. Details

4.3.1. Initialization

We use NASS CDL data from 2001 to 2011 for initializing the land cover in the first two years of simulation and compare with the simulation results in the remaining years. NASS CDL is a raster data set at 30 meter spatial resolution. It contains land cover information of specific crop types including corn and soybean. We use the majority zonal operator in ArcGIS 10.0 to aggregate the pixel-specific crop type into the CLU level.

4.3.2. Input Data

The ABM inputs include prices and land parcel information. The price file contains annual time series of prices for crops (corn, soybean, switchgrass, corn stover), fertilizers (N, P, K), and fuels (diesel, LPG-liquid petroleum gas). The parcel information file contains land parcel specific data about acreage, CSR, yield potential for switchgrass, and land cover types for the previous two years. These data were used to construct net returns for each field in the watershed for each crop rotation [6]. In addition, a parameter file characterizing farmer agents' attitudes towards biofuel crop production (as in Table 1) is a necessary input into this model. The price data was based on United States Department of Agriculture's (USDA's) Economic Research Service, U.S. Energy Information Administration and Iowa State University Extension's budgets [38,40,41], the CSR data was obtained from the soil database ISPAID, the historical landcover was constructed using the CDL, and the attitudes of farmers were parameterized from a land use and attitudes survey conducted in 2010 [36,37].

4.3.3. Submodels

As mentioned, there are two parts of the decision making processes. In Part I, for current iteration year, the farm based objective function is:

$$maximize\ P = \sum_{i=1}^{n} (S_i - C_i) \tag{1}$$

where i is the land parcel index, n is the total number of land parcels within a farm, P is the profit, S is revenue from crop sale, and C is crop production cost. Equation (1) illustrates that the farm agents' objective is to maximize farm profit each year. With decision variables included, Equation (1) can be expressed as:

$$maximize\ P = \sum_{i=1}^{n} \sum_{j=1}^{m} f\left(c_{ij} X_{ij}\right) \tag{2}$$

where c_{ij} is the objective function coefficient corresponding to the jth variable for land parcel i, and X_{ij} is the jth decision variable for land parcel i, n is the total number of land parcels within a farm, and m is the total number of decision variables. As illustrated in Equation (2), the farm profit maximization for corn and soybean is a linear programming problem, which is scripted in the mathematical programming software (AIMMS) and implemented as an ABM using C++.

After the implementation of Part I, the decision variable values and the resulting profit rates from the AIMMS optimization component are passed to the decision-tree based algorithm in Part II. In this part, profit rates for biofuel crops (corn with stover harvest, and switchgrass) are evaluated with the agent-specific risk premiums and compared to commodity crop profit rates obtained from the Part I algorithm. The biofuel crop is chosen for a land parcel when the risk premium for biofuel crops is met and expected profits exceed those of traditional crops. Empirical parameters characterizing farmers' risk premiums about biofuel crops are incorporated in the decision-tree based algorithm. The parameters are derived from the land use survey and applied in the land use decision rule for switchgrass and corn stover production. The parameters and the sampling strategy of parameter values derived from the land use survey database are described in Table 1 and the Appendix B.

5. Data and Simulation Settings

5.1. Agricultural Landowners and Operators Survey

A survey of agricultural landowners and operators in the Clear Creek watershed was conducted in 2010 to obtain information about current farming practices and farmers' willingness to participate in conservation practices and biofuel crop production [36,37]. The survey questions were sent by mail to all non-urban landowners and agricultural operators (about 998) within the watershed. Responses were received from 397 of them (response rate 41.1%). A survey database was developed based on those responses. The database covers multiple topics including personal information, farm characteristics, and farmers' information sources and their attitudes towards watershed conservation and growing biofuel crops. For this study, we are interested in questions about (1) farm size; (2) minimum net profit rates that farmers require from biofuel crop production; (3) the acreage on which farmers would plant, harvest, and market corn stover and/or switchgrass; and (4) the proportion of corn stover harvested if such profits were realized (Table A1 in the Appendix B).

5.2. Price Scenarios and Simulation Settings

Price inputs to the model include commodity prices of corn, soybean, corn stover, switchgrass, fertilizers (N, P, K) and fuels (LPG and diesel). In this study, we run the ABM simulations with four different input datasets for commodity and biofuel crop prices to investigate: (1) the land use pattern resulting from profit maximization (Simulation Set I); (2) the sensitivity of land use pattern to commodity crop prices (Simulation Set II); (3) the impacts of biofuel crop prices on land use patterns given the risk associated with biofuel crop production (Simulation Set III, and Simulation Set IV).

5.2.1. Historical Market Prices

Simulation Set I is based on historical market prices from 2003 to 2011 (Table 2). The corn and soybean prices are the average prices in the calendar year (ISU cash corn and soybean prices) [42]. From 2003 to 2006, corn prices are replaced with the target prices of the counter-cyclical payments for corn. The fertilizer prices are the average U.S. farm prices of selected fertilizers in March or April published by USDA's Economic Research Service. The LPG price is the weekly Iowa propane residential price in the middle of March and the diesel price is the price of Midwest No. 2 diesel retail sales by all sellers. The data is published by the U.S. Energy Information Administration.

5.2.2. Price Scenarios for Corn and Soybean

Simulation Set II (Table 3) is based on price scenarios for 360 combinations of corn prices and soybean prices. By confining the price ranges within the historical records and the ten year baseline projections for U.S. agricultural markets by the Food and Agricultural Policy Research Institute, University of Missouri (FAPRI-MU) [43], we determine 24 levels of corn prices and 15 levels of soybean to corn price ratios (Table 3). The 24 levels of corn price start from 360 (24 × 15) combinations of corn price and the price ratio are used for generating price inputs into the model. For the

360 price input files, the other prices including switchgrass, corn stover, fertilizers (N, P, and K), and fuels (diesel and LPG) are held constant. Switchgrass and corn stover are not taken into account and so their prices are considered as zero. Prices of fertilizers and fuels are the average values of FAPRI-MU projections.

5.2.3. Price Scenarios for Switchgrass and Corn Stover

Simulation Set III (Table 3) is based on 16 levels of switchgrass prices. By referring to the projected prices of warm season grasses and corn stover for biofuel markets by FAPRI-MU [44], we determine 16 levels of switchgrass prices from 58 $/metric ton to 224 $/metric ton with an interval of 11 $/metric ton. The price ratio of switchgrass to corn stover is fixed as 1.31 according to the average value in FAPRI-MU (2011). The other price inputs for corn, soybean, fertilizers and fuels are fixed as the average prices for Year 2012 to 2021 in the FAPRI-MU projections. In Simulation Set III, simulations are run for each of the 16 price levels with the 30 samples of the empirical parameter set. The model is run for a total of 480 (16 × 30) simulations. The last set of simulations (Simulation Set IV) (Table 3) uses the same price settings as Simulation Set III except that the price ratio of switchgrass to corn stover varies from 1.1 to 2.1 with interval of 0.2 (a total of 6 levels) Therefore, in Simulation Set IV, price scenarios are 96 (16 × 6) combinations of switchgrass and corn stover prices. Similarly to Simulation Set IV, simulations are run for each of the 96 combinations with 10 samples of the empirical parameter set. The model is run for totally 960 (96 × 10) simulations.

Table 2. Historical market prices (corn, soybean switchgrass and corn stover prices in $/metric ton, fertilizer prices in $/kg, fuel prices in $/liter, soybean to corn price ratio is dimensionless).

Year	Corn	Soybean	Switch Grass	Corn Stover	N	P	K	Diesel	LPG	Soybean to Corn Ratio
2003	102.36 *	223.38	0	0	0.41	0.27	0.18	0.38	0.29	2.18
2004	103.54 *	281.06	0	0	0.42	0.29	0.20	0.44	0.28	2.71
2005	103.54 *	216.03	0	0	0.46	0.33	0.27	0.56	0.33	2.09
2006	103.54 *	203.91	0	0	0.57	0.36	0.30	0.75	0.39	1.97
2007	132.68	285.84	0	0	0.58	0.46	0.31	0.73	0.40	2.15
2008	188.19	417.00	0	0	0.83	0.88	0.62	1.16	0.56	2.22
2009	150.00	369.60	0	0	0.75	0.70	0.94	0.57	0.43	2.46
2010	151.97	362.26	0	0	0.55	0.56	0.56	0.80	0.46	2.38
2011	234.65	458.88	0	0	0.83	0.70	0.66	1.06	0.49	1.96

* Corn prices are replaced with the target prices of the counter-cyclical payments according to the Farm Security and Rural Investment Act of 2002.

Table 3. Price levels of alternative scenarios (corn and soybean prices in $/metric ton, switchgrass and corn stover prices in $/metric ton, fertilizer prices in $/kg, fuel prices in $/L, price ratios are dimensionless).

Simulation Set	Number of Combinations	Corn	Soybean	Soybean: Corn Price Ratio	Switchgrass	Corn Stover	SWG: Stvr Price Ratio	N	P	K	Diesel	LPG
II	360	Start = 102.36 To = 283.46 Step = 7.9 (24 levels)	Corn price × SC price ratio (15 levels)	Start = 1.8 To = 3.2 Step = 0.1 (15 levels)	0	0		0.89	0.80	0.76	1.03	0.48
III	16	186.6	419.94	2.25	Start = 58.41, To = 223.71, Step = 11 (16 levels)	SWG price/SS price ratio (16 levels)	1.31	0.89	0.80	0.76	1.03	0.48
IV	96	186.6	419.94	2.25	Start = 58.41, To = 223.71, Step = 11 (16 levels)	SWG price/SS price ratio (6 levels)	Start = 1.1, To = 2.1, Step = 0.2 (6 levels)	0.89	0.80	0.76	1.03	0.48

6. Results and Discussions

6.1. Model Verification

The yield drag incorporated in the ABM model on the basis of agronomic data effectively precludes the choice of continuous corn for the simulated farmers. Thus, farmers are *de facto* constrained to think in terms of corn-soybeans or corn-corn-soybeans in the model. In practice farmers may not behave as if the yield drag matters, which is verified by the crop rotation pattern comparison below.

6.1.1. Crop Rotation Pattern

Figure 2 compares the simulated crop rotations with the NASS CDL based crop rotations using graphs that summarize the land area percentage and the CLU count percentage corresponding to the maximum number of continuous crop years during the 11-year time period. In Figure 2a, we see that about 74% of the simulated land is in corn-corn-soybean and about 26% of land is in corn-soybean during any of the 11 years. In reality, however, less land is in corn-corn-soybean (about 39%) while about the same percentage of land (25%) is in corn-soybean; and there are some fields (about 36%) which were in continuous corn for three or more years. For soybean (Figure 2b), the simulated and real landscapes are very similar: Single-year soybean is predominant (more than 80%). This indicates that in reality farmers do not necessarily follow the rotation patterns of corn-corn-soybean or corn-soybean, and that the continuous corn years could be longer than two, but soybean is grown almost always only in rotation with corn.

Assuming the NASS CDL is reality, we can conclude the inclusion of yield drag in the model and the likely underestimate of the effect of yield drag by actual farmers causes the model to underestimate the amount of continuous corn, because simulated farmers are in practice restricted in their options. For example, after two years of corn a simulated farmer will plant soybeans even if relative crop prices suggest that corn is the more profitable crop because of the effects of yield drag on productivity. Yield drag is, however, hard for actual farmers to ascertain because they do not run long term controlled experiments in their fields, many factors change simultaneously from year to year, and yield drag is affected by tillage and weather. Unfortunately, though there is anecdotal evidence [45,46] that farmers may underestimate yield drag, there are no peer-reviewed studies that compare perception and reality on this issue.

Overall though, the model gives a reasonable approximation of the behavior of farmers in the watershed for the study period. For example, the average acreage of cropland in corn according to the CDL dataset was 56.7% and the simulated average using the one year planning horizon was 52.2%. Again, the underestimation of corn acreage is related to the yield drag issue.

93

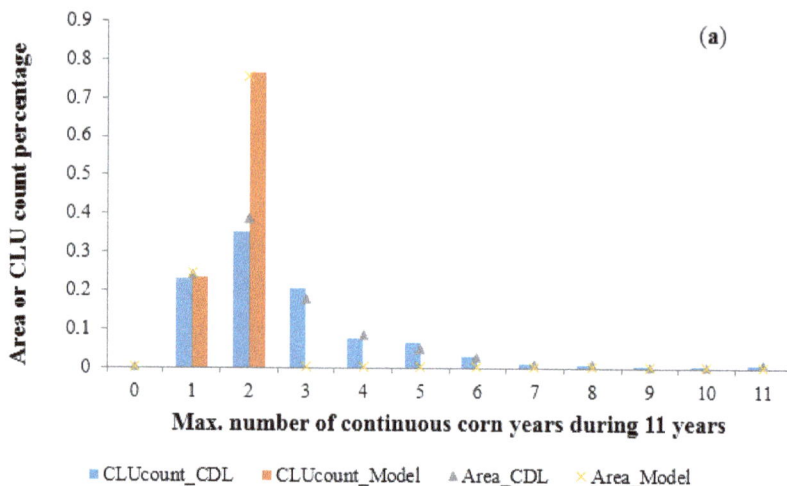

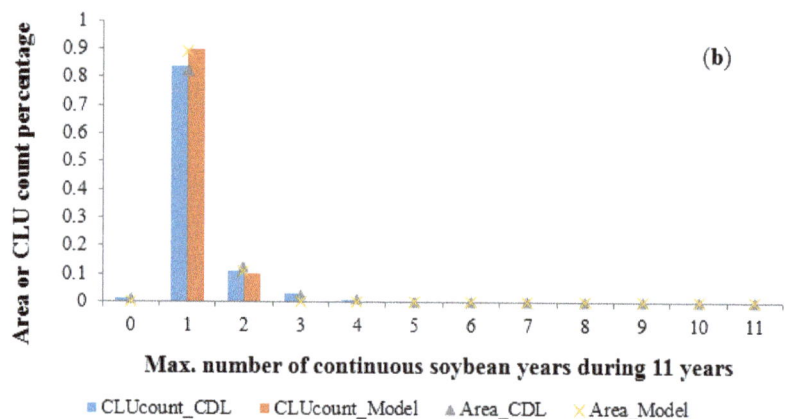

Figure 2. Maximum number of continuous corn (**a**); and soybean (**b**) years during an 11-year time period.

6.1.2. Field and Farm Scale Statistics

At the field scale, box plots are made for the cost and profit rates of corn and soybean throughout the 11 simulation years (Figure 3a–d). Generally, the simulated cost and profit rates increase through the years, and corn (Figure 3a,c) has higher cost and profit rates than soybean (Figure 3b,d). The box heights are generally larger in Figure 3c than Figure 3d, which indicates that the spatial variability (across fields) of the simulated cost rate of corn is much higher than soybean. The reason is that nitrogen fertilizer is only applied for corn and the N application rate varies across different corn fields depending on the physical characteristics of land (e.g., yield potential). At the farm scale, a box plot is made for the simulated profit rate over all simulation years (Figure 3e). The simulated average net farm income rates from 2008 to 2011 is about 694 $/ha, which is comparable to the overall average of typical cash rents from 2008 to 2011 for corn and soybean fields (respectively 460, 477, 484, and 541 $/ha) in the corresponding agricultural district (District 6) in Iowa [47]. The lower cash rent is likely due to the fact that cash rent is generally built as the average of recent past net returns and therefore lags.

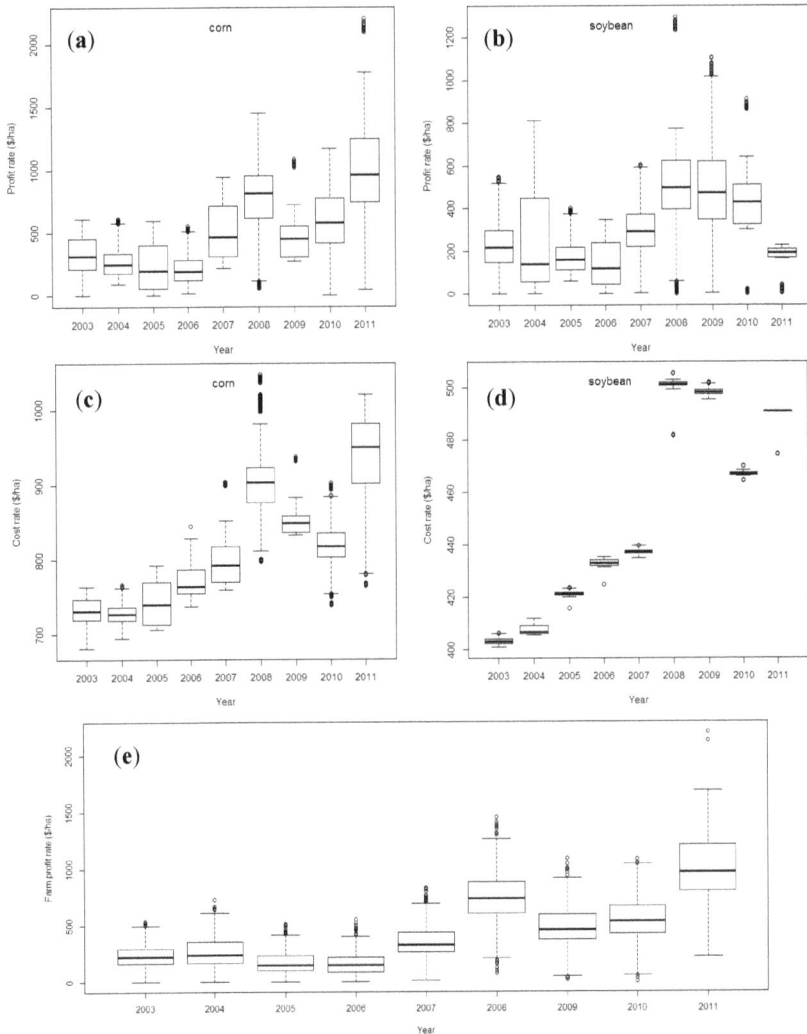

Figure 3. Box plots of simulated profit rate ((a) corn; (b) soybean); and cost rate ((c) corn; (d) soybean); and of farm-level profit rate (e).6.2. Model Results: Corn and Soybean Price Scenarios (Simulation Set II).

With the second set of simulation results, we plotted watershed-scale crop area percentages (Figure 4a,b) and fertilizer rates (Figure 4c,d) against changes in corn price and soybean to corn price ratio. As illustrated by Figure 4a, the percent of land in corn increases as corn price increases and/or as the ratio of soybean to corn price decreases. The percent of land in soybean changes with the opposite trends (Figure 4b). Corn and soybean lands stabilize at 50/50 within the corn price range of about 181.1–283.5 $/metric ton and the range of soybean to corn price ratio of about 2.8–3.2. Beyond this range, the percent of the landscape in corn and soybean is very sensitive to changes in corn price and/or soybean to corn price ratio.

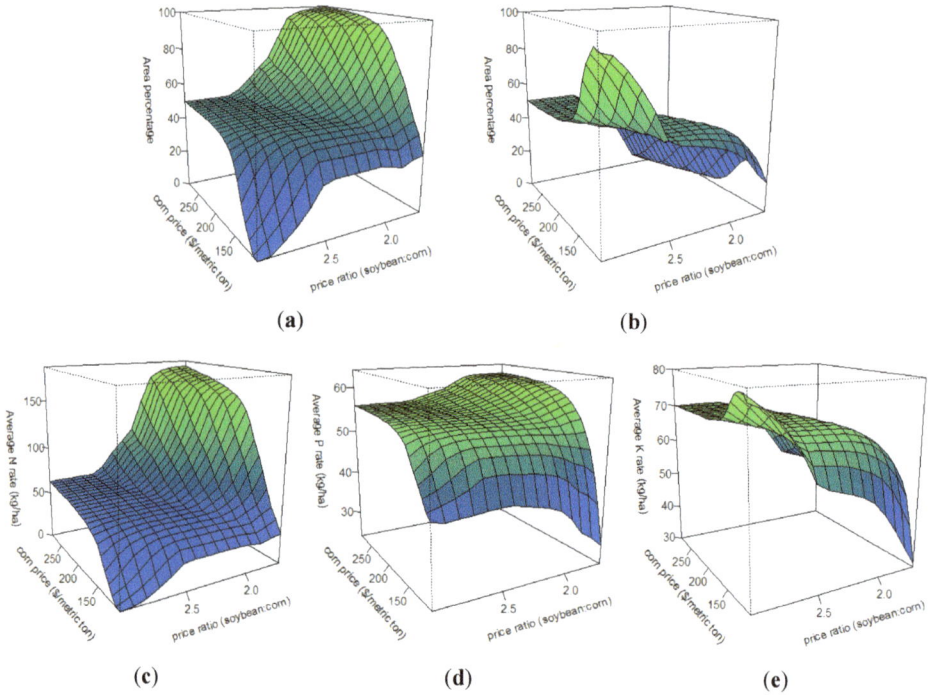

Figure 4. Simulated crop area percentage ((**a**) corn; (**b**) soybean); and fertilizer rates averaged over the whole watershed ((**c**) N; (**d**) P; (**e**) K) for 360 combinations of corn prices and soybean to corn price ratios.

According to Figure 4c, nitrogen responds to the changes in corn price and soybean to corn price ratio in a very similar way to corn acreage since N fertilizer is only applied for corn. The shape of the P surface is more similar to the corn land percentage surface than to the soybean, except that the lowest value of the average P application rate occurs when corn price is at the lowest (181.1 $/metric ton) and soybean to corn price ratio is at the lowest (2.8). This is because the average P application rate is related to both corn and soybean planting but dominated by corn since corn requires more P input than soybean. Figure 4e characterizes the response of K application rate to changes in corn price and the soybean to corn price ratio. The K surface has a shape more similar to that of the percent of the land in soybean than to corn, except that the lowest value of the average K application rate occurs at the same position as P. This is reasonable because the average K application rate is related to both corn and soybean planting but dominated by soybean since soybean requires more K input than corn. Within similar ranges (181.1–283.5 $/metric ton corn prices, and 2.8–3.2 of soybean to corn price ratio), the average N, P, and K rates stabilize respectively at around 65, 56, and 71 kg/ha.

6.2. Model Results: Switchgrass and Corn Stover Price Scenarios

6.2.1. Simulations with *vs.* without Land Use Survey Information Included (Simulation Set III)

We run the model with positive switchgrass prices as defined in Simulation Set III under two scenarios:

Scenario 1. The agricultural landowners and operators survey information is not included in the model, and agents do not appreciate that there is additional risk in planting perennial crops for a new market. In this scenario farmer agents follow the same rule of farm profit maximization as they do in Simulation Set I (profit maximization unaffected by farmer risk perception);

Scenario 2. Information from the farmer survey is incorporated into the model and used to parameterize the perceived risks associated with planting a new perennial crop. In this scenario, agent-specific constraints are incorporated into the model which stipulate the minimum expected economic return required before a farmer would consider biofuel crops and the maximum percent of their farm they would allocate to biofuel crop production if such returns could be realized.

As illustrated in Figure 5, the incorporation of farmer attitudes about risk has a significant impact on land use. Considerably more land is allocated to switchgrass and corn stover in scenario 1 (profit maximization, triangles in Figure 5) than in scenario 2 (constrained by perceived risk, circles in Figure 5). Even with 30 simulations for each of switchgrass price levels, the variations are low (in Figure 5, the dispersion of 30 circles is small compared to the discrepancy between the circles and triangles). Figure 5a–d shows that when perceived risk is modeled, the simulated crop composition starts to change at switchgrass price of $88.16/metric ton (switchgrass to corn stover ratio fixed at 1.31) and above: Corn (no stover) (Figure 5a) and soybean (Figure 5b) acreages start to decrease and switchgrass (Figure 5c) and corn (stover) (Figure 5d) start to increase. The crop area percentage curves approach a stabilization stage at switchgrass prices of about 132–165 $/metric ton. The crop composition stabilizes at about 32% corn (no stover), 38% soybean, 22% switchgrass, and 8% corn (stover).

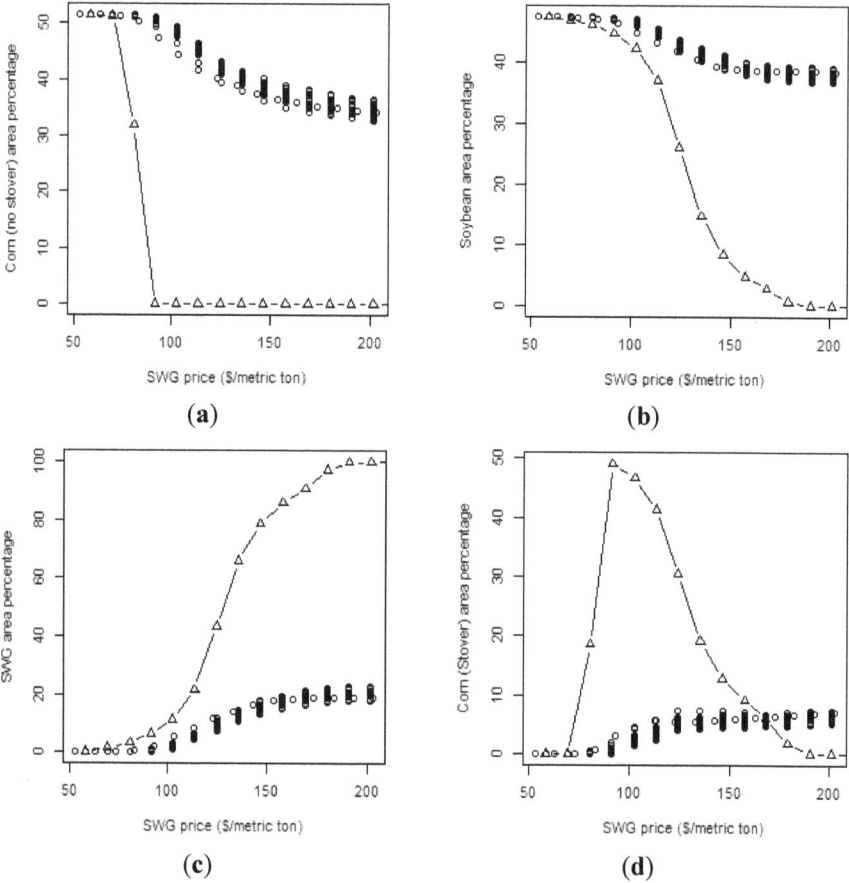

Figure 5. Crop area percentages ((a) corn—no stover; (b) soybean; (c) switchgrass; (d) corn—stover) from 36 simulations (36 SWG price levels and fixed switchgrass to corn stover price ratio of 1.31) on the basis of with (dots) *vs.* without (triangles in dashed line) land use survey information included.

Similarly, there is a large discrepancy in simulated watershed level fertilizer application rates between scenario 1 and 2 (Figure 6). The variation of the fertilizer rates with scenario 2 given 30 random samples from the empirical distribution is small. Under scenario 2 the simulated N and K curves are sigmoid shaped while the P curve mirrors the sigmoid shape. As the switchgrass price increases within a certain range, the average N rate and K rate increase while the average P rate decreases.

This is reasonable because P rate for switchgrass is relatively low compared to corn (stover or no stover) and soybean.

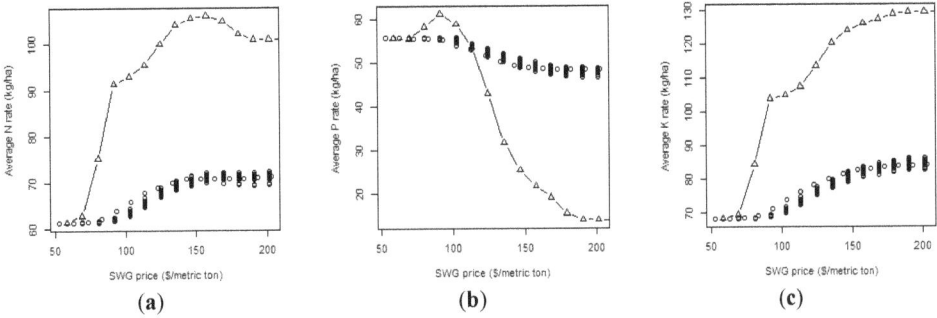

(a) (b) (c)

Figure 6. Watershed-averaged fertilizer rates ((a) N; (b) P; and (c) K) from 36 simulations (36 SWG price levels and fixed switchgrass to corn stover price ratio of 1.31) on the basis of with (dots) *vs.* without (triangles in dashed line) land use survey information included.

The two runs in Simulation Set II correspond to the commodity crop price settings in Simulation Set III, corn price is 186.6 $/metric ton and the price ratio of soybean to corn is 2.25. The two runs are (1) corn price is 181.1 $/metric ton, and the soybean to corn price ratio is 2.4; and (2) corn price is 188.98 $/metric ton, and the soybean to corn price ratio is 2.24. Considering the two runs in Simulation Set II as baseline scenarios, the watershed-average N rate in Simulation Set III (ranging from 62.77 to 71.14 kg/ha) is higher than in the baseline (between 60.53 and 62.77 kg/ha). So is the watershed-average K rate in Simulation III (ranging from 67.25 to 84.07 kg/ha) compared to the baseline (68.37 kg/ha). The watershed-average P rate in Simulation III (ranging from 47.08 to 56.04 kg/ha) is lower than in the baseline (56.04 kg/ha). These results show that biofuel crop markets may potentially result in more N and K inputs and less P inputs into the watershed. This is understandable because (1) in the baseline scenario, corn-soybean rotation only needs N input in corn years, while once enrolled in switchgrass planting, it requires N input every year; (2) switchgrass requires more K input than corn and soybean do. Since switchgrass potentially has beneficial effects of reducing nutrient and sediment runoff [48,49] while commodity crops management and harvesting corn stover may have adverse effects [50], it would therefore be very interesting to investigate the impacts of potential biofuel crop markets on watershed water quality in a further study.

6.2.2. Impacts of Corn Stover and Switchgrass Price (Simulation Set IV)

In Simulation Set IV, both the switchgrass price and the price ratio of switchgrass to corn stover vary. Switchgrass price varies from 58.41 $/metric ton to

223.71 $/metric ton, and the price ratio of switchgrass to corn stover varies from 1.1 to 2.1. For each combination of the two variables, we run the simulation 10 times and plotted the average of the 10 as a response "surface" (Figures 7 and 8).

Figure 7c,d shows that within the price ranges in Simulation Set IV, switchgrass occupies at most about 22%–24% of the watershed and corn stover occupies at most 10%. These numbers are consistent with those in Simulation Set III. In Figure 7a–d, at any fixed ratio of switchgrass: Corn stover (1.1–2.1), the crop area percent responds similarly to the curves represented by circles in Figure 6a–d except for corn (stover). At any fixed ratios greater than about 1.4, the corn (stover) area percentage curve does not stabilize within the switchgrass price range from 58.41 to 223.71 $/metric ton. Instead, it still keeps increasing even at the very high switchgrass price of 223.71 $/metric ton.

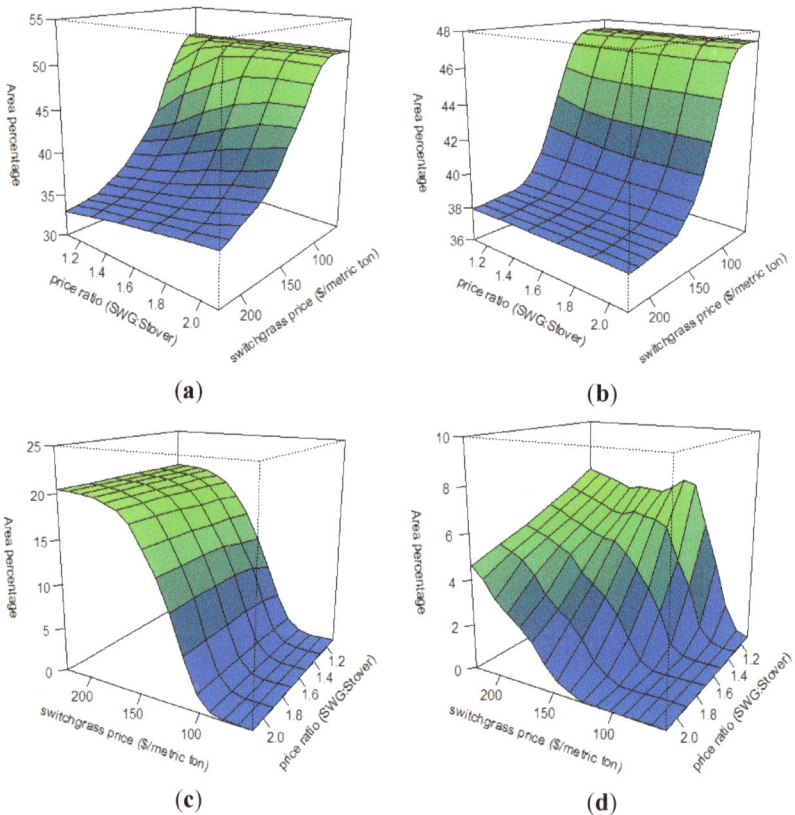

Figure 7. Crop area percentages ((a) corn—no stover; (b) soybean; (c) switchgrass; (d) corn—stover) from 960 simulations (16 levels of switchgrass price by 6 levels of switchgrass to corn stover price ratio, average of the 10 random samples based simulations for each combination).

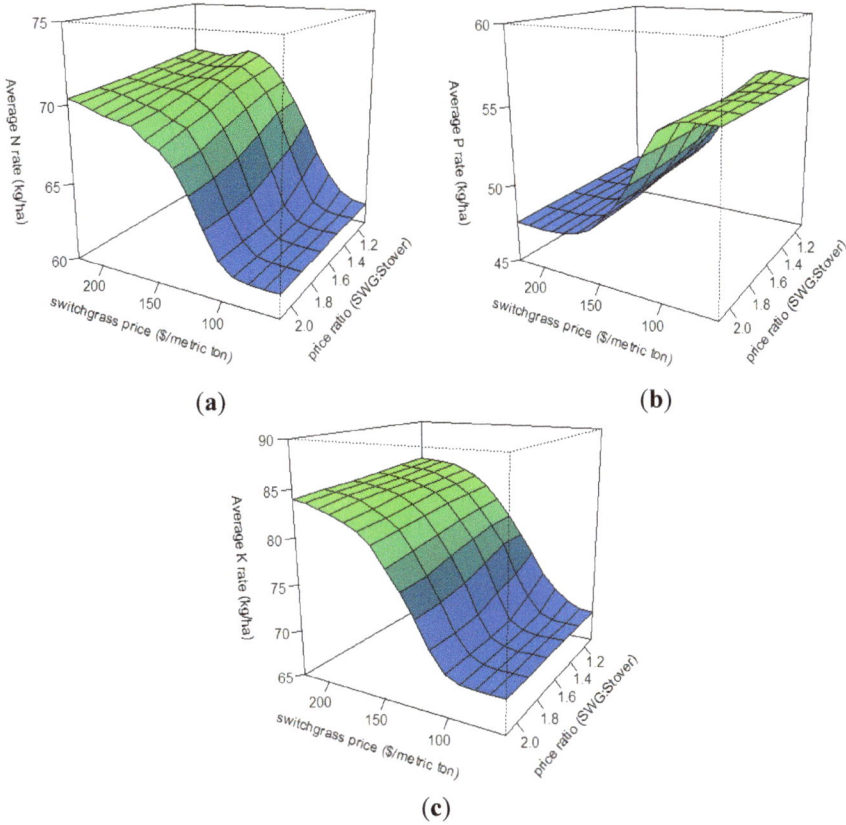

Figure 8. Watershed-averaged fertilizer rates ((**a**) N; (**b**) P; and (**c**) K) from 960 simulations (16 levels of SWG price by 6 levels of switchgrass to corn stover price ratio, average of the 10 random samples based simulations for each combination).

Overall, the crop area percentage in Figure 4a,b (corn and soybean scenarios) responds more rapidly to the change of price ratio than in Figure 7a–d (switchgrass and corn stover scenarios). We may conclude that the price ratio of switchgrass to corn stover does not influence the crop composition as much as the price ratio of soybean to corn.

In Figure 8a–c, we plotted watershed-averaged N, P, and K rates in response to the change of switchgrass prices and switchgrass to corn stover price ratios. Similarly to Figure 6a–c, the watershed-averaged N and K rates increase as switchgrass prices increase but the watershed-averaged P rate decreases as switchgrass prices increase. N, P, and K rates stabilize, respectively, at 67.25–73.98 kg/ha, 44.83–49.32 kg/ha, and 81.82–87.43 kg/ha when switchgrass price reaches 132.24–165.3 $/metric ton and higher. The fertilizer rates in the switchgrass-corn stover scenarios (Figure 8a–c) respond more rapidly to the change of switchgrass price than to the change of price ratio. At fixed level of switchgrass price, fertilizer rates hardly respond to the change of switchgrass to corn stover ratio within the range of 1.1 to 2.1.

7. Conclusions

The major findings in this research include: (1) discrepancies exist between simulated and satellite-derived land acreages and crop rotation patterns, with implications in terms of farmers' estimation on crop yield drag effects; (2) the simulated biofuel crop land acreage response surface starts to plateau at prices beyond $150/ton for switchgrass; and (3) simulated biofuel crop land acreages and fertilizer application rates in response to alternative crop market scenarios differ significantly depending on whether or not the risk averse behavior of farmers towards biofuel crop production is considered. These findings help define the decision space for future biofuel production, and provide insights for different stakeholders, such as agricultural policy makers who are concerned about commodity and biofuel crop land use and inputs of fertilizer into the agricultural system. Since currently no crop insurance is available for biofuel crops, our results also indicate the importance of creating such a program if biofuel crops are to be promoted.

The simulated corn and soybean land percentages respond to market prices (commodity crop, fertilizers, and fuel) more strongly than would be suggested by satellite imagery. The simulations strictly follow the rotation pattern of corn-soybean or corn-corn-soybean while according to the satellite data, about 36% of the crop land were in more than two-year continuous corn. The discrepancy between the simulations and satellite data suggests that farmers are underestimating the yield drag associated with continuous corn. The topic of farmers' yield drag perception and estimation *versus* agronomic evidence is worth further investigation and verification since misperceptions can result in non-optimal economic and environmental consequences.

The simulated corn and soybean land acreage and fertilizer rates responses to corn price change and/or soybean to corn price ratio changes are realistic. Those responses stabilize within 181.10–283.46 $/metric ton and the range of soybean to corn price ratio of about 2.61–2.99: Corn and soybean area percentages stay 50/50; average N, P, and K rates stabilize at respectively about 65.01, 56.04, and 70.61 kg/ha.

Given farmers' attitudes towards biofuel crop production, large scale cellulosic biofuel crop production is likely to require some mechanism to reduce risk for farmers. Our results indicate that if risk for planting switchgrass is eliminated and farmers only follow a farm-profit maximization rule, given a switchgrass to corn stover price ratio of 1.31, switchgrass would occupy the whole watershed when its price is about 179.63 $/metric ton and higher. However, in the perceived risk case, switchgrass would occupy at most about 22%–24% of the watershed when switchgrass price is about 132.24–165.3 $/metric ton or higher, and corn stover would occupy at most about 10%. Correspondingly, average N, P, and K rates stabilize respectively at 67.25–73.98 kg/ha, 44.83–49.32 kg/ha, and 81.82–87.43 kg/ha when switchgrass price reaches 132.24–165.3 $/metric ton and higher. The 2014 Farm bill eliminates crop subsidies, though temporary establishment subsidies for dedicated biomass feedstocks were maintained in the Energy title of the bill under the Biomass Crop Assistance Program (up to five years for a maximum of $111.2/ha). However, this is unlikely to suffice. Some type of subsidized crop insurance would also have to be developed.

Acknowledgments: This study was supported by the National Science Foundation Cyber-Enabled Discovery and Innovation Program (award 0835607), the National Science Foundation Dynamics of Coupled Natural and Human Systems Program (award 1114978), and the IIHR-Hydroscience and Engineering at the University of Iowa.

Author Contributions: Deng Ding designed the ABM, wrote the model scripts, analyzed the data and wrote the manuscript. David Bennett supervised the research, participated in designing the model, reviewed the manuscript, and provided suggestions and edits. Silvia Secchi provided data and citation sources about the economic models adapted in the model, reviewed the manuscript and provided suggestions and edits.

Conflicts of Interest: The authors declare no conflict of interest.

Appendix

Appendix A. Crop Yield Drag Coefficients and Fertilizer Rates

The effects of crop rotation, tillage type, and nitrogen fertilizer rates (corn only) are accounted for when calculating crop yields. Crop yield drag coefficients that capture the rates at which the crop productivity declines as a function of continuous planting, are derived from corn and soybean yield functions [7] and represented in Figure A1. Given the impact of the previous year's crop, continuous corn (CC) corresponds to lower yield than corn in annual rotation with soybean (SC). Soybean yields in the present year are influenced by the previous two years crops. Among the crop rotation series (corn-corn-soybean: CCS, corn-soybean-soybean: CSS, soybean-corn-soybean: SCS), CCS corresponds to the highest soybean yield, followed by SCS, and then CSS. The impacts of tillage types on crop yields are taken into account for both corn and soybean. Among the three tillage types (conventional tillage, mulch tillage, and no tillage), the crop with conventional tillage has the highest yield and no till has the lowest. Instead of using a quadratic function [7] to characterize corn yield response to nitrogen, ten different levels of nitrogen rates and corresponding corn yield drag coefficients are sampled from the response curve. This implementation reduces the optimization problem from a nonlinear into a linear programming problem.

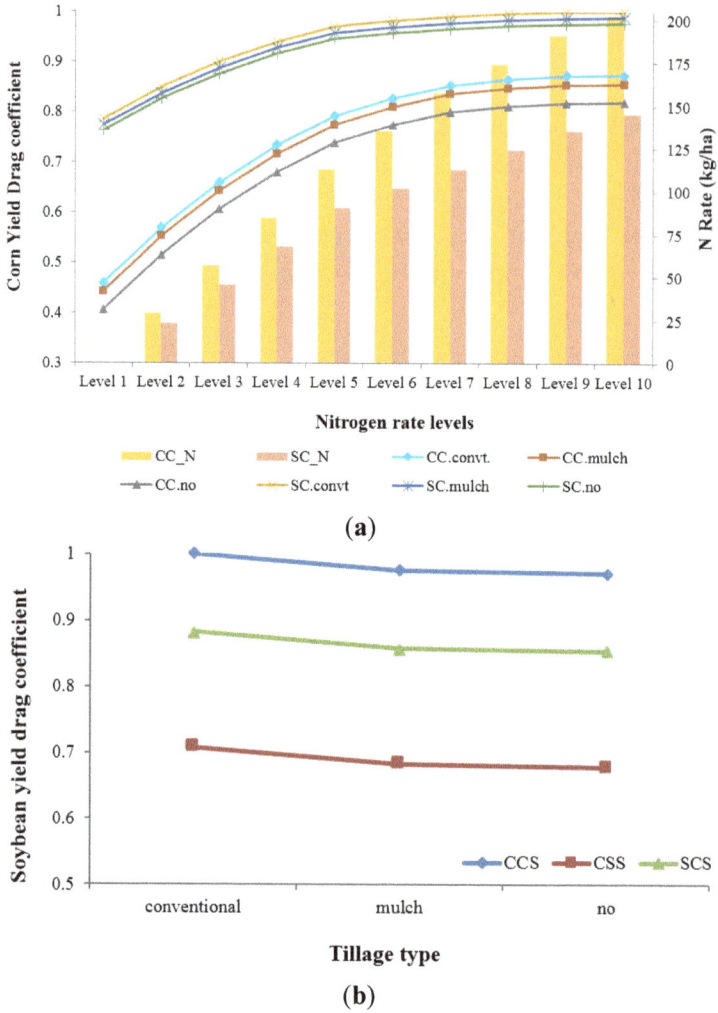

Figure A1. Corn yield drag coefficients (**a**) for different rotation, tillage, and N fertilizer rate levels, and soybean yield drag coefficients; (**b**) for different rotation types: Corn is denoted as "C" and soybean as "S", the last letter in a rotation pattern means the current crop (e.g., CCS means corn-corn-soybean, and soybean is in the current year).

In the model, it is assumed that soybean does not need nitrogen fertilizer. The fertilizer rates of phosphorus (P) and potassium (K) are fixed for soybean: Respectively, 44.83 kg/ha and 84.07 kg/ha. According to Iowa State University Extension [38,51], corn after corn requires lower fertilizer input of phosphorus (P) and potassium (K) than corn after soybean. For corn after corn, the P rate is 61.65 kg/ha and the K rate is 50.44 kg/ha; for corn after soybean, the rates are 67.25 kg/ha and 56.04 kg/ha.

Switchgrass is a perennial with an establishment period. Therefore, we assume that the switchgrass starts to be harvested from the second year. Switchgrass yields depend on soil type. We scale the tall introduced grasses yield (TIGRSYLD) in the Iowa Soil Properties and Interpretations Database (ISPAID) into the data range from 4.48 to 14.35 metric ton/ha to represent the switchgrass yield for different soil types. According to Duffy (2008) [40], in the first year when switchgrass is established nitrogen fertilizer is not required, but 33.63 kg/ha of P and 44.83 kg/ha of K are applied. 112.09 kg/ha of N is applied from the second year on to compensate for the harvest of switchgrass. The amount of P and K applied depends on the amount of switchgrass biomass harvested. For each ton of switchgrass that is harvested, 0.88 kg of P and 10.34 kg of K are applied.

Corn stover yield is estimated as the product of the portion of corn stover harvested and corn yield (we assume that dry weight of corn grain is equal to corn stalk weight). Additional amounts of N, P, and K fertilizers are applied to compensate for the nutrients removed from the soil system by corn stover harvest. 9.07 kg of N, 2.68 kg of P, and 11.34 kg of K are applied per ton of dry matter stover harvest [41].

Appendix B. Empirical Parameters and Parameterization

To generate samples from the agricultural landowners and operators survey database for the eight parameters above, we performed the following analysis on the survey responses. Firstly, we considered the respondents who self-identified as farmers. Concerning the parameter *Type*, the farmer respondents were classified into the four types based on whether they responded to Question 48 and/or 56 (Table A1). Secondly, we performed one-way ANOVA test on the farm sizes (*Area*) of the four types of farmers. The results showed that there is sufficient evidence (significance level: 0.001) to reject the hypothesis that the farm size means are all equal among the four farmer types. Thirdly, we analyzed the correlation among the other parameters (Table A2). The results do not support the conclusion that *Percent1 SWG* and *Profit SWG* are correlated for Type 2 and 3 farmers. Neither is there evidence to suggest that *Percent1 Stvr* and *Profit Stvr* are correlated for *Type* 1 and 3 farmers. But for both *Type* 2 and 3 farmers, *Percent1 SWG* and *Percent2 SWG* are significantly correlated. Similarly, *Percent1 Stvr* and *Percent2 Stvr* are significantly correlated for *Type* 1 and 3 farmers. For *Type* 3 farmers, *Profit SWG* and *Profit Stvr* are significantly correlated as are *Percent1 SWG* and *Percent1 Stvr*.

Table A1. Land use survey questions adopted for model parameterization.

Question #	Question
3	Do you consider yourself a farmer?
18c	Number of acres you own that you farmed in 2009
18d	Number of acres you leased or rented to other people to farm in 2009
18e	Number of acres you rented from others to farm in 2009
18f	Number of acres in the Conservation Reserve Program (CRP) in 2009
48	What is the minimum net profit per acre you would need to get in order to consider marketing corn stover?
49	If you could get that profit per acre, how many acres of corn stover would you consider harvesting?
50	If you supplied corn stover to a bio-refinery, would you prefer to harvest 30%, 50%, or 70% of the corn stover in your fields?
51	If you could get a net profit 50% higher than what you indicated in question 48, how many acres of corn stover in total would you consider harvesting?
56	What is the minimum net profit per acre you would need to get in order to consider growing switchgrass?
57	If you could get that profit per acre, how many acres of switchgrass you consider planting?
58	If you could get a net profit 50% higher than what you indicated in question 55, how many acres of switchgrass in total would you consider growing?

Based on the results of this analysis, we used the following strategy to empirically draw 30 samples using the bootstrap sampling method. First, farmer respondents and farmer agents are respectively classified into five groups based on the land acreages they reported ($\leq$40.47, 40.47–121.41, 121.41–202.34, 202.34–303.51, and >303.51 ha). Second, for all agents in a certain group, the sample of parameter *Type* is drawn from the empirical distribution of *Type* for farmers in the corresponding group. Third, the remaining parameter values are generated for farmer agents in each group and of each type. For *Type* 1 agents, parameter values of *Profit Stvr*, *Percent1 Stvr* and *Portion Stvr* are independently generated; parameter values of *Percent2 Stvr* are associated with *Percent1 Stvr*. For *Type* 2 agents, parameter values of *Profit SWG*, *Percent1 SWG* are independently generated; parameter values of *Percent2 SWG* are associated with *Percent1 SWG*. For *Type* 3 agents, parameter values for *Profit Stvr*, *Percent1 Stvr*, and *Portion Stvr* are independently generated; parameter values of *Profit SWG* are associated with *Profit Stvr*; *Percent2 Stvr*, *Percent1 SWG* and *Percent2 SWG* are associated with *Percent1 Stvr*.

Table A2. Pearson's r (*p*-value) between the parameters characterizing farmers' attitudes.

	Type	Profit SWG	Percent1 SWG	Percent2 SWG	Profit Stvr	Percent1 Stvr	Percent2 Stvr	Portion Stvr
Profit SWG	1 or 2	1	0.4145 (0.1407)	na *	na	na	na	na
	3	1	−0.2488 (0.1560)	na	0.8471 (2.03×10^{-11})	na	na	na
Percent1 SWG	1 or 2		1	0.9687 (4.18×10^{-11})	na	na	na	na
	3		1	0.8885 (1.06×10^{-12})	na	0.6822 (3.31×10^{-6})	na	na
Percent2 SWG	1 or 2			1	na	na	na	na
	3			1	na	na	na	na
Profit Stvr	1 or 2				1	0.1918 (0.4768)	na	−0.0525 (0.8526)
	3				1	−0.1714 (0.3248)	na	−0.0648 (0.7033)
Percent1 Stvr	1 or 2					1	0.8897 (9.01×10^{-6})	na
	3					1	0.8952 (1.76×10^{-13})	na
Percent2 Stvr	1 or 2						1	na
	3						1	na
Portion Stvr	1 or 2							1
	3							1

* No Data.

References

1. O'Neal, M.R.; Nearing, M.A.; Vining, R.C.; Southworth, J.; Pfeifer, R.A. Climate change impacts on soil erosion in Midwest United States with changes in crop management. *Catena* **2005**, *61*, 165–184.
2. Lettenmaier, D.P.; Hooper, E.R.; Wagoner, C.; Faris, K.B. Trends in stream quality in the continental United States, 1978–1987. *Water Resour. Res.* **1991**, *27*, 327–339.
3. Yadav, V.; Malanson, G.P.; Bekele, E.; Lant, C. Modeling watershed-scale sequestration of soil organic carbon for carbon credit programs. *Appl. Geogr.* **2009**, *29*, 488–500.
4. Bekele, E.G.; Lant, C.L.; Soman, S.; Misgna, G. The evolution and empirical estimation of ecological-economic production possibilities frontiers. *Ecol. Econ.* **2013**, *90*, 1–9.
5. Secchi, S.; Gassman, P.W.; Jha, M.; Kurkalova, L.; Kling, C.L. Potential water quality changes due to corn expansion in the Upper Mississippi River Basin. *Ecol. Appl.* **2011**, *21*, 1068–1084.
6. Secchi, S.; Kurkalova, L.; Gassman, P.W.; Hart, C. Land use change in a biofuels hotspot: The case of Iowa, USA. *Biomass Bioenergy* **2011**, *35*, 2391–2400.

7.   Kurkalova, L.A.; Secchi, S.; Gassman, P.W. Corn stover harvesting: Potential supply and water quality implications. In *Handbook of Bioenergy Economics and Policy*; Khanna, M., Scheffran, J., Zilberman, D., Eds.; Springer-Verlag: New York, NY, USA, 2010; pp. 307–323.

8.   Lee, K.H.; Isenhart, T.M.; Schultz, R.C.; Mickelson, S.K. Nutrient and sediment removal by switchgrass and cool-season grass filter strips in central Iowa, USA. *Agrofor. Syst.* **1998**, *44*, 121–132.

9.   Sarkar, S.; Miller, S.A.; Frederick, J.R.; Chamberlain, J.F. Modeling nitrogen loss from switchgrass agricultural systems. *Biomass Bioenergy* **2011**, *35*, 4381–4389.

10.   Parker, D.C.; Manson, S.M.; Janssen, M.A.; Hoffmann, M.J.; Deadman, P. Multi-agent systems for the simulation of land-use and land-cover change: A review. *Ann. Assoc. Am. Geogr.* **2003**, *93*, 314–337.

11.   Bennett, D.A.; Tang, W.; Wang, S. Toward an understanding of provenance in complex land use dynamics. *J. Land Use Sci.* **2011**, *6*, 211–230.

12.   Matthews, R.; Gilbert, N.G.; Roach, A.; Polhill, J.; Gotts, N.M. Agent-based land-use models: A review of applications. *Landsc. Ecol.* **2007**, *22*, 1447–1459.

13.   Robinson, D.T.; Brown, D.G.; Parker, D.C.; Schreinemachers, P.; Janssen, M.A.; Huigen, M.G.A.; Wittmer, H.; Gotts, N.M.; Promburom, P.; Irwin, E.G.; *et al.* Comparison of empirical methods for building agent-based models in land use science. *J. Land Use Sci.* **2007**, *2*, 31–55.

14.   Smajgl, A.; Brown, D.G.; Valbuena, D.; Huigen, M.G.A. Empirical characterisation of agent behaviours in socio-ecological systems. *Environ. Model. Softw.* **2011**, *26*, 837–844.

15.   Manson, S.M. Challenges in evaluating models of geographic complexity. *Environ. Plan. B Plan. Design* **2007**, *34*, 245–260.

16.   Matthews, R. The people and landscape model (palm): Towards full integration of human decision-making and biophysical simulation models. *Ecol. Model.* **2006**, *194*, 329–343.

17.   Matthews, R.; Selman, P. Landscape as a focus for integrating human and environmental processes. *J. Agric. Econ.* **2006**, *57*, 199–212.

18.   Epstein, J.M. *Generative Social Science: Studies in Agent-Based Computational Modeling*; Princeton University Press: Princeton, NJ, USA, 2006.

19.   Lansing, J.S.; Kremer, J.N. Emergent properties of Balinese water temple networks: Coadaptation on a rugged fitness landscape. *Am. Anthropol.* **1993**, *95*, 97–114.

20.   Le, Q.B.; Park, S.J.; Vlek, P.L.G. Land Use Dynamic Simulator (LUDAS): A multi-agent system model for simulating spatio-temporal dynamics of coupled human-landscape system 2. Scenario-based application for impact assessment of land-use policies. *Ecol. Inform.* **2010**, *5*, 203–221.

21.   Le, Q.B.; Park, S.J.; Vlek, P.L.G.; Cremers, A.B. Land-Use Dynamic Simulator (LUDAS): A multi-agent system model for simulating spatio-temporal dynamics of coupled human-landscape system. I. Structure and theoretical specification. *Ecol. Inform.* **2008**, *3*, 135–153.

22.   Bousquet, F.; Le Page, C. Multi-agent simulations and ecosystem management: A review. *Ecol. Model.* **2004**, *176*, 313–332.

23. Sengupta, R.; Lant, C.; Kraft, S.; Beaulieu, J.; Peterson, W.; Loftus, T. Modeling enrollment in the Conservation Reserve Program by using agents within spatial decision support systems: An example from southern Illinois. *Environ. Plan. B Plan. Des.* **2005**, *32*, 821–834.

24. Scheffran, J.; BenDor, T. Bioenergy and land use: A spatial-agent dynamic model of energy crop production in Illinois. *Int. J. Environ. Pollut.* **2009**, *39*, 4–27.

25. Ng, T. Response of Farmers' Decisions and Stream Water Quality to Price Incentives for Nitrogen Reduction, Carbon Abatement, and Miscanthus Cultivation: Preditions Based on Agent-Based Modeling Coupled with Water Quality Modeling. Ph.D. Dissertation, University of Illinois at Urbana-Champaign, Urbana-Champaign, IL, USA, 2010.

26. Ng, T.; Eheart, J.W.; Cai, X.M.; Braden, J.B. An agent-based model of farmer decision-making and water quality impacts at the watershed scale under markets for carbon allowances and a second-generation biofuel crop. *Water Resour. Res.* **2011**, *47*.

27. Schreinemachers, P.; Berger, T. Land use decisions in developing countries and their representation in multi-agent systems. *J. Land Use Sci.* **2006**, *1*, 29–44.

28. Berger, T.; Schreinemachers, P. Creating agents and landscapes for multiagent systems from random samples. *Ecol. Soc.* **2006**, *11*, 19.

29. Castella, J.C.; Verburg, P.H. Combination of process-oriented and pattern-oriented models of land-use change in a mountain area of Vietnam. *Ecol. Model.* **2007**, *202*, 410–420.

30. Torii, D.; Bousquet, F.; Ishida, T.; Trébuil, G.; Vejpas, C. *Using Classification Learning in Companion Modeling Multi-Agent Systems for Society*; Lukose, D., Shi, Z., Eds.; Springer: Berlin/Heidelberg, Germany, 2009; Volume 4078, pp. 255–269.

31. Brown, D.G.; Robinson, D.T. Effects of heterogeneity in residential preferences on an agent-based model of urban sprawl. *Ecol. Soc.* **2006**, *11*, 46.

32. Huigen, M.G.A. First principles of the mameluke multi-actor modelling framework for land use change, illustrated with a Philippine case study. *J. Environ. Manag.* **2004**, *72*, 5–21.

33. Castillo, D.; Saysel, A.K. Simulation of common pool resource field experiments: A behavioral model of collective action. *Ecol. Econ.* **2005**, *55*, 420–436.

34. Irwin, E.G.; Bockstael, N.E. Interacting agents, spatial externalities and the evolution of residential land use patterns. *J. Econ. Geogr.* **2002**, *2*, 31–54.

35. Brown, D.G.; Robinson, D.T.; An, L.; Nassauer, J.I.; Zellner, M.; Rand, W.; Riolo, R.; Page, S.E.; Low, B.; Wang, Z.F. Exurbia from the bottom-up: Confronting empirical challenges to characterizing a complex system. *Geoforum* **2008**, *39*, 805–818.

36. Druschke, C.G.; Secchi, S. The impact of gender on agricultural conservation knowledge and attitudes in an Iowa watershed. *J. Soil Water Conserv.* **2014**, *69*, 12.

37. Varble, S.; Druschke, C.G.; Secchi, S. An examination of growing trends in land tenure and conservation practice adoption: Results from a farmer survey in Iowa. *Environ. Manag.* **2015**.

38. Iowa State University Extension. *Nitrogen Fertilizer Recommendations for Corn in Iowa*; Leopold Center: Ames, IA, USA, 1997; Available online: http://www.iasoybeans.com/advancenewsletter/PDF/ADV15_0611_4_PM1714.pdf (accessed on 1 October 2015).

39. Grimm, V.; Berger, U.; DeAngelis, D.L.; Polhill, J.G.; Giske, J.; Railsback, S.F. The ODD protocol: A review and first update. *Ecol. Model.* **2010**, *221*, 2760–2768.

40. Duffy, M. *Estimated Costs for Production, Storage and Transportation of Switchgrass*; Iowa State University Extension: Ames, IA, USA, 2008; Available online: https://www.extension.iastate.edu/agdm/crops/pdf/a1-22.pdf (accessed on 1 October 2015).

41. Edwards, W. *Estimating a Value for Corn Stover*; Iowa State University Extension: Ames, IA, USA, 2014; Available online: https://www.extension.iastate.edu/agdm/crops/pdf/a1-70.pdf (accessed on 1 October 2015).

42. Johanns, A.M. *Iowa Cash Corn and Soybean Prices*; Iowa State University Extension and Outreach: Ames, IA, USA, 2015; Available online: http://www.extension.iastate.edu/agdm/crops/pdf/a2-11.pdf (accessed on 1 October 2015).

43. Food and Agricultural Policy Research Institute. *U.S. Baseline Briefing Book: Projections for Agricultural and Biofuel Markets*; University of Missouri: Columbia, MO, USA, 2015; Available online: http://www.fapri.missouri.edu/wp-content/uploads/2015/03/FAPRI-MU-Report-01-15.pdf (accessed on 1 October 2015).

44. Food and Agricultural Policy Research Institute. *Competition for Biomass among Renewable Energy Policies: Liquid Fuels Mandate versus Renewable Electricity Mandate*; University of Missouri: Columbia, MO, USA, 2011; Available online: http://www.fapri.missouri.edu/wp-content/uploads/2015/02/FAPRI-MU-Report-11-11.pdf (accessed on 1 October 2015).

45. Davidson, D. *Corn on Corn: How Much Yield is Enough*; Progressive Farmer: Birmingham, AL, USA, 2007.

46. Smith, P. *The Pros and Cons of Going corn-on-Corn*; Fields of Facts: Brandon, MS, USA, 2013; Available online: http://agfax.com/2013/09/23/the-pros-and-cons-of-going-corn-on-corn/ (accessed on 1 October 2015).

47. Edwards, W.; Johanns, A.M. *Cash Rental Rates for Iowa 2012 Survey*; Iowa State University Extension and Outreach: Ames, IA, USA, 2013; Available online: https://www.extension.iastate.edu/agdm/wholefarm/pdf/c2--10_2012.pdf (accessed on 1 October 2015).

48. McLaughlin, S.B.; Walsh, M.E. Evaluating environmental consequences of producing herbaceous crops for bioenergy. *Biomass Bioenergy* **1998**, *14*, 317–324.

49. Lee, K.H.; Isenhart, T.M.; Schultz, R.C. Sediment and nutrient removal in an established multi-species riparian buffer. *J. Soil Water Conserv.* **2003**, *58*, 1–8.

50. Mann, L.; Tolbert, V.; Cushman, J. Potential environmental effects of corn (*Zea mays* L.) stover removal with emphasis on soil organic matter and erosion. *Agric. Ecosyst. Environ.* **2002**, *89*, 149–166.

51. Iowa State University Extension. *A General Guide for Crop Nutrient and Limestone Recommendations in Iowa*; Iowa State University Extension and Outreach: Ames, IA, USA, 2013; Available online: http://store.extension.iastate.edu/Product/A-General-Guide-for-Crop-Nutrient-and-Limestone-Recommendations-in-Iowa-PDF (accessed on 1 October 2015).

Why Don't More Farmers Go Organic? Using A Stakeholder-Informed Exploratory Agent-Based Model to Represent the Dynamics of Farming Practices in the Philippines

Laura Schmitt Olabisi, Ryan Qi Wang and Arika Ligmann-Zielinska

Abstract: In spite of a growing interest in organic agriculture; there has been relatively little research on why farmers might choose to adopt organic methods, particularly in the developing world. To address this shortcoming, we developed an exploratory agent-based model depicting Philippine smallholder farmer decisions to implement organic techniques in rice paddy systems. Our modeling exercise was novel in its combination of three characteristics: first, agent rules were based on focus group data collected in the system of study. Second, a social network structure was built into the model. Third, we utilized variance-based sensitivity analysis to quantify model outcome variability, identify influential drivers, and suggest ways in which further modeling efforts could be focused and simplified. The model results indicated an upper limit on the number of farmers adopting organic methods. The speed of information spread through the social network; crop yields; and the size of a farmer's plot were highly influential in determining agents' adoption rates. The results of this stylized model indicate that rates of organic farming adoption are highly sensitive to the yield drop after switchover to organic techniques, and to the speed of information spread through existing social networks. Further research and model development should focus on these system characteristics.

Reprinted from *Land*. Cite as: Schmitt Olabisi, L.; Wang, R.Q.; Ligmann-Zielinska, A. Why Don't More Farmers Go Organic? Using A Stakeholder-Informed Exploratory Agent-Based Model to Represent the Dynamics of Farming Practices in the Philippines. *Land* **2015**, *4*, 979–1002.

1. Introduction

Acreage under organic farming methods is increasing globally, as government support for organic farmers and market demand for organic products grows [1]. A major driving force behind both of these trends is the recognition of the negative impacts of chemically-intensive farming methods on the environment and human health [2,3]. Approximately one-third of global land under organic production is located in the developing world [1]. In many developing nations, agricultural production is dominated by smallholder farmers, who often lack access to crop insurance or inexpensive credit. Some have argued that organic agriculture can benefit smallholder farmers by eliminating their reliance on expensive, fossil fuel-derived chemical inputs [4]. Organic farming therefore could make smallholder farmers more resilient to input price shocks, which are a significant source of insecurity for them [5]. This argument is bolstered by research from the United States, which indicates that organic farmers spend less on inputs than conventional farmers [6]. Policy makers, extension agents, and non-governmental organizations that serve developing-world farmers need a clear understanding of farmers' motivations and challenges in converting to organic agriculture, so that efforts to promote organic farming will be maximally effective.

Few organic adoption studies to date have portrayed a dynamic and complex decision environment, rather than a static snapshot of farmer decision-making [7]. Farmers' demographic characteristics, their economic motivations, their concern for the environment and for their families' health, all interact in complex and heterogeneous ways as they consider their options and choose their production method. Moreover, farmers' choices may be different year to year, depending on the decision context. They may choose to convert to organic, and then return to conventional methods when their situation changes. Most previous studies have not taken advantage of modeling tools that could portray this type of decision-making in dynamic contexts, such as agent-based modeling (ABM).

ABM has been used extensively in studying land-use conversion decisions, and has provided insights into how decision/environment feedback loops operate to produce non-intuitive outcomes [8–15]. Another advantage of ABM is its ability to represent the dynamics of social influences and information propagation through social networks via peer-to-peer, or "word-of-mouth", communications [16,17]. Previous studies have found that a farmer's source of information about farming and organic techniques influences his or her decision to adopt organic methods [18]. Peer-to-peer sharing through social networks can be an important source of information for farmers [19]. Modeling the spread of information about organic agricultural techniques through social networks could therefore provide insight into how best to promote these technologies and support farmer adoption.

ABM has been successfully applied to agricultural systems all over the world [20–22], to study the complex interactions between farmers, global and regional crop markets, and biophysical (especially hydrological and crop-soil) systems [8,13,20,23,24]. Agricultural applications vary from new practice adoption [8,14,20,24–27], changes in agricultural production and its viability [14,25,28], the impact of different decision making practices on agricultural land use/land cover change [14,23,25,28],evaluation of landscape structure [13,28], farmers' imitative behavior [8,29], and explicit analysis of agricultural policies [8,20,30]. For example, Berger [8] uses the concept of innovation diffusion to study various farming production alternatives and their influence on local hydrology in Chile. Happe *et al.* [30] present an ABM for agricultural policy analysis of different farm structures in Germany, Schreinemachers *et al.* [31] build an ABM to analyze the diffusion of greenhouse agriculture in Thailand, and Evans *et al.* [25] develop an empirically-rich ABM to explore the transition from shifting cultivation to rubber production in Laos. Many researches stress the usefulness of ABM to represent the diversity of farmer decision-making [13,20,27,30].

Of particular interest to our study are ABM applications that simulate organic farming adoption [26,27]. For example, Kaufmann *et al.* [27] develop an agent-based model of agricultural decision making that utilizes the theories of planned behavior and innovation diffusion, coupled with a survey-informed social network sub-model, to evaluate the diffusion of organic farming in Latvia and Estonia. The goal of their modeling exercise is to test how the economic changes are intertwined with agent-agent interactions in the formation of beliefs concerning the transition to organic farming. They conclude that mere social influence is not sufficient when modeling conversion from traditional to organic farming. Without exogenous economic factors, in the form of subsidies, *"organic farmers remain organic, and conventional farmers remain conventional"* (p. 2589). In a comparable study, Deffuant *et al.* [26] develop an ABM to model organic farming in a selected region of France. The novelty of their approach lies in introducing an auxiliary *institution* agent that evaluates the farm potential and assists the farmer in the decision to go organic. Similarly to Kaufmann *et al.* [27], they conclude that organic farming adoption results from complex interactions between economic and social processes.

In this paper, we define ABM as a simulation environment composed of heterogeneous computational entities (called agents) that represent Philippine smallholder farmers, for the purpose of exploring group dynamics around organic agriculture adoption. The agents are situated in a common agricultural environment of an upland paddy rice system, where organic farming practices are being actively promoted by a local non-governmental organization (NGO). Agents' decision making to adopt and maintain organic agriculture is constrained by limited access to information about the optimal farming strategies and the economic resources available. The agents are driven by their individual goals and social behaviors, and they constantly adapt to changing agro-economic and ecological conditions [8,12,32]. Previous agricultural ABM research suggests that, to account for a fuller complexity of farmer decision making and provide room for experimentation aimed at sustainable resource use, agent behavior should be informed by both normative science (economics) and social science that more realistically represents the actual resource-use decision making [33]. Thus, in our model, the socioeconomic micro-decisions of agents, strengthened by the interactions among them, generate macro-structures of the system including the level of organic farming adoption. The goal of this structure is to determine how landscape-scale patterns of adoption emerge from individual farmer decision-making.

2. Study Area and Research Questions

The Philippines, like many countries in the developing world, has passed legislation supporting organic agriculture, citing its benefits to human health and the environment [34]. Acreage under organic agriculture has risen steeply in recent years, from approximately 3500 hectares in 2004 to 52,400 hectares in 2009 [35]. This represents only 0.6% of the country's 9.2 million hectares of cultivated land area. The Philippine commitment to organic agriculture is nonetheless significant, given the country's history of promoting chemically intensive technologies to raise yields of rice, a staple crop [36]. Achieving self-sufficiency in rice production is another important goal of the Philippine government (the country consistently ranks as one of the top five rice importers) [37]. This necessitates the country increasing rice yields to the levels of some of its more productive Asian neighbors [38]. Achieving both higher rice yields and greater amounts of land under organic production will require investments in programs that target smallholder farmers, who constitute the majority of rice producers in the Philippines.

We chose Negros Island, in the central Philippines, as a study site (Figure 1). The island is located in the country's agricultural belt; major crops include sugarcane, maize, rice and coconut. While the rice-growing areas just north of Manila are known as the country's rice bowl, rice lands on Negros are also highly productive, according to the Philippine Bureau of Agricultural Statistics (www.bas.gov.ph); three harvests per year are standard.

The Negros Institute for Rural Development (NIRD) is an internationally funded and locally governed NGO, which has been promoting organic agricultural techniques in an upland rice producing area on Negros since 1999. The municipality of Canlaon, where NIRD is located, is therefore an ideal location for studying the spread of organic technology adoption. According to NIRD staff, some farmers in the area have adopted organic techniques and used them consistently, while a larger number of farmers converted to organic methods but then went back to conventional methods after some time. This presents an intriguing research question which we addressed in our model: Why do some farmers on Negros go organic, while others do not? Put another way, what are the driving factors determining the rate of organic adoption over time which Negros has experienced?

114

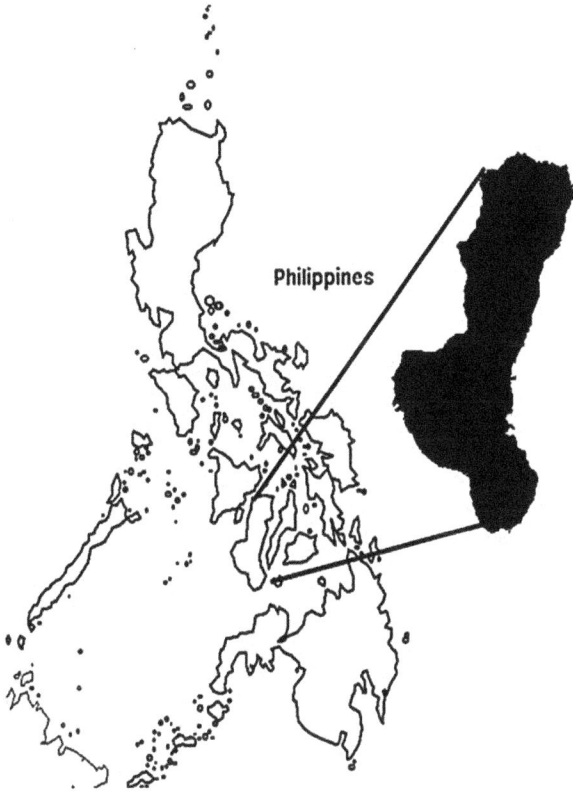

Figure 1. Map of the Philippines showing location of Negros Island where the farmer focus groups were conducted in an upland rice-growing area.

3. Data Collection

Focus Group Discussions and Model Parameterization

We collected both qualitative and quantitative data from the study site to parameterize the model. We organized focus groups in November, 2010, to assess the state of organic agriculture adoption in the Canlaon region, and to identify barriers to adoption. Because NIRD introduced organic methods to this region and conducts periodic workshops and trainings in organic techniques, they are in contact with most of the organic farmers in the area. Ten farmers who had practiced organic methods for at least the past three years were able to attend the focus group sessions, and we constructed groups of similar size to represent conventional farmers (eight participants) and farmers who had tried organic farming but had returned to conventional methods (ten participants). Organic farmers were over-represented in this sample compared to their presence on the landscape, because the purpose of

115

the focus group exercise was to examine the motivations and decision processes of farmers who choose to go organic.

The groups were asked to discuss the questions: (1) Why do you farm using your chosen method (organic/conventional)? (2) What are the benefits of using your chosen farming method? and (3) What are the challenges of using your chosen farming method? A note-taker was present in each group to record key concepts as the discussions took place. Afterward, the three groups were brought together in a common forum to share their observations and ask one another questions. These sessions were recorded and coded using three categories (Farm Characteristics, Organic Adoption, and Challenges) and fifteen sub-categories (farm size; family size; years farming; general history (any information about the farm not covered under other categories); markets (where farm products are sold); off-farm income; livestock; fertilizer use; history using organic; perceived benefits of organic; characteristics of organic (aspects of organic technologies that farmers discussed, other than benefits); pest management; yield; knowledge of organic; labor). All conversations took place in Cebuano-Visayan, the local language. The model decision structure represented in Figure 2 was developed to reflect, as closely as possible, the decision-making process around organic conversion described to us by farmers. Both organic and conventional farmers agreed on the importance of an "experimentation" period for organic conversion, in which they would grow organically on one portion of land to test the efficacy of this method. In addition, all farmers agreed that, in order to farm organically, one has to know how to use organic methods and be able to bear the initial cost of changeover from conventional to organic methods.

From June 2010 through December 2010 (the length of one cropping season), NIRD field staff assisted the authors in collecting information on crop yields and farm size from the farmers who participated in the focus groups. A total of twenty-eight farmers participated in this data collection (ten organic farmers, eight conventional farmers, and ten mixed-method farmers). "Mixed-method" refers to farmers who at some point have tried organic farming, but do not do so consistently and may have reverted to conventional farming at the time of data collection. This sample was not intended to be statistically representative for the larger population of rice farmers; rather, the information was to be used to develop the farm characteristics represented in the model.

The average farm size of the participants was 0.9 hectares (0.63 ha for the organic group; 0.85 ha for the conventional group, and 1.1 ha for the mixed-methods group). This is smaller than the 2.8 ha average farm size for the island [39]. Given the highly unequal distribution of farm acreage between large-scale sugarcane plantations and smallholder rice and vegetable farms on Negros Island, the average farm size of the study participants represents the landholdings of small-scale farmers [40]. The average yield for one cropping season reported by the organic group was 3.0 MT/ha;

for the conventional farmers it was 5.4 MT/ha; and for the mixed-methods farmers it was 4.5 MT/ha. We believe the organic yields were lower for two reasons: first, some of the organic farmers reported that access to organic fertilizers was a problem, and they may not have been adding adequate levels of nutrients to ensure yields. Secondly, some of the farmers who converted to organic methods in the recent past may have been suffering from the initial yield loss at changeover (see below for more information). However, we would caution against using one cropping season's data from a limited sample of farmers to depict an overall pattern. For this reason, we used a wider yield distribution drawn from regional datasets to parameterize our model simulation.

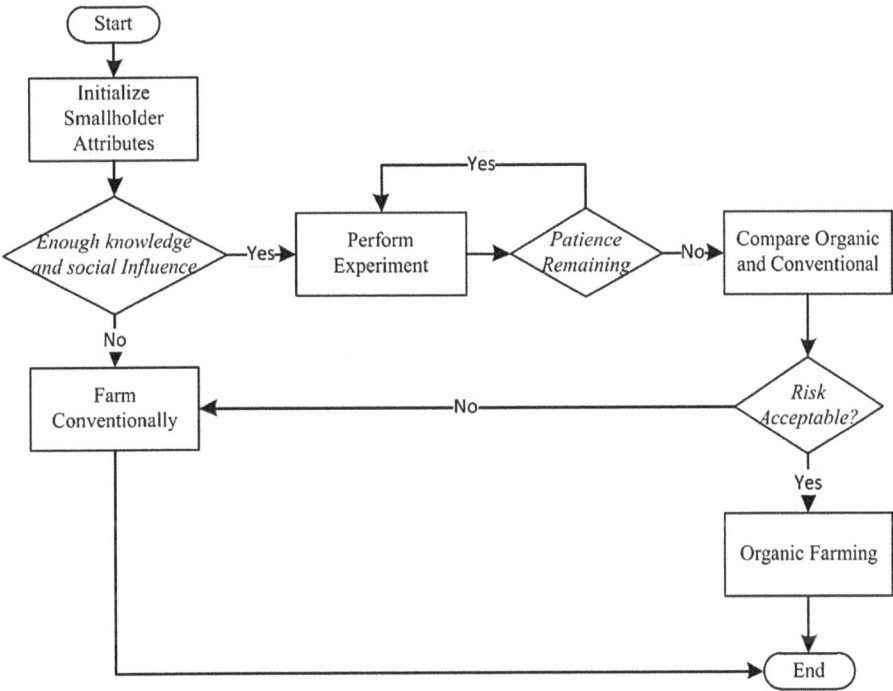

Figure 2. The decision-making process used by farmer agents in the agent-based modeling (ABM). In order to experiment with organic agriculture (plant half of their land using organic methods), farmers must have knowledge of organic techniques. Once they begin an experiment with organic techniques, they wait a certain number of rice seasons to compare their yields under organic production with their yields under conventional production. The number of seasons they are willing to wait is designated by "patience", a characteristic of each farmer agent that describes his/her dedication to organic production. After the finished experiment, they evaluate both methods while considering the risk imposed by external climatic and economic conditions. Then they chose which method to adopt.

Among organic farmers who participated in the focus group, concern for the environment and for their family's health were agreed to be key reasons for choosing organic methods. Participants also noted that organic methods reduce input costs and therefore improve net profit. In addition, ducks and fish can be integrated into paddy rice production when organic methods are used, while chemical herbicides and pesticides are harmful to these animals. The incorporation of animal agriculture/aquaculture with rice production was clearly seen as desirable by organic farmers, who mentioned the enhanced productivity that this integration affords, and the weed and pest-controlling characteristics of ducks. These observations are consistent with literature studies of paddy rice agriculture elsewhere in Asia [41]. Price premium for organic rice was not a determinant in the choice of organic methods for focus group participants, because the farmers do not have access to customers who might be willing to pay a premium.

Farmers who used conventional techniques, or who had switched between organic and conventional techniques, were not as strongly devoted to their method of production as the organic farmers. In fact, most of the farmers in these groups stated that they try to incorporate some organic techniques, such as mixing animal manure with chemical fertilizer, or using organic pest control methods. The most significant barriers to organic farming mentioned were labor requirements (mainly for composting and weeding), lack of nutrient inputs (particularly for farmers without animals), and lack of information about organic techniques. Some conventional farmers, but not all, believed that high-yielding varieties of rice (HYVs) were preferable to traditional varieties for their consistent high yields, and these HYVs require chemical pest control. Farmers also stated that they are sometimes forced to use chemical inputs by creditors, who view these inputs as necessary to guarantee their return on investment. Although we didn't explicitly model this requirement as part of the simulation, we noted that without access to credit, most farmers would not be able to produce. This statement is therefore well worth further examination in the context of promoting organic agriculture.

4. Methods

4.1. Agent-Based Model

The ABM simulates a virtual farming area containing 2500 farmer agents who must choose between organic and conventional farming methods over the course of 100 cropping seasons (approximately 33 years). We chose the number 2500 because this is a reasonable approximation of the number of rice farmers that live in the Canlaon region; this number therefore represents the "upper limit" of organic diffusion, if all farmers were to adopt organic technologies [39]. Each farmer agent represents a farm household, and all agents operate in a virtual "social space" (a concept which is relevant for the social network learning described below), but the model is not spatially explicit. We built the model using Repast Simphony software and programmed it using Java language. The farmer agents all have the same decision-making rule (see Figure 2), but since their attributes are different, their decision outcomes are also different. This decision-making framework was developed based on the focus group discussions, and reflects the considerations of the farmers as described in the groups.

4.1.1. Farmer Agent Specification

The literature on organic farming adoption indicates that economic benefits can be a significant driver in farmers' decisions to convert to organic methods [42]. Several studies have also pointed to farmer characteristics that influence whether farmers choose to convert to organic methods. These include age, gender and education of the producer [18,43]; farmers' social and ecological values [44,45]; sources of information about farming [18]; and, possibly, the desire to become more resilient to input price shocks or climatic shocks [46]. All of these empirical observations were used during agent conceptualization of our ABM of organic farming, together with the focus group results.

The agents were assigned values for land area, animal ownership, and number of people in the household (corresponding to household food demand and labor availability), taken from the range of observed values for these variables in the field and in Philippine agricultural statistics, and assuming a uniform distribution across the range [47] (See Table 1 for range of values assigned to agent characteristics; more details about the model and its parameterization can be found in Supplementary Section 1). We used uniform distributions for all our input variables because the small sample of participants that served as the source of input to our agent attributes was insufficient to generate more complex empirical probability density functions. The rationale behind this distribution is that it requires at most two parameters (*i.e.*, upper/lower bounds) identified from data. Initial capital assets were also randomly distributed across a uniform range, but after the start of model simulation capital assets were calculated for each time step based on yield.

We constructed a stylized social network for the farmer agents to represent the ways in which farmers influence their social connections and share information about organic techniques. The structure of the network was informed by qualitative data (*i.e.*, how farmers described receiving information). The farmer agents were assigned values for "social reach", the unitless maximum distance between nodes which, in this case, were connected farmers (*i.e.*, the extent of influence). Any two farmers within each other's social reach were connected by a link. Each link creates a dyad and each farmer in the dyad is a social connection to the other. These links were the channels through which the model passed information about organic methods; farmers could not experiment with organic techniques unless they were connected to a farmer who already used them. The links represent the farmer's social influence on his/her networks. The values of social reach were initially set at 5, and then drawn from a uniform distribution during our computational experiment of sensitivity analysis, because we did not collect information on the structure of the social network at the field site [48].

Table 1. Factors (parameters) used in the model with their respective probability distributions (U—continuous uniform, D—discrete uniform).

Input Factor	Definition and Units	Distribution
Area of land (a)	Hectares	U = (0.25, 10)
Labor availability (la)	People in household	D = {1, 2, 3, ... , 8}
Social reach (sr)	Distance in agent's social space (unitless)	U = (5, 25)
Social influence (si)	Density of social connections among agents	D = {1, 2, 3, ... , 30}
Animal ownership (ao)	Proportion of farmers owning animals (unitless)	U = (0, 1)
Patience (p)	Number of farming seasons (3 seasons/year)	D = {1, 2, 3, ... , 15}
Risk taken (r)	Attitude towards risk related with the adoption of organic farming (unitless)	U= (0, 1)
Cost of conventional fertilizer (ccf)	Philippine Pesos per kilo	U = (40, 400)
Cost of organic fertilizer (cof)	Philippine pesos per kilo	U = (40, 400)
Average cost of other conventional input (coci)	Philippine pesos per hectare	U = (3000, 7500)
Average cost of other organic input (cooi)	Philippine pesos per hectare	U = (1100, 2900)
Average cost of labor (cl)	Philippine pesos	U = (6000, 13,000)
Average price of rice (p)	Philippine pesos per kilo	U = (7.44, 17.86)
Average yield (y)	Kilos per hectare	U = (1200, 6000)
Organic fertilizer threshold (oft)	Hectares	U = (0,10)
Land area for one labor (lal)	Hectares/person	U = (0.1,10)

Farmer agents were initially randomly distributed in unitless social space. To initialize the model, we set up the network structure by applying Hamil and Gilbert's methods [48]. A virtual grid space is created with the dimension of 100 × 100. Therefore, the space is divided into 10,000 grid cells. Then 2500 agents are created, and each of them occupies a cell. Each cell can only contain one agent. The values of the social attributes are initialized following the distributions mentioned above. After the attributes are set, agents are connected based on the values of social reach. Any two agents a and b are connected if Equation (1) is satisfied.

$$distance\,(a, b) = \sqrt{(x_a - x_b)^2 + (y_a - y_b)^2} \leq min(social\ reach\,(a)\,, social\ reach\,(b)) \quad (1)$$

The model was "seeded" with between 5 and 10 organic farmer agents. This is representative of the pioneer smallholder farmers in the Canlaon region who learned organic farming through NIRD when the NGO first was established there. Figure 3 depicts the social network for one model run.

Farmer agents were also assigned a value for the attribute "patience", and a value for "risk attitude". The patience attribute represented the number of rice cropping seasons a farmer was willing to experiment with organic methods

120

before comparing his or her net profit under organic farming with net profit under conventional methods. "Patience" therefore represented all of the values a farmer might hold that might prompt him or her to consider organic farming, even if it was not initially as productive as conventional farming.

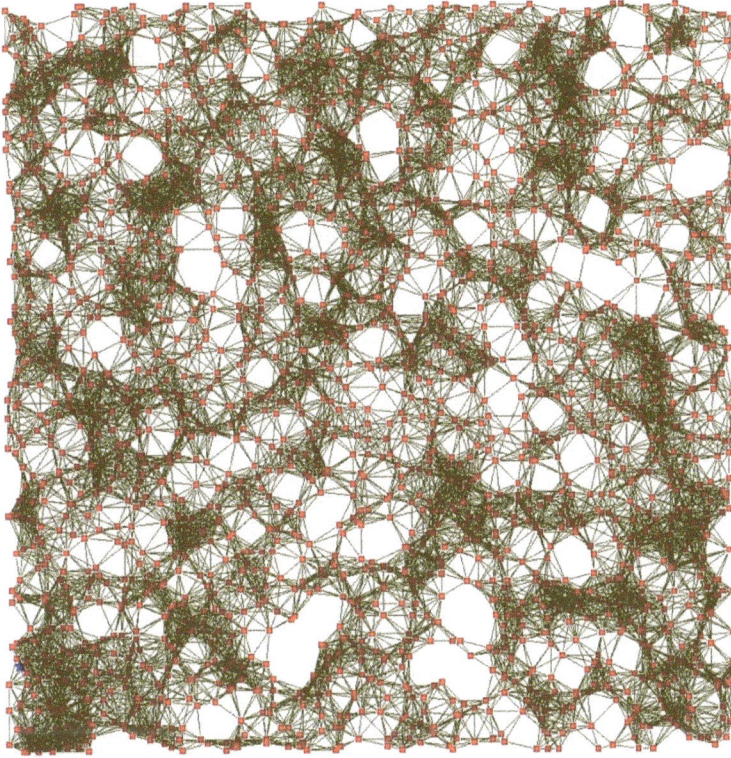

Figure 3. Social network depicted at the end of a model run with the "social reach" variable set at 20. The agents are distributed randomly in social space; circles represent farmer agents, and lines represent social connections between the farmer agents.

"Risk attitude" was an index assigned to each farmer agent reflecting their willingness to try a new farming method after their field experiment (in this case, organic farming). This index varied between 0 and 1 for each agent, and was compared to a randomly generated "risk index" between 0 and 1 for each farming season, representing the time-dependent environmental riskiness of trying a new technology (encompassing, for example, weather and economic conditions, political instability, *etc.*). Notice that "risk attitude" was only used to determine which farming method to choose when a given farmer agent expected profit from both organic farming and conventional farming. If a farmer had a relatively low risk attitude, indicating he/she was risk taking, he/she would be willing to try new technologies when expected profits were both positive. Conversely, if a farmer had a relatively high risk attitude, indicating risk aversion, he/she would almost never be willing to try organic farming. The qualitative and quantitative evidence from the field support this depiction of conversion to organic methods as a significant risk. Studies suggest an initial yield loss associated with organic conversion, which represents a risk to a farmer's income stream and household food supply [4].

4.1.2. Decision Algorithm

Before the start of each time step, a farmer agent must decide whether to experiment with organic farming (Figure 2). This can only take place if the farmer is connected to other farmer agents with adequate knowledge of organic techniques. The model assumes that this knowledge is disseminated through the social network, so if a farmer agent is connected to a certain number of organic farmers or farmers experimenting with organic techniques, he/she is able to begin an experiment. We assumed the number, which is called "influence threshold", could be drawn from a discrete uniform distribution between 1 and 8, representing a minimum and maximum number of people a farmer might reasonably turn to for farming advice. Once the condition was satisfied, the farmer agent would begin his/her experiment. This means that he/she plants half of his/her land using the organic techniques and half using the conventional techniques.

Input costs were calculated for each farmer agent using a given farming technique at each model time step. These costs included the cost of fertilizer, the cost of other inputs, and the cost of labor (if labor provided by the household was not sufficient). To simulate the fluctuation of input costs across the model run time, the amount in Philippine pesos assigned to labor, fertilizer and input costs for a given time step was drawn randomly from a uniform distribution based on data collected in the field and from the Philippine Bureau of Agricultural Statistics (see Table 1). Conventional farmers must purchase all of their fertilizer, but organic farmers may derive some of their fertilizer from composting agricultural waste (such as manure and rice straw), and purchase organic fertilizer for their remaining needs. We assumed that farmers with at least one hectare of land and one animal were able to produce all of their fertilizer on-farm, up to 10 hectares (this threshold was based on empirical observations, and tested with the sensitivity analysis described below). The amount of fertilizer applied was assumed to be up to 1.5 times the average nitrogen application for the region recorded in 2010 agricultural statistics (organic and conventional fertilizer types are adjusted by their respective nitrogen contents), based on what the farmer-agent could afford.

Yield was calculated as a function of the nitrogen application for either conventional or organic methods, using a Michaelis-Menten equation applied to rice yields and fertilizer application rates from provincial statistics:

$$yield = (land\ area \times 1.0045^{seasoncount} \times 4800 \times fertilizer\ amount)/(fertilizer + 6.1) \quad (2)$$

where *fertilizer amount* is the maximum nitrogen application rate a farmer can afford, and *seasoncount* is the model time step [49]. Data from the region indicate that conventionally grown rice yields increase at a rate of 0.45% per season, and this annual increase was simulated in the model. When a farmer agent switched their land to organic methods (or planted half of their land in organic methods as an experiment), the modeled yield initially dropped by 50% before recovering at a rate of 5% per season to the yields described by the Michaelis-Menten equation above. Therefore, after approximately 10 seasons (or three years), yields for a farmer who had converted to organic methods would be no different than if he/she had continued to farm conventionally. This initial yield drop represents a high estimate of initial yield losses reported in field experiments by organic farmers, due to the need for soil organic matter and soil chemistry to recover post-chemical fertilizer application [4].

A farm household consumes part of the rice (enough to feed the household members during the cropping season) and sells the remainder, if any. The price of rice was also drawn randomly from a uniform distribution based on recent farmgate prices reported by the Bureau of Agricultural Statistics (see Table 1). The sale of the rice provides the capital input necessary for the rice season that follows. It is important to note that organic rice is sold at the same price as conventionally grown rice in our modeled environment, because farmers in the Canlaon region do not have access to markets at which organic rice might fetch a premium due to transportation constraints.

As mentioned above, each farmer agent was randomly assigned a value for the variable "patience", which represents the length of the experiment he/she conducts before comparing the net profit from conventional farming methods to the net profit from organic farming methods and choosing the method that is more profitable. In the model, this number ranged from one to 15 farming seasons (based on the farmer focus groups, we thought five years would be the maximum amount of time a farmer would experiment with an organic technology). The more "patience" an agent has, the more likely he/she is to choose organic farming, because once the yields have recovered from the initial loss caused by switching to organic methods, input costs tend to be lower for organic farmers, who can produce inputs on-farm rather than buying them.

Based on the presented specification, our model can be placed somewhere on the continuum between the highly abstract agricultural ABM that result in stylized simulations [29] and empirically-rich ABM equipped with microeconomic mechanisms and complex agricultural markets [20,30]. This mixed approach, combining observational and simulated data in computational experimentation, has been successfully applied in other agricultural ABM [14,23].

4.2. Exploring Model Outcome Variability with Sensitivity Analysis

Sensitivity analysis of agricultural ABM is rarely undertaken. Notable examples include Happe *et al.* [30] and Schouten *et al.* [50] where regression-based metamodels are developed to evaluate the influence of the uncertain model inputs on outputs. In their study, Schouten *et al.* [50] also employ a simple one-parameter-at-a-time (OAT) sensitivity analysis and compare it with the regression metamodel.

The ABM described above is an example of a dynamic model that emulates a complex agricultural system. It is, therefore, imperative to apply a simulation procedure appropriate to complex nonlinear models, which would account for input and output variability as well as the potential input interactions. Neither regression nor OAT meet these requirements. Crosetto *et al.* [51], Gomez-Delgado and Tarantola [52], Chu-Agor *et al.* [53], and Ligmann-Zielinska *et al.* [11] among others, postulate the use of Monte Carlo simulation that incorporates global sensitivity analysis (GSA) as a part of computational experiments. GSA is a method of experimentation in which the variability of model results is quantified based on simultaneous sampling of the whole set of input variables, which are then examined individually and in combinations [54].

The most comprehensive method of GSA is based on model-independent output variance decomposition, in which model outcome variability (represented using variance V) is apportioned to various model inputs [54,55], so that the underlying causes of variable outcomes can be explicitly identified. The procedure starts from generating M samples of input variable values using a selected experimental design. The model is then executed for each sample $m \in M$ and the result is recorded. These output values form a distribution which can be summarized using descriptive statistics like mean and variance (V). GSA then uses output variance decomposition which partitions V based on the contribution of each input variable to V. This partitioning (aka decomposition) is accomplished by estimating the conditional variances of every input variable k. By calculating the ratio of conditional variance due to k to the total V, we obtain a first order sensitivity index (Sk). If a given variable k has a relatively high value of Sk, its single influence on output variability is substantial. To express the interactions among variables, which are ubiquitous in complex system models, we also compute a total effect index for every k (STk), which accounts for all higher-order effects of inputs [56]. Consequently, variables with relatively low values of the (Sk,STk) pairs are deemed unimportant in shaping output variability and, in the consecutive experiments, can be set to constant values contributing to model simplification. In addition, the (Sk,STk) pairs provide valuable information on the mechanisms that affect the dynamics of the model and, consequently, can serve as quantitative indicators explaining model drivers. For details on (Sk,STk) calculation the reader is referred to Saltelli *et al.* [54].

To calculate the (Sk,STk) pairs we employed the quasi-random Sobol experimental design described in Saltelli (2002), which proved to be the most effective in approximating the values of sensitivity indices [57]. To compute the indices, we used SimLab open source software (http://ipsc.jrc.ec.europa.eu/?id=756).

4.3. Computational Experiments

Following the quasi-random experimental design mentioned above, we employed Monte Carlo simulation and executed the model M times, where M was set to 17,408 runs based on the procedure described in Ligmann-Zielinska and Sun [58]. At the end of each simulation, we recorded the number of agents who adopted organic farming (N) and compared that to the total number of farmers (T) in the model using a simple ratio N/T, referred to as RATIO in the following sections. Given that the focus of our experimentation is organic farming adoption, we set the number of farmers to 2500 (*i.e.*, T = 2500), with a time step of one cropping season, and ran the model for 100 seasons. The results were analyzed in two ways. First, we compiled the RATIO values into a probability distribution and summarized it using descriptive statistics (including average and variance). Second, we performed variance decomposition of the RATIO distribution, and calculated Sk and STk for each of the 16 input variables (a total of 32 indices plus an interaction index) to investigate the overall model behavior and determine its critical drivers. The interpretation of the pairs of indices (Sk, STk) followed the procedure described in Ligmann-Zielinska and Sun [58].

5. Results

Agent-Based Model

By the end of the model simulation, all farmer agents had the opportunity to experiment with organic techniques, as the knowledge of organic agriculture spread throughout the social network. Experimentation concluded by rice season 23, at which point all farmer agents selected a farming type. The number of organic adopters in the model leveled out after experimentation in the model run using default parameter values (Table 1, Supplementary Section 1) (Figure 4). This ratio of organic farmers to total farmers was robust to the seasonal variation in labor, fertilizer, and input prices, as well as model-simulated seasonal variation in yields, as indicated by the fact that after the initial decision period, the ratio of organic to total farmers did not change significantly. This result is qualitatively similar to the patterns of adoption reported to us by the NGO operating in the Canlaon region, and by the farmers who participated in the focus group. Both NGO workers and farmers confirmed that, after an initial period when many farmers seemed to be adopting organic methods, the adoption rate leveled off and most farmers did not switch to organic despite learning about organic methods. This observation is similar to the conclusion derived from the study by Deffuant *et al.* [26] which suggests that, while sympathetic to organic agriculture, farmers did not convert to organic even though they were exposed to positive messages from the public.

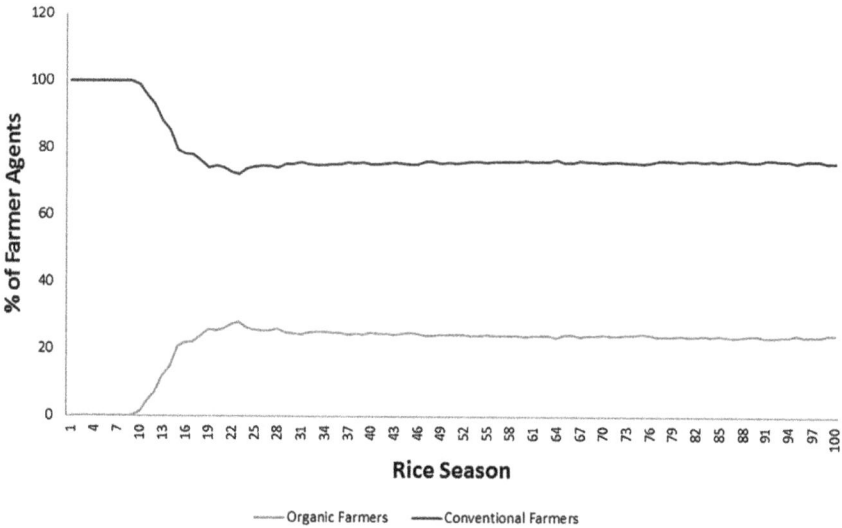

Figure 4. Model output depicted number of organic farmers and number of conventional farmers over the 100 cropping seasons simulated in the model, at default parameter values (see supplementary for description).

The box plot in Figure 5 summarizes the distribution of RATIO. The Monte Carlo simulations result in RATIO ranging from 0% adoption of organic farming to the maximum of 60% organic farming adoption. The mean adoption of organic farming equals 31% of the total agent population, with a substantial level of variability (std = 33%), which we analyze in the following section.

Sensitivity Analysis: Variance Decomposition of Model Results

The results of variance decomposition of the ratio of organic farmers to total farmer agents (RATIO) are depicted in Figure 6. As mentioned above, model sensitivity to different inputs is analyzed separately for their individual influence

(Sk) and their total (STk) influence. With the first order sensitivity index (Sk) we look for important input variables that, if fixed independently, would reduce the variance of RATIO the most. In other words, the variables with relatively high S values have the most impact on the variability of the adoption of organic farming. In our experimentation, the social reach (defined in Section 4.1.1) scores highest ($Ssr = 20\%$ of total variance), followed by yield ($Sy = 15\%$), and area of land under cultivation ($Sa = 12\%$).

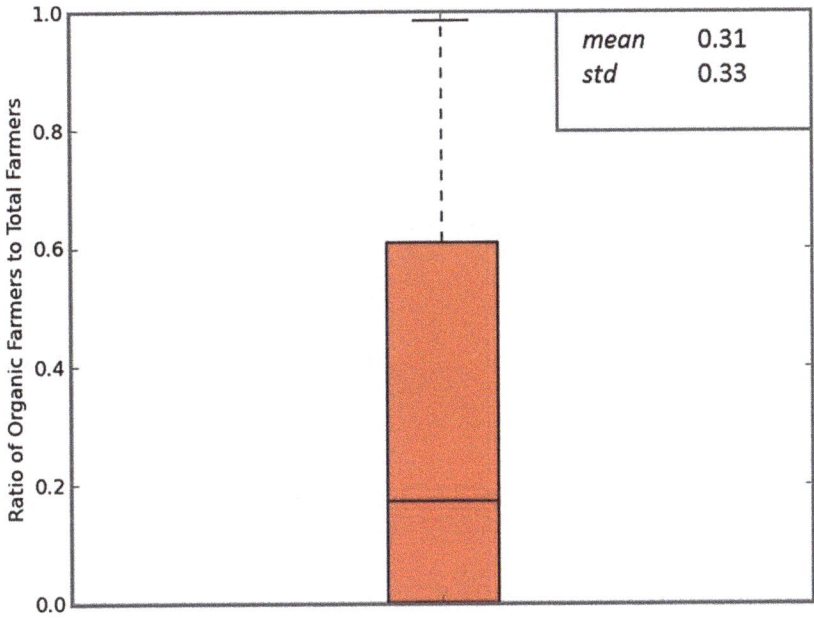

Figure 5. Distribution of the ratio of organic farmer agents to total farmer agents at the end of the model run (100 cropping seasons), over all Monte Carlo sensitivity simulations. This ratio ranges from 0% of farmer agents adopting organic methods to a maximum of 60% of farmer agents adopting organic methods, with a mean adoption rate of 31%.

The model's nonlinear behavior is expressed by the relatively high level of variable interactivity, which is derived from the percentage sum of all first order indices. In our simulation, this sum amounts to 71% of RATIO variability (Figure 6). Therefore, only 71% of the variance in organic farming adoption rates can be explained by analyzing the variables in isolation. The remaining 29% of V is attributed to variable interdependence, which is expressed using the total effect index (STk).

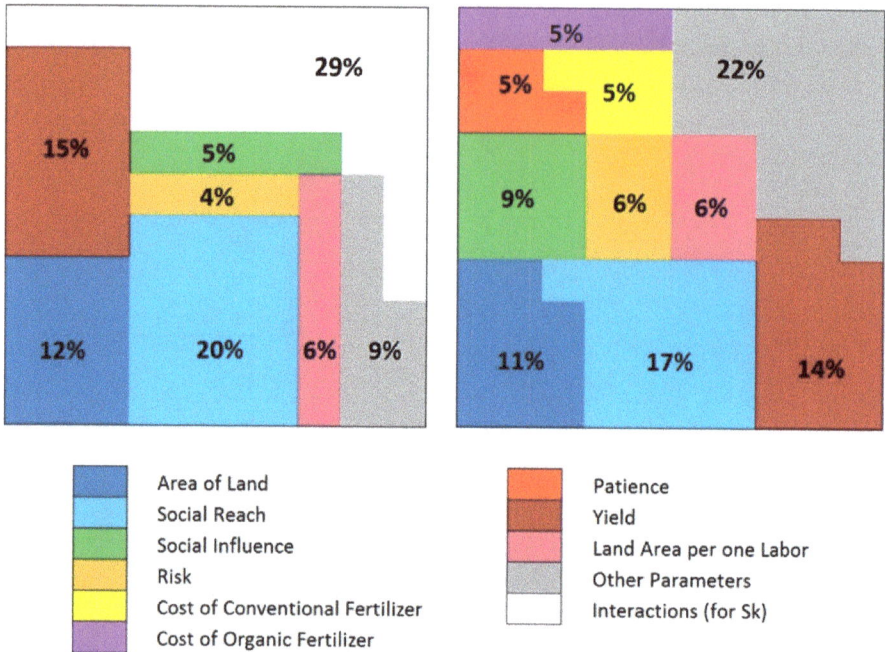

First Order Index Sk

Total Effect Index STk

29%

15% 5%

4%

12% 20% 6% 9%

5%

5% 5%

22%

9% 6% 6%

11% 17% 14%

Area of Land
Social Reach
Social Influence
Risk
Cost of Conventional Fertilizer
Cost of Organic Fertilizer

Patience
Yield
Land Area per one Labor
Other Parameters
Interactions (for Sk)

Figure 6. Variance decomposition of variable effects on model output (ratio of organic-adopting farmer agents to total farmer agents). Model sensitivity to different input variables is analyzed separately for their first order influence (Sk) and their total effects (STk) influence, in which variable interactions are explicitly quantified. The variables with relatively high S values have the most impact on the variability of the proportion of farmer agents who adopt organic methods.

Over half of the variables (9 out of 16) contribute substantially to the complex behavior of the model. The highest interaction effect can be observed for social reach (STsr = 17%), yield (STy = 14%), area of farmland (STa = 11%), and influence threshold *i.e.*, the density of social connections of any given agent (STsi = 9%). Consequently, not only do these four variables individually contribute to the variability of the RATIO, but they also have a strong influence when analyzed as a group.

Other variables that somewhat contribute to RATIO variability include the cost of fertilizer (both conventional, where STccf = 5%, and organic, where STcof = 5%), labor availability (STlal = 6%), attitude to risk taking (STr = 6%), and patience (STp = 6%).

6. Discussion

There appears to be an upper limit on the number of farmers choosing to go organic that is fairly robust to fluctuations in prices and the external risk environment. The results of the sensitivity analysis, which reveal how influential the various model variables are in determining the ratio of organic adopting farmers to total farmers, gives us insight into why this limit exists. The sensitivity analysis suggests that the decision to shift from traditional to organic farming is mainly driven by yield, and by imitative behavior, that is, by the number of farmer agents that have already adopted organic farming, as indicated by the strong effects of social reach and influence threshold on the organic adoption ratio (Figure 6). Consequently, our ABM is a classic example of simulating innovation diffusion, where the decision made by agent (A) is influenced by the decisions of proximal agents in A's social space [59–61]. The fact that social reach and influence threshold account for 26% of output variance corroborates our postulate about the role of diffusion of innovation in explaining organic farming adoption in our ABM.

Particularly, social reach and influence threshold affect the rate at which farmers "learn" about organic farming, a prerequisite to their experimentation with organic technologies. This is important in the model context because of the assumption that conventional yields rise over time. The more rice cropping seasons that go by before a farmer agent learns about organic technologies and is able to experiment with them, the higher the baseline conventional yields to which he/she compares organic yields—therefore, the less likely it is that he/she will choose to switch to organic. When yields are higher, the loss of 50% of yield during switchover is a greater absolute yield loss compared to when yields are lower. For example, if conventional yields are one ton/ha, upon switchover organic yields would be 0.5 ton/ha. Given the lower costs of organic farming, the farmer might be able to make up this 0.5 ton/ha difference between organic and conventional yields in net profit by going organic. However, if yields are 5 tons/ha, the yield loss upon switchover would be 2.5 tons/ha—a greater profit gap to make up.

We can also conclude that the adoption decision is dependent on the land area an agent farms. Land area constraints on organic farming adoption were discussed in the farmer focus groups, and the model's sensitivity to this variable confirms that farmers with limited land may not be able to generate the nutrients they need to use as inputs for organic production. However, the model was not sensitive to the "organic threshold", or the amount of land required to produce on-farm organic fertilizer. This provides evidence for rejecting our hypothesis that only farmers who can produce their own fertilizer on-farm will be able to use organic methods. Interestingly, none of the variables that would seem to have a direct influence on the variability of organic farming (like cost of organic fertilizer and cost of other organic inputs) drive the switch from conventional to organic farming, suggesting that cost comparisons between farming methods may not comprise a farmer's main motivation for adopting organic.

Patience and fertilizer costs play particularly intriguing roles in determining organic adoption rates. When treated independently, these variables have negligible influence on the adoption of organic farming (their Sk values equal zero). However, their influence becomes much more pronounced when analyzed in interactions with other variables, suggesting that they play a complex intertwined role in the decision to switch cultivation to organic farming. This suggests that value orientations towards promoting family health and health of the environment, which were prominent topics of discussion in the focus groups, do not by themselves determine whether a given farmer will choose to adopt organic methods. Rather, knowledge of organic methods and the ability to produce organic fertilizer may be pre-requisites for a farmer putting his or her pro-health or pro-environment values into action. Farmers who have not been able to practice organic methods may nonetheless be interested in integrated pest management techniques, and this was a major topic of discussion in the focus group sessions [62].

Based on the sensitivity analysis we can also suggest the direction of ABM simplification. In particular, animal ownership, labor availability, crop price, labor costs, the amount of land needed to produce organic fertilizer on-farm (organic threshold) and other organic and conventional costs proved to be unimportant in shaping the variability of the RATIO statistics. Consequently, an equivalent simplified model would have these variables set to constant average values, reducing the dimensionality of the model from 16 to 9 uncertain inputs. This simplified model would be equivalent to its predecessor in that it would maintain the variability of RATIO (*i.e.*, with V unchanged) [11,54]. This property can be of value in policy analysis where a certain level of result variability is required to account for any unanticipated events (surprises) in the target system. At the same time, a model with a smaller number of variables can be easier for stakeholders and decision-makers to understand. The drawback of our model simplification is that it is dependent on the type and form of the output variable. Every time we substitute RATIO with a different output of interest (e.g., we introduce different categories of organic farmers based on their land management practices or household income), we should revisit the initial ABM formulation and perform a separate variance-based model simplification.

Implications and Limitations

Quantitative data on organic adoption over time, which would allow for more extensive validation of the model, is not yet available from this region. However, these model results are consistent with anecdotal observations collected by NIRD and expert opinion of in-region farmers—for example, the NGO has observed that, despite a long-term presence in the region, the number of farmers who have decided to convert to organic methods remains limited. Similarly, the model depicted nearly all farmers experimenting with organic methods after encountering these methods through their social network, but only a minority of the farmer agents ended the model run by adopting organic. This confirms the model's usefulness as a scoping exercise that can shed light on where future research and farmer support activities should be focused.

We can suggest that further research would be warranted into the social network among farmers in this region, and farmers' means of sharing information, since social reach and influence threshold were both found to be significant in determining rates of organic adoption. Our model simulation used a uniform, random social network to represent information sharing, which obviously does not reflect the network topology in a typical community [27]. Moreover, there is no feedback effect from farmers observing organic experiments which have been abandoned—it is possible that if a significant number of these occurred in their social network, they might be more reluctant to try organic agriculture in the first place. However, it is worth noting that this dynamic was not mentioned in the focus groups as a factor in determining organic experimentation.

An interesting implication of the model output seems to be that, as conventional yields rise due to improved crop varieties and other variables, at some point it may be "too late" for farmers to consider going organic, because conventional yields would be high enough that the 50% loss occasioned by organic switchover would represent too large of a yield gap for the farmer. The significance of the yield variable in determining organic adoption rates confirms this interpretation of the model. Therefore, if organic adoption is desirable in a given area, informing interested farmers about organic techniques relatively quickly (possibly through strategic use of social networks), while minimizing the initial yield losses from switching to organic methods and boosting organic yields as quickly as possible to compete with conventional yields (for example, through intensive soil fertility amendments), would seem to be the best set of strategies. To return to the original research question proposed by our NGO partner NIRD, we seem to have identified two major drivers of organic adoption: yield and information sharing through social networks. Additional research on how these variables change over time should help to confirm this determination.

We did not collect detailed demographic data on the farmers who participated in the focus groups, which limited the degree to which we were able to assess how organic adoption rates may differ by gender, age, years spent farming, *etc*. It is likely that these demographic variables also affect the nature of the social networks in which farmers are embedded. The topology of the social networks represented in the model was therefore totally under-developed, and warrants further exploration in a more sophisticated modeling framework. In addition, our data collection was limited to one rice cropping season by the time and budget available for this study. Given that we are simulating a dynamic system, observing empirical trends over multiple seasons would give us more insight into whether the behavior simulated in the model is similar to the behavior of the real-world system. One of the reasons organic adoption is being promoted worldwide is for resilience to climate change-induced drought and other extremes in precipitation and temperature patterns [63]. An improved version of this model would include such shocks, in order to observe their effects on organic adoption and yields compared to conventional yields. In addition, economic shocks such as restriction to credit or large spikes in prices (larger than we modeled here) could shed light on the relative resilience of organic techniques compared to conventional techniques.

Another shortcoming of our model is its lack of spatial information. Clearly, many aspects of farm production are highly dependent on that farm's location in the landscape, and one would expect organic adoption to be affected by spatial characteristics such as farm slope, proximity to roads and water sources, soil characteristics, *etc.* We hope to include spatial aspects in a future version of this model.

Importantly, we were using the ABM as an exploratory, rather than a predictive, model [64,65]. Given the complex and uncertain nature of the simulated upland farming system, and the paucity of data from the region, we believe a traditional quantitative model that seeks to make predictions about systemic outcomes is inappropriate [66]. Rather, our goal was to shed light on the dynamic aspects of the system described in questions 1 and 2 in the introduction, so that we might target future research and modeling efforts. This approach is philosophically different from the traditional use of statistical or optimization techniques to design a "best" model which produces a "right" answer [66], but it is no less powerful. Systems modelers have argued for decades that an optimization approach to a complex system with high levels of uncertainty can lead to incomplete consideration of system drivers at best [67], and misleading or erroneous conclusions at worst [68]. A dynamic simulation of a complex system, as undertaken here, can be used for theory development and hypothesis testing [69]. With more empirical data with which to calibrate and validate the model, however, this model could be used to make more targeted projections to inform policy in a given location.

7. Conclusions

We built a stylized agent-based model to explore the reasons behind low adoption rates of organic agriculture in a productive rice-growing region in the upland Philippines, where organic agriculture has been promoted for fourteen years. The model was run for multiple input variable sets that denote different socioeconomic, behavioral, and ecological characteristics of the farmers and the agricultural system. Given the uncertainty of the system, we explored the variability of the results using a variance-based sensitivity analysis framework to identify the core drivers of the decision to adopt organic farming. We found that the speed of information spread through the social network was highly influential in agents' decisions to adopt organic agriculture, because the longer it takes for them to hear about organic techniques and experiment with them, the more progress conventional varieties and techniques achieve in boosting yields, leading to opportunity costs for farmers who convert to organic. Land area is a constraint to organic adoption, because farmers with small fields are not able to generate on-farm organic inputs that are sufficient to maintain yields. The model revealed a high degree of complexity in farmers' decision-making, with interactive effects between decision variables explaining one-third of the variation in organic adoption rates. Yield improvements and information spread through social networks appear to be the major drivers of organic adoption, warranting further research and modeling attention, and perhaps programmatic targeting by our partner NGO. In addition, more research on farmers' decision-making processes is warranted, as these processes are complex and driven by interactions between economic, social and ecological factors. Continued interaction with farmers in the region will allow us to gain more insight into their decision-making environment and constraints.

Acknowledgments: We thank the Center for Advanced Studies in International Development at Michigan State University, which funded this research, and the Negros Oriental Institute for Rural Development.

Author Contributions: Laura Schmitt Olabisi oversaw the overall development of this project and is the lead writer of this paper. Laura Schmitt Olabisi performed the field interviews with stakeholders and processed the data. Ryan Qi Wang was the primary developer of the model and performed its calibration. Ryan Qi Wang is also the lead author of the Methods section and the Supplementary. Laura Schmitt Olabisi and Arika Ligmann-Zielinska conceived and designed the experiments. Ryan Qi Wang performed the experiments. Arika Ligmann-Zielinska performed and wrote the sensitivity analysis sections.

Conflicts of Interest: The authors declare no conflict of interest.

References

1. Willer, H.; Kilcher, L. (Eds.) *The World of Organic Agriculture: Statistics & Emerging Trends 2009*; International Federation of Organic Agriculture Movements (IFOAM): Bonn, Germany, 2009.
2. Horrigan, L.; Lawrence, R.S.; Walker, P. How sustainable agriculture can address the environmental and human health harms of industrial agriculture. *Environ. Health Perspect.* **2002**, *110*, 445–456.
3. Pimentel, D. Green revolution agriculture and chemical hazards. *Sci. Total Environ.* **1996**, *188* (S1), S86–S98.

4. Mendoza, T.C. Evaluating the benefits of organic farming in rice agroecosystems in the Philippines. *J. Sustain. Agric.* **2004**, *24*, 93–115.

5. Eakin, H. Smallholder maize production and climatic risk: A case study from Mexico. *Clim. Change* **2000**, *45*, 19–36.

6. Uematsu, H.; Mishra, A.K. Organic farmers or conventional farmers: Where's the money? *Ecol. Econ.* **2012**, *78*, 55–62.

7. Burton, M.; Rigby, D.; Young, T. Modelling the adoption of organic horticultural technology in the UK using duration analysis. *Aust. J. Agric. Resour. Econ.* **2003**, *47*, 29–54.

8. Berger, T. Agent-based spatial models applied to agriculture: A simulation tool for technology diffusion, resource use changes and policy analysis. *Agric. Econ.* **2001**, *25*, 245–260.

9. Parker, D.C.; Manson, S.M.; Janssen, M.A.; Hoffman, M.J.; Deadman, P. Multi-agent systems for the simulation of land-use and land-cover change: A review. *Ann. Assoc. Am. Geogr.* **2003**, *93*, 314–337.

10. An, L.; Linderman, M.; Qi, J.; Shortridge, A.; Liu, J. Exploring complexity in a human-environment system: An agent-based spatial model for multidisciplinary and multiscale integration. *Ann. Assoc. Am. Geogr.* **2005**, *95*, 54–79.

11. Ligmann-Zielinska, A.; Kramer, D.B.; Cheruvelil, K.S.; Soranno, P.A. Using uncertainty and sensitivity analyses in socioecological agent-based models to improve their analytical performance and policy relevance. *PLoS ONE* **2014**, *9*, e109779.

12. Acosta-Michlik, L.; Espaldon, V. Assessing vulnerability of selected farming communities in the Philippines based on a behavioural model of agent's adaptation to global environmental change. *Glob. Environ. Change* **2008**, *18*, 554–563.

13. Valbuena, D.; Verburg, P.; Veldkamp, A.; Bregt, A.K.; Ligtenberg, A. Effects of farmers' decisions on the landscape structure of a Dutch rural region: An agent-based approach. *Landsc. Urban Plan.* **2010**, *97*, 98–110.

14. Bert, F.E.; Podesta, G.P.; Rovere, S.L.; Menendez, A.N.; North, M.; Tatara, E.; Laciana, C.E.; Weber, E.; Toranzo, F.R. An agent based model to simulate structural and land use changes in agricultural systems of the argentine pampas. *Ecol. Model.* **2011**, *222*, 3486–3499.

15. Huang, Q.; Parker, D.C.; Sun, S.P.; Filatova, T. Effects of agent heterogeneity in the presence of a land-market: A systematic test in an agent-based laboratory. *Comput. Environ. Urban Syst.* **2013**, *41*, 188–203.

16. Goldenberg, J.; Libai, B.; Muller, E. Talk of the network: A complex systems look at the underlying process of word-of-mouth. *Mark. Lett.* **2001**, *12*, 211–223.

17. Allsop, D.T.; Bassett, B.R.; Hoskins, J.A. Word-of-Mouth research: Principles and applications. *J. Advert. Res.* **2007**, *47*, 398–411.

18. Burton, M.; Rigby, D.; Young, T. Analysis of the determinants of adoption of organic horticultural techniques in the UK. *J. Agric. Econ.* **2008**, *50*, 47–63.

19. Conley, T.; Udry, C. Social learning through networks: The adoption of new agricultural technologies in Ghana. *Am. J. Agric. Econ.* **2001**, *83*, 668–673.

20. Schreinemachers, P.; Berger, T. An agent-based simulation model of human-environment interactions in agricultural systems. *Environ. Model. Softw.* **2011**, *26*, 845–859.

21. Matthews, R.B.; Gilbert, N.G.; Roach, A.; Polhill, J.G.; Gotts, N.M. Agent-based land-use models: A review of applications. *Landsc. Ecol.* **2007**, *22*, 1447–1459.

22. Kaye-Blake, W.; Li, F.Y.; Martin, A.M.; McDermott, A.; Rains, S.; Sinclair, S.; Kira, A. *Multi-Agent Simulation Models in Agriculture: A Review of Their Construction and Uses*; Lincoln University: Christchurch, New Zealand, 2010.

23. Bithell, M.; Brasington, J. Coupling agent-based models of subsistence farming with individual-based forest models and dynamic models of water distribution. *Environ. Model. Softw.* **2009**, *24*, 173–190.

24. Matthews, R. The People and Landscape Model (PALM): Towards full integration of human decision-making and biophysical simulation models. *Ecol. Model.* **2006**, *194*, 329–343.

25. Evans, T.P.; Phanvilay, K.; Fox, J.; Vogler, J. An agent-based model of agricultural innovation, land-cover change and household inequality: The transition from swidden cultivation to rubber plantations in Laos PDR. *J. Land Use Sci.* **2011**, *6*, 151–173.

26. Deffuant, G.; Huet, S.; Bousset, J.P.; Henriot, J.; Amon, G.; Weisbuch, G. Agent based simulation of organic farming conversion in Allier département. In *Complexity and Ecosystem Management: The Theory and Practice of Multi-Agent Systems*; Janssen, M.A., Ed.; Edward Elgar Publishers: Chelten-Ham, UK, 2002; pp. 158–187.

27. Kaufmann, P.; Stagl, S.; Franks, D.W. Simulating the diffusion of organic farming practices in two New EU Member States. *Ecol. Econ.* **2009**, *68*, 2580–2593.

28. Bakker, M.M.; van Doorn, A.M. Farmer-specific relationships between land use change and landscape factors: Introducing agents in empirical land use modelling. *Land Use Policy* **2009**, *26*, 809–817.

29. Gotts, N.M.; Polhill, J.G. When and how to imitate your neighbours: Lessons from and for FEARLUS. *J. Artif. Soc. Soc. Simul.* **2009**, *12*, 2.

30. Happe, K.; Kellermann, K.; Balmann, A. Agent-based analysis of agricultural policies: An illustration of the agricultural policy simulator AgriPoliS, its adaptation, and behavior. *Ecol. Soc.* **2006**, *11*, 49.

31. Schreinemachers, P.; Berger, T.; Sirijinda, A.; Praneetvatakul, S. The diffusion of greenhouse agriculture in northern Thailand: Combining econometrics and agent-based modeling. *Can. J. Agric. Econ.* **2009**, *57*, 513–536.

32. Valbuena, D.; Verburg, P.; Veldkamp, A.; Bregt, A.K.; Ligtenberg, A. An agent-based approach to explore the effect of voluntary mechanisms on land use change: A case in rural Queensland, Australia. *J. Environ. Manag.* **2010**, *91*, 2615–2625.

33. Jager, W.; Janssen, M.A.; De Vries, H.J.M.; De Greef, J.; Vlek, C.A.J. Behaviour in commons dilemmas: Homo economicus and Homo psychologicus in an ecological-economic model. *Ecol. Econ.* **2000**, *35*, 357–379.

34. Congress of the Philippines. An Act Providing for the Development and Promotion of Organic Agriculture in the Philippines and for Other Purpose, in Republic Act 10068. Fourteenth Congress: Manila, Philippines, 2009.

35. Lesaca, P.R.A. Organic Agriculture in the Philippines: Going Back to Basic. *BAR Digest* *2012*; 13; February; 2014; Available online: http://www.bar.gov.ph/organic-agriculture (accessed on 10 January 2015).

36. David, C.C. Agriculture. In *The Philippine Economy: Development, Policies, and Challenges*; Balisacan, A.M., Hill, H., Eds.; Oxford University Press: New York, NY, USA, 2003; pp. 175–218.

37. National Economic and Development Authority (NEDA). *Medium-Term Philippine Development Plan, 2004–2010*; NEDA: Manila, Philippines, 2004.

38. Herdt, R.W.; Barker, R.; Rose, B. *The Rice Economy of Asia*; International Rice Research Institute: Manila, Philippines, 1985; Volume 2.

39. Collado, P.M.G.; Tia, M.E.; del Prado, D.G.L.; Taguibolos, G.B.; Lipio, G.M., Jr. *Characteristics of Farm Holdings: Evidence from the Philippines' Census of Agriculture*; Southeast Asian Regional Center for Graduate Study and Research in Agriculture: Los Baños, Philippines, 2013.

40. Schmitt, L. Developing and applying a soil erosion model in a data-poor context to an island in the rural Philippines. *Environ. Dev. Sustain.* **2009**, *11*, 19–42.

41. Muramoto, J.; Hidaka, K.; Mineta, T. Japan: Finding opportunities in the current crisis. In *The Conversion to Sustainable Agriculture: Principles, Processes and Practices*; Gliessman, S.R., Rosemeyer, M., Eds.; CRC Press: Boca Raton, FL, USA, 2010; pp. 273–302.

42. Klonsky, K.; Greene, C. Widespread adoption of organic agriculture in the US: Are market-driven policies enough? In Proceedings of the American Agricultural Economics Association Annual Meeting, Providence, RI, USA, 24–27 July 2005.

43. D'Souza, G.; Cyphers, D.; Phipps, T. Factors affecting the adoption of sustainable agricultural practices. *Agric. Resour. Econ. Rev.* **1993**, *221*, 59–165.

44. Best, H. Environmental concern and the adoption of organic agriculture. *Soc. Nat. Resour.* **2010**, *23*, 451–468.

45. Mzoughi, N. Farmers adoption of integrated crop protection and organic farming: Do moral and social concerns matter? *Ecol. Econ.* **2011**, *7*, 1536–1545.

46. Cauwenbergh, N.V.; Biala, K.; Bielders, C.; Brouckaert, V.; Franchois, L.; Garcia Cidad, V.; Hermy, M.; Mathijs, E.; Muys, B.; Reijnders, J.; *et al.* SAFE—A hierarchical framework for assessing the sustainability of agricultural systems. *Agric. Ecosyst. Environ.* **2007**, *1202*, 229–242.

47. Philippine Bureau of Agricultural Statistics. *Area Planted/Harvested of Crops*; Report 2007; Philippine Bureau of Agricultural Statistics: Manila, Philippines, 2007.

48. Hamill, L.; Gilbert, N. Social circles: A simple structure for agent-based social network models. *J. Artif. Soc. Soc. Simul.* **2009**, *12*, 3.

49. Datta, S.K.D. Improving nitrogen fertilizer efficiency in lowland rice in tropical Asia. *Fertil. Res.* **1986**, *91*, 71–186.

50. Schouten, M.; Verwaart, T.; Heijman, W. Comparing two sensitivity analysis approaches for two scenarios with a spatially explicit rural agent-based model. *Environ. Model. Softw.* **2014**, *54*, 196–210.

51. Crosetto, M.; Tarantola, S.; Saltelli, A. Sensitivity and uncertainty analysis in spatial modelling based on GIS. *Agric. Ecosyst. Environ.* **2000**, *81*, 71–79.

52. Gomez-Delgado, M.; Tarantola, S. GLOBAL sensitivity analysis, GIS and multi-criteria evaluation for a sustainable planning of a hazardous waste disposal site in Spain. *Int. J. Geogr. Inf. Sci.* **2006**, *20*, 449–466.

53. Chu-Agor, M.L.; Muñoz-Carpena, R.; Kiker, G.; Emanuelsson, A.; Linkov, I. Exploring vulnerability of coastal habitats to sea level rise through global sensitivity and uncertainty analyses. *Environ. Model. Softw.* **2011**, *26*, 593–604.

54. Saltelli, A.; Ratto, M.; Andres, T.; Campolongo, F.; Cariboni, J.; Gatelli, D.; Saisana, M.; Tarantola, S. *Global Sensitivity Analysis: The Primer*; Wiley-Interscience: Chichester, UK, 2008; p. 304.

55. Lilburne, L.; Tarantola, S. Sensitivity analysis of spatial models. *Int. J. Geogr. Inf. Sci.* **2009**, *23*, 151–168.

56. Homma, T.; Saltelli, A. Importance measures in global sensitivity analysis of nonlinear models. *Reliab. Eng. Syst. Saf.* **1996**, *52*, 1–17.

57. Saltelli, A.; Annoni, P.; Azzini, I.; Campolongo, F.; Ratto, M.; Tarantola, S. Variance based sensitivity analysis of model output. Design and estimator for the total sensitivity index. *Comput. Phys. Commun.* **2010**, *181*, 259–270.

58. Ligmann-Zielinska, A.; Sun, L. Applying time dependent variance-based global sensitivity analysis to represent the dynamics of an agent-based model of land use change. *Int. J. Geogr. Inf. Sci.* **2010**, *24*, 1829–1850.

59. Tarde, G. *The Laws of Imitation*; The Mershon Company Press: Rahway, NJ, USA, 1903.

60. Ryan, B.; Gross, N.C. The diffusion of hybrid seed corn in two Iowa communities. *R. Soc.* **1943**, *8*, 15–24.

61. Rogers, E.M. *Diffusion of Innovations*; Free Press: New York, NY, USA, 2003.

62. Heong, K.L.; Escalada, M.M. A comparative analysis of pest management practices of rice farmers in Asia. In *Pest Management of Rice Farmers in Asia*; Heong, K.L., Escalada, M.M., Eds.; International Rice Research Institute: Los Baños, Philippines, 1997; pp. 227–242.

63. Borron, S. *Building Resilience for an Unpredictable Future: How Organic Agriculture Can Help Farmers Adapt to Climate Change*; United Nations Food and Agriculture Organization: Rome, Italy, 2006.

64. Bankes, S. Exploratory modeling for policy analysis. *Oper. Res.* **1993**, *41*, 435–449.

65. Weaver, C.P.; Lempert, R.J.; Brown, C.; Hall, J.A.; Revell, D.; Sarewitz, D. Improving the contribution of climate model information to decision making: The value and demands of robust decision frameworks. *Wiley Interdiscip. Rev.: Clim. Change* **2013**, *4*, 39–60.

66. Bankes, S.; Lempert, R.; Popper, S. Making computational social science effective: Epistemology, methodology, and technology. *Soc. Sci. Comput. Rev.* **2002**, *20*, 377–388.

67. Downey, E.; Brill, J. The use of optimization models in public-sector planning. *Manag. Sci.* **1979**, *25*, 413–422.

68. Meadows, D. *Thinking in Systems: A Primer*; Chelsea Green: White River Junction, VT, USA, 2008.
69. Peck, S.L. Simulation as experiment: A philosophical reassessment for biological modeling. *Trends Ecol. Evol.* **2004**, *19*, 530–534.

An Approach for Simulating Soil Loss from an Agro-Ecosystem Using Multi-Agent Simulation: A Case Study for Semi-Arid Ghana

Biola K. Badmos, Sampson K. Agodzo, Grace B. Villamor and Samuel N. Odai

Abstract: Soil loss is not limited to change from forest or woodland to other land uses/covers. It may occur when there is agricultural land-use/cover modification or conversion. Soil loss may influence loss of carbon from the soil, hence implication on greenhouse gas emission. Changing land use could be considered actually or potentially successful in adapting to climate change, or may be considered maladaptation if it creates environmental degradation. In semi-arid northern Ghana, changing agricultural practices have been identified amongst other climate variability and climate change adaptation measures. Similarly, some of the policies aimed at improving farm household resilience toward climate change impact might necessitate land use change. The heterogeneity of farm household (agents) cannot be ignored when addressing land use/cover change issues, especially when livelihood is dependent on land. This paper therefore presents an approach for simulating soil loss from an agro-ecosystem using multi-agent simulation (MAS). We adapted a universal soil loss equation as a soil loss sub-model in the Vea-LUDAS model (a MAS model). Furthermore, for a 20-year simulation period, we presented the impact of agricultural land-use adaptation strategy (maize cultivation credit *i.e.*, maize credit scenario) on soil loss and compared it with the baseline scenario *i.e.*, business-as-usual. Adoption of maize as influenced by maize cultivation credit significantly influenced agricultural land-use change in the study area. Although there was no significant difference in the soil loss under the tested scenarios, the incorporation of human decision-making in a temporal manner allowed us to view patterns that cannot be seen in single step modeling. The study shows that opening up cropland on soil with a high erosion risk has implications for soil loss. Hence, effective measures should be put in place to prevent the opening up of lands that have high erosion risk.

Reprinted from *Land*. Cite as: Badmos, B.K.; Agodzo, S.K.; Villamor, G.B.; Odai, S.N. An Approach for Simulating Soil Loss from an Agro-Ecosystem Using Multi-Agent Simulation: A Case Study for Semi-Arid Ghana. *Land* **2015**, *4*, 607–626.

1. Introduction

Erratic rainfall is a major challenge facing agricultural practice in the semi-arid regions of West Africa. The quality as well as the amount of land and water resources accessible for agriculture and other climate-dependent sectors such as forestry and fisheries are affected by climate change [1,2]. Farmers are changing their agricultural practices and devising ways to modify livelihoods in light of the changing climate and other multiple stresses. In some cases, the changes could be considered actual or possible successes in adapting to climate change. It could also be just coping, or it may be considered maladaptation where they create environmental degradation [3]. In semi-arid northern Ghana, changing agricultural practices (e.g., crop diversification) have been identified amongst other climate variability and climate change adaptation measures [4–6]. Various policy instruments have been introduced to enhance farmers' resilience towards the impact of climate change, for example, fertilizer subsidies, farm credit, training in alternative sources of livelihood, *etc.* Some of these policies might require change in the agricultural land use/cover.

Soil is directly linked to many ecosystem services, hence conserving the soil will preserve and maintain the availability of these ecosystem services, such as food production, water filtration, carbon storage, *etc.* Soil loss is a process caused by erosion and its prepositional power [7]. The combination of climate, steep slopes, and inappropriate land use/cover patterns triggers soil erosion [8]. Various human activities, for example, population growth, removal of forest, land cultivation, overgrazing, and higher demands for firewood often cause soil erosion [9]. Soil loss may result in a decline in soil fertility and a decrease in the volume of reservoirs and water bodies due to siltation. When productivity of soil is reduced, the outputs derived from renewable natural resource systems of the biosphere are affected [10]. Soil carried by erosion also moves pesticides, soil nutrients, and other harmful chemicals into water bodies as well as ground water resources [11,12]. Soil erosion is also a channel through which carbon is lost from the ecosystem [13], hence the implication for greenhouse gas emissions. In Africa, decreases in productivity due to soil loss have been estimated to be between 2% and 40%, with an average of 8.2% for the whole continent [14], and about 19% of the reservoir storage volumes of Africa are silted [15]. In Ghana, about 30%–40% of the total land area, most of which is concentrated in the northern, drier part of the country, is experiencing some form of land degradation. The soils of northern Ghana are erodible due to low organic matter content, in the range of 1.8%–3.2% [16,17].

In this part of the world, soil loss due to agricultural land use change has not been adequately addressed. Agriculture is a primary source of livelihood in the semi-arid northern Ghana [18], and human decision-making will play a vital role when it comes to agricultural land use change (ALUC). We cannot ignore the heterogeneity of farm households (agents) when addressing issues on land use/cover change, especially when livelihood is primarily dependent on land. Multi-agent simulation (MAS) modeling is a data-demanding modeling approach and soil erosion/soil loss study is a resource-demanding field of study. These may have contributed to fewer applications of MAS model in the domain of soil erosion as compared to other fields. This paper therefore presents an approach for simulating soil loss from an agro-ecosystem using a multi-agent simulation (MAS) model. We simulated the impact of agricultural land use change adaptation strategy (maize cultivation credit-maize credit scenario) on soil loss and we compared the impact with the baseline scenario, i.e. business-as-usual.

2. Methods

2.1. Study Area

This study was conducted in the Vea catchment (Figure 1) in the Upper East Region (UER) of Ghana. The region is located in the northeast corner of Ghana between latitudes 10°30′ and 11°8′ North and longitudes 1°15′ West and 0°5′ East. The UER, together with the Upper West Region and Northern Region, constitute the three regions of northern Ghana. The region is bordered by Burkina Faso in the north and Togo to the east. Most parts of the region belong to the West African semi-arid Guinea Savannah [19]. The region covers a total land area of 8842 km^2 and this represents about 3.7% of the total area of Ghana [20]. In the 2010 national census report [18], the UER of Ghana has a population of about 1,046,545 habitants (~48.4% male and 51.6% female), which constitutes about 4.2% of the total population of Ghana. The average household size in the region is 5.8 persons per household, rural locality is about 79%, and about 70% of the economically active population (ages 15 years and above) are involved in agricultural activities [18].

Rainfall in the region is mono-modal and the peak of the rainy season is around July–September. The average annual rainfall is about 1044 mm and this is suitable for a single wet season crop [21]. About 60% of the annual rain falls between July and September. The wet period in the region is relatively short and is further marked by variations in the arrival time, duration, and intensity of rainfall [21]. The annual temperature is around 28–29 °C, whereas the absolute minimum temperature is around 15–18 °C [22].

The region has experienced a series of climate change impacts, such as a shift in seasons and irregular climatic conditions. The real problem for farmers in the northern part of Ghana is the unreliability of rainfall caused by inter-annual variability of both the total amounts and distribution of rainfall [23]. In the study area, rainfall is a key underlying factor influencing farmers' agricultural land use change options [24]. Erratic rainfall makes agricultural planning very difficult and is one of the principal sources of risk for rain-fed agriculturalists in the Sahel [23].

141

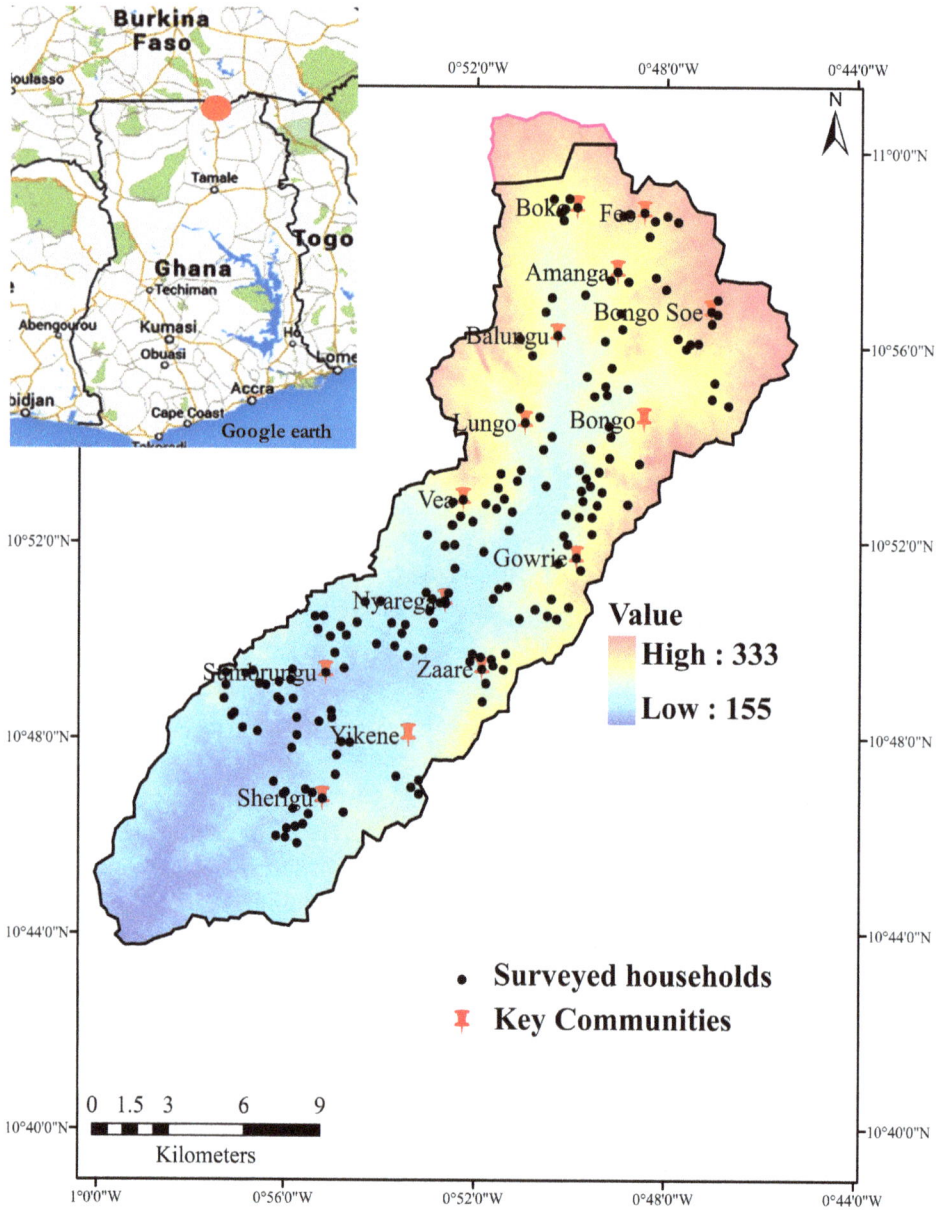

Figure 1. Elevation map of Vea catchment showing the locations of sampled farm households. The upper pink boundary represents the Burkina Faso section of the catchment.

2.2. Household Agricultural Land Use Choice

Seven main categories of agricultural land-use choices were identified (Table 1) from the household survey. They include traditional cereals (guinea corn culture, late millet culture, mixed-traditional culture), groundnut (monoculture groundnut, mixed-culture groundnut), rice, and maize.

Table 1. Agricultural land-use choice classes.

	Sub-Category/Description	Code
1	Traditional cereals culture, where Guinea corn (GC) is main crop	GC_CULT
2	Traditional cereals culture, where Late millet (LM) is main crop	LM_CULT
3	Traditional cereals culture, where there is an equal ratio of GC and LM	MIX_TRAD_CULT
4	Groundnut in a mixture of other crops	MIX_GNUT
5	Groundnut in a mono culture	MONOGNUT
6	Rice is the main crop.	RICE
7	Maize is the main crop.	MAIZE

2.3. Model Description: Vea-LUDAS

The Vea-LUDAS model (Figure 2) adapted the framework of Land-Use Dynamic Simulator (LUDAS) [25]. The Vea-LUDAS model is mainly based on the existing versions of LUDAS models [25–28]. The new feature of this version of the LUDAS model (*i.e.*, Vea-LUDAS) is the incorporation of soil loss, which was parameterized in the context of the Vea catchment in the Upper East Region of Ghana.

LUDAS is a MAS model that was first applied to an upland watershed of about 90 km^2 in central Vietnam. LUDAS was first applied by Le *et al.* [25] because of the heterogeneous nature of biophysical conditions, the diverse livelihood patterns of local farming households, and the need to formulate policies balancing nature conservation and economic development purposes. The description of the Vea-LUDAS model using the ODD protocol (overview, design concept, and details) [29,30] is presented in Appendix A. The ODD protocols of the Vea-LUDAS model followed similar steps to other versions of the LUDAS model. Vea-LUDAS model programming and simulation was carried out in NetLogo [31].

The human (household) agent and environmental (landscape) agent are the two agents in the Vea-LUDAS model and each of these agents has numerous state variables. Human agents are represented in the model as farm households (*i.e.*, household agent) and each farm household has its spatial location, hence it can be identified with respect to its position. The state variables of human agent are household characteristics (age of household head, household size, household labor, household dependency ratio), human-plot characteristics (land holding per capita, rain-fed land holding, land area cultivated for different crops, household proximity to plots, river and irrigation area), household financial characteristics (income per capita, income from rain-fed crop). Landscape agents consist of biophysical spatial raster layers and other variables in the form of GIS-raster layers. Landscape agent is also referred to as patch and this includes biophysical features (land cover, elevation, upslope contributing area, wetness index, and soil texture components) and proximity features (plot distance to river and plot distance to irrigation area).

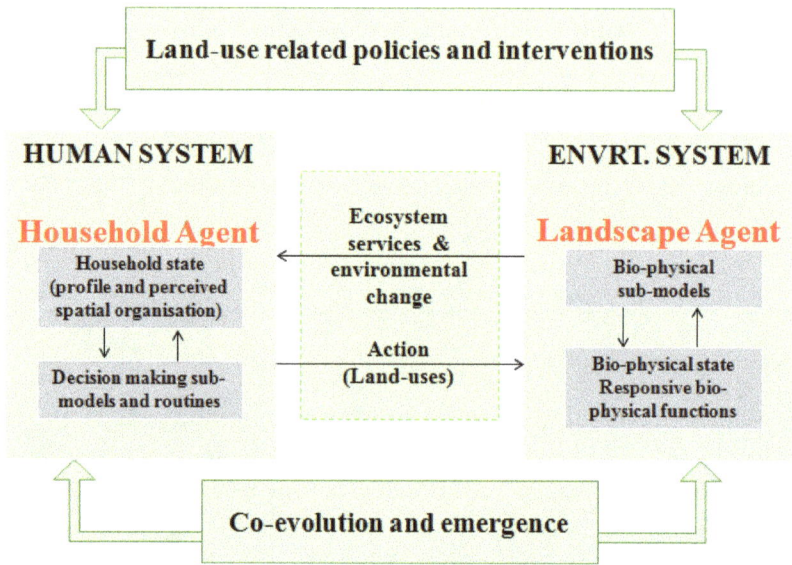

Figure 2. Vea-LUDAS framework. Adapted from [25].

2.3.1. Key Sub-Model Adapted for This Study: Soil Loss Sub-Model

The Universal Soil Loss Equation (USLE) [32] was adapted for a soil loss estimation sub-model (Figure 3) in the Vea-LUDAS model. The USLE (Equation (1)) has been used extensively to estimate soil loss and it has also found usage in Africa. Kaolinite is the dominant clay in soils of West Africa, thus permitting the use of USLE [33]. The soil loss estimation sub-model was embedded inside the landscape module of the Vea-LUDAS model. The erosivity (R), erodibility (K), slope factor (LS), and cover factor (C) layers were imported into Vea-LUDAS. The C layer is linked to land-use/cover layer. As the farm households make their cropping decisions in terms of agricultural land-use, the C layer updates and soil loss is determined through the following equation:

$$A = R \times K \times LS \times C \times P \qquad (1)$$

where A = Mean annual soil loss (t·ha^{-1}·yr^{-1}), R = Rainfall/runoff erosivity (MJ mm·ha^{-1}·h^{-1}·yr^{-1}), K = Soil Erodibility (t·h·MJ^{-1}·mm^{-1}), LS = Slope length and steepness factor (Unitless), C = Cover and management factor (Unitless), P = Conservation/support practice (Unitless).

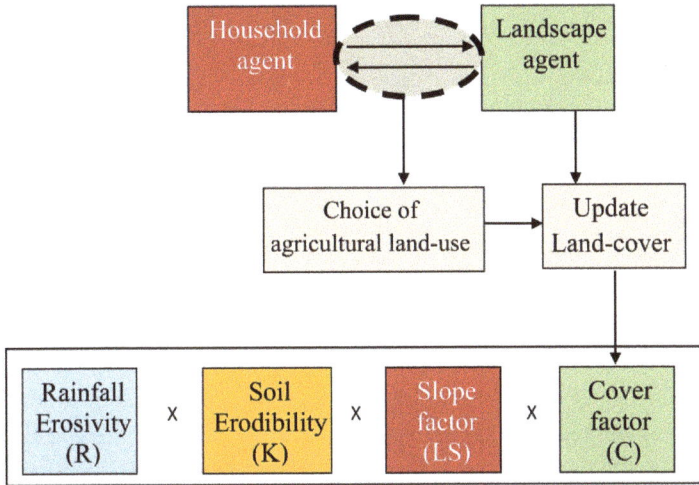

Figure 3. Soil loss sub-model.

2.3.2. Variable Specification for Soil Loss Estimation

Rainfall Erosivity Factor (R)

This factor gives an indication on how erosive the rainfall is. The conventional method of estimating rainfall erosivity is the use of Erosion Index-EI_{30} [32]. However, this is difficult to obtain in developing countries where continuous data availability has been a major challenge. Hence, several other methods have been developed in different parts of the world, for example, [34–38]. In some locations in West Africa, the relationship between annual rainfall and erosivity (Equation (2)) was tested and verified by [35] with 20 rainfall recording station in Cote d'Ivoire, Burkina Faso, Senegal, Niger, and Chad, excluding stations located around the mountains as well as near the sea. The Fournier index (FI) [34] has also been used to estimate rainfall erosivity, but has been improved upon with the modified version *i.e.,* Modified Fournier index (MFI) (Equation (3)) [36]. We generated the MFI for the study area using time series rainfall data provided by the Ghana Meteorological Service Department. MFI was determined for the rainy season period of each year (April to October) and the average for the years was used. We obtained the estimated monthly Erosivity (Ri) using Equation (4).

$$R = [(0.5 \pm 0.05) \ P] \tag{2}$$

where R = Rainfall erosivity, P = Annual rainfall

$$MFI = \sum_{i=1}^{12} Pi^2/P \qquad (3)$$

where MFI = Modified Fournier index, i = Months, Pi = average rainfall in month i (mm), P = Annual rainfall.

$$Ri = a + b \ (MFI) \qquad (4)$$

where Ri = Monthly erosivity, MFI = Modified Fournier Index, and a (21) and b (1.96) are site-specific empirical constants.

Soil Erodibility (K-value)

The K-value represents the soil loss per unit of EI_{30} as measured in the field on a standard plot with a length of 22 m and 9% slope [32]. There are three popular methods [32,39,40] used in the estimation of erodibility [41]; soil particles play an essential role in all cases. A soil erodibility nomograph was developed by Wischmeier *et al.* [42] to read K-value. In using the nomograph, % silt content, % sand content, % organic matter, soil structural class, and soil permeability are required. In Williams *et al.* [39], the fine sand, silt, clay, and organic carbon content of the soil were used to estimate soil erodibility. In a data-scarce environment, an alternative method for estimating soil erodibility, *i.e.*, ERFAC-K (Equation (5)), was proposed by Geleta [43]. In deriving the ERFAC-K, soil particles of different ratios, such as (i) silt to clay, (ii) silt to sand, and (iii) silt to sand and clay were compared with the measured K-value, and the highest coefficient of correlation (0.88) was obtained using the silt to sand and clay ratio [43]. Furthermore, soil characteristics from FAO soil database [44] were tested and a correlation coefficient of (0.82) was obtained [43]. Hence, the ERFAC-K method was adapted for the estimation of K-value as follows:

$$ERFAC\text{-}K = a \left[\frac{\% \ Silt}{\% \ Sand \ + \ \%Clay} \right]^b \qquad (5)$$

where ERFAC-K = Proposed alternative soil Erodibility factor, % Silt = % silt content of the soil, % Clay = % clay content of the soil, % Sand = % sand content of the soil, a = 0.32, and b = 0.27.

Slope Length and Steepness Factor (LS)

Slope length and steepness are usually combined in USLE. The LS-factor represents the ratio of soil loss on a given slope length and steepness to soil loss from a 22.1 m slope length and a steepness of 9% under otherwise identical conditions [45].

LS factor can be calculated in various ways, for example [32,46–48]. According to Van der Knijff et al. [49], the LS equation (Equation (6)) described in Moore et al. [46,48] has the advantage over the original equation [32] because it uses specific contributing area as a slope length estimate, and this is more amenable to three-dimensional landscapes. We therefore used the method described in [49] for the estimation of the LS factor:

$$LS = m + 1 \left[\frac{As}{22.13} \right]^m \left[\frac{SinB}{0.0896} \right]^n \qquad (6)$$

where A_s is upslope contributing area, B is the slope in degrees, and m and n are empirical exponents.

C- and P-Factor

C-factor is the ratio of soil loss from land cropped under specific conditions to the corresponding loss from clean-tilled continuous fallow [32]. P-factor describes the erosion conservation practice put in place. The value of C-factor depends on vegetation type, stage of growth, and cover percentage [8]. C-factor is the most important conditional factor, and if vegetation cover is uninterrupted, erosion and runoff are small despite the erosivity of the rainfall, slope steepness, and soil instability [35]. C-factor can be estimated on the field by comparing soil loss on clean-tilled, continuous fallow with other types of land-use/cover [50]. A normalized vegetation index has also been used to estimate crop factor, for example, [51,52]. The study area is primarily agriculture based, and agriculture constitutes the main source of livelihood. Very few studies we are aware of have looked at the C-factor for different crop types in West Africa, for example [35]. In Roose [35], C-factors for different crops were presented based on the yield of crops, but the study did not provide the standard yield used in the estimation of C-factor. However, Henao and Baanante [53] summarized the C-factor for some selected cover types in Africa (Table 2). Hence we adapted C-factors presented in [53].

Table 2. Crop Cover and Management Factor for selected crops [53].

Cover Type	Cover and Management Factor (C)
Millet and sorghum	0.3–0.9
Cotton	0.5–0.7
Groundnuts	0.4–0.8
Cowpea	0.2–0.4
Maize	0.4–0.7
Rice (paddy)	0.3–0.5
Bare land	0.8–1.0

2.4. Scenario Exploration

Two scenarios were tested in this study, namely (i) Baseline (BS) and (ii) maize credit scenario (MCS). BS describes the business-as-usual situation whereby the behavior of agents on the ground is that there is no policy intervention. On the other hand, MCS operates on the grounds that credit is offered to farmers for maize cultivation. The concept of MCS arose because in the northern part of Ghana, maize has been identified as an agricultural land-use adaptation practice [4,24,54]. Also, a program promoting maize cultivation was observed in the study area. For example, in the block farm program, farmers are provided with support to enable them to improve their production, and they pay back the credit in kind at the time of harvest. Maize is one of the target crops under the block farm program [55]. Hence, this study opted for maize as an agricultural land-use change option influenced by credit. The choice of household agents to accept maize cultivation followed the maize credit adoption sub-model (Figure 4). This sub-model adapted the decision-making sub-model for willingness to accept payment for ecosystem services in [28,56] by following a process-based decision [56,57]. The sub-model is linked with the crop decisions of the household agent. At each time step and with respect to preferences coefficient generated using binary logistic regression, the sub-model randomly determines the probability of whether a household will accept maize credit to cultivate maize; otherwise the household uses the choice probability of his land holding. A yearly household increment of 1.2% for the study area [18] was specified in the model.

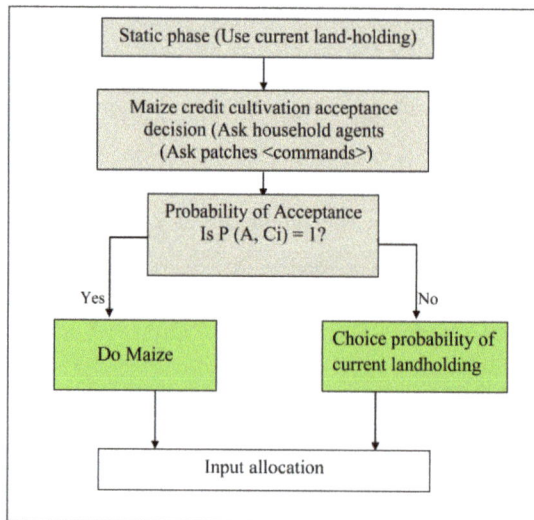

Figure 4. Maize credit adoption sub-model. Adapted from [28,56].

3. Results

3.1. Soil Loss Estimation Parameters

Rainy season (April–October) rainfall erosivity between 1976 and 2012 (Figure 5) ranged between 414.9 and 701.1 MJ·mm·ha^{-1}·h^{-1}·yr^{-1}. LS-factor and soil erodibility factor are presented in (Figure 6). LS-factor ranged between from 10^{-8} and 3.49% (Figure 6a) and the soil erodibility factor ranged from 0.026 to 0.035 t·h·MJ^{-1}·mm^{-1} (Figure 6b). The highest value for soil erodibility was obtained in the fluvisol, while the least values were obtained from the lixisols.

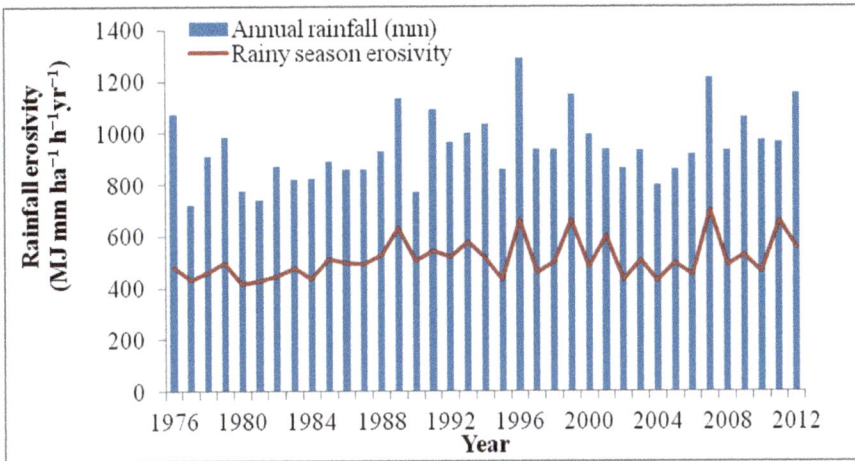

Figure 5. Rainfall erosivity.

3.2. Agricultural Land-Use Change

The change in area cultivated for different crops between year 1 and year 20 under the two scenarios is presented in Figure 7. Under the BS, the steady increase in the land area cultivated for different crops was attributed to the 1.2% yearly household increment specified in the model. In the case of MCS, the change in area cultivated for different crops was linked to the 1.2% yearly household increment specified in the model, as well as the influence of maize adoption rate (influenced by maize cultivation credit) at the expense of other agricultural land uses.

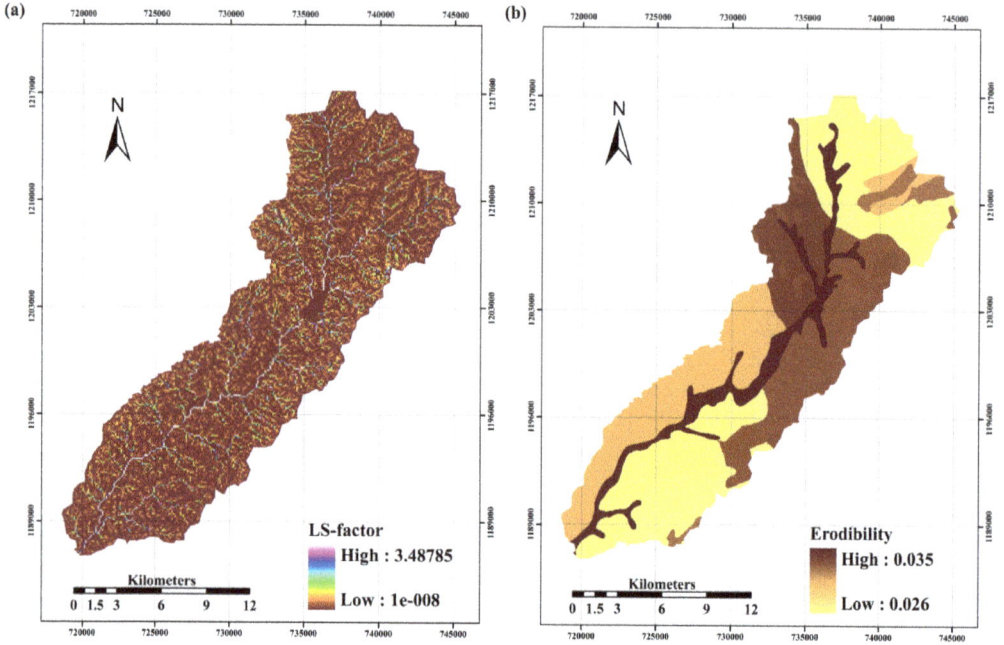

Figure 6. (**a**) LS-Factor %; (**b**) soil erodibility factor-t·h·MJ^{-1}·mm^{-1}.

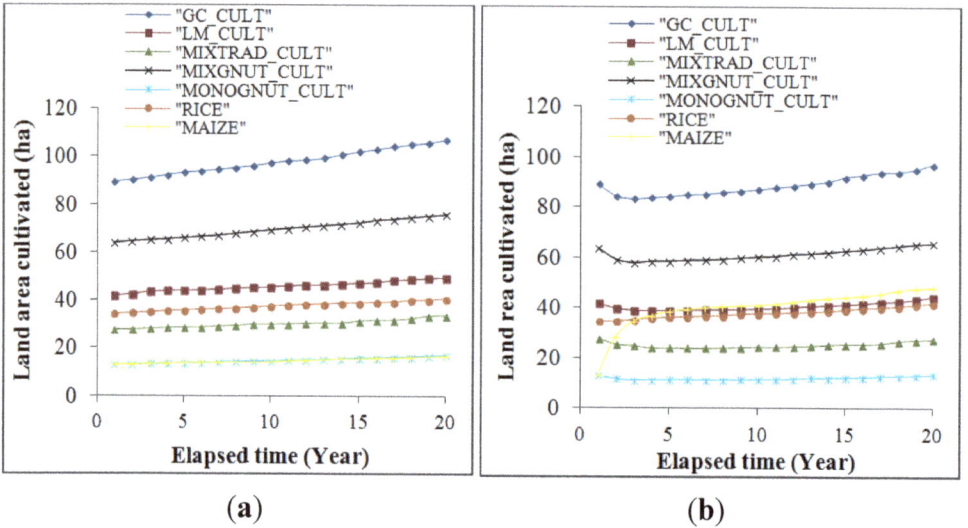

Figure 7. Simulated agricultural land-use change under the (**a**) BS and (**b**) MCS scenarios.

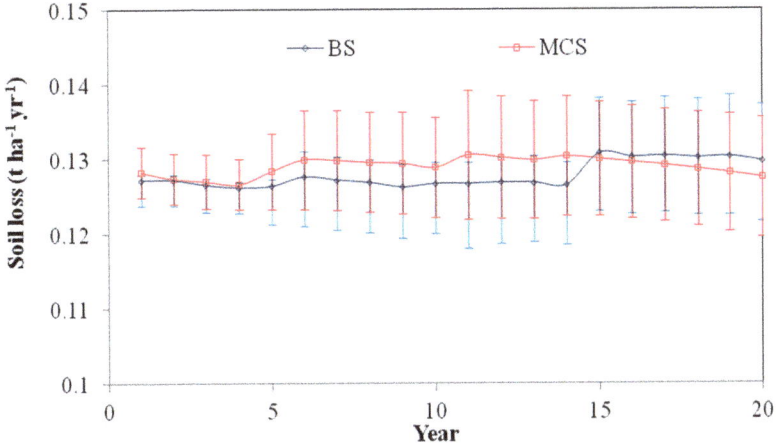

Figure 8. Simulated annual soil loss ($p < 0.05$; confidence interval 95%).

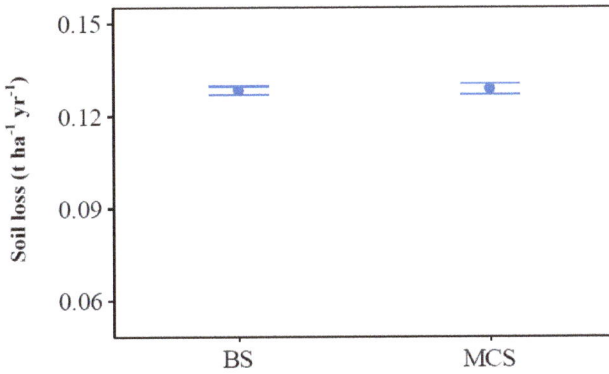

Figure 9. Average annual soil loss ($p < 0.05$; confidence interval 95%).

3.3. Soil Loss

The impact of MCS on soil loss in comparison with BS showed a mixed pattern. Between year 4 and 15, simulated annual soil loss was higher under MCS as compared to BS. On the other, between year 15 and 20, simulated annual soil loss was higher under BS as compared to MCS (Figure 8). There was however no statistical difference ($p < 0.05$) (Figure 9) in the average simulated annual soil loss under MCS and BS.

4. Discussion

The application of MAS model for research has shown a tremendous increase in the last two decades [58], and this cuts across several disciplines. A key strength of the MAS model is the ability to clearly simulate the implications of human decision-making processes [59,60].

In Ghana, rainfall model simulation results have showed more uncertainty compared to temperature model simulation [61]. The uncertainty in the rainfall will influence erosivity, which is the rainfall indicator for soil loss estimation. Similarly, inter-seasonal variation in rainfall will also influence rainfall erosivity and this in turn will influence the soil loss. In semi-arid Ghana, rain-fed agriculture predominates and rainfall plays a very important role in influencing agricultural land-use choice and alternatives [24]. An increase in spatial patterns of rainfall has been reported in the study area. For example, close-by locations that usually have similar rainfall patterns are now experiencing varying patterns. This situation has implications on soil loss because the cover factor will vary at different times of the year. However, this study did not consider inter-seasonal and spatial variation in rainfall and cover between different locations in the study area due to limited availability of widespread rainfall data across the study area. Furthermore, the primary farming practice in the study area is subsistence and the majority of the farmers are small holders, hence the difficulty in collecting data on the sowing and harvesting time for each type of crop cultivated. As a result of data scarcity, we settled for an alternative method of estimating the K-factor due to data limitation, which has been a huge challenge for research in Sub-Saharan Africa.

In Roose [33], C-factor is described as the most important conditional factor influencing soil loss. On the other hand, the mixed pattern observed in soil loss under the BS and MCS is associated with the fact that as farm households clear new land for crop cultivation, the type of crop cultivated on the land is not the only factor contributing to the soil loss; the influence of other biophysical characteristics (e.g., erodibility and topographic factors) of the newly cleared land also counts. This also points out the importance of incorporating farm household crop decision-making into soil loss estimation in the agro-ecosystem. It is well known that soil loss is also driven by various human activities, such as overgrazing, higher demands for fire, *etc.* However, this study only captured the implications of agricultural land-use change on soil loss.

5. Conclusions

Land-use change is second to fossil fuel burning in terms of contribution to greenhouse gas emissions. It has been reported that in the coming years, the contribution of land-use change to climate change will increase considerably. Africa will contribute significantly to the projected 9 billion people by 2050, and this implies an increase in the demand for land resources. Therefore, Africa might play a major role in future climate change. Soil erosion has an important impact on the loss of carbon from the soil into the atmosphere. We presented an approach for simulating soil loss from an agro-ecosystem using a multi-agent simulation model (the Vea-LUDAS model). Following a process-based decision approach, we simulated the impact of maize cultivation credit (maize credit scenario) on agricultural land-use change and subsequently the impact on soil loss. This impact was compared with the baseline scenario, *i.e.*, business-as-usual. The Vea-LUDAS model has shown its potential to explicitly simulate soil loss from an agro-ecosystem. The temporal modeling suitability of the Vea-LUDAS model and the incorporation farm household decision-making allowed us to view patterns that cannot be seen in single step modeling. Although there was no statistical difference in the soil loss under the two tested scenarios, the simulation shows that converting high erosion risk soil to cropland has implications for soil carbon loss (*i.e.*, climate change), which we propose to apply in areas with high erosion risk soils. Consequently, policy should be elaborated to prevent further land degradation of high erosion risk soils. Furthermore, sufficient infrastructure needs to be put in place so that reliable climatic data will be available and accessible. This is important so that farmers can have reliable information on expected weather patterns, thus enabling them to effectively plan their cultural practices and not having bare soil during the period of higher rainfall erosivity.

Acknowledgments: We thank the West African Science Service Center on Climate Change and the Adapted Land Use (WASCAL) Graduate Research Program (GRP) for providing research funds. Appreciation also goes to the field assistants and the WASCAL team. We also thank the anonymous reviewers for their valuable comments and suggestions.

Author Contributions: Biola K. Badmos was involved in the design of research, data collection and supervision of data collection, data analysis, modeling, and writing of the article. Sampson K. Agodzo contributed to the supervision of data collection and writing of the article. Grace B. Villamor contributed to the design of research, data collection and supervision of data collection process, data analysis, modeling, and writing of the article. Samuel N. Odai contributed to the supervision of data collection and writing of the article.

Conflicts of Interest: The authors declare no conflict of interest.

Appendix A. ODD Protocol for Vea-LUDAS

In this section, we describe the Vea-LUDAS using the ODD (overview, design concept, and details) protocol [29,30]. The Vea-LUDAS adopts/follows most of the functionalities with other LUDAS models [25–28].

A1. Overview (O)

Purpose: This study applied the Vea-LUDAS model to assess the impact of maize cultivation credits on agricultural land-use change and farm household livelihood in Vea catchment, Upper East Region of Ghana.

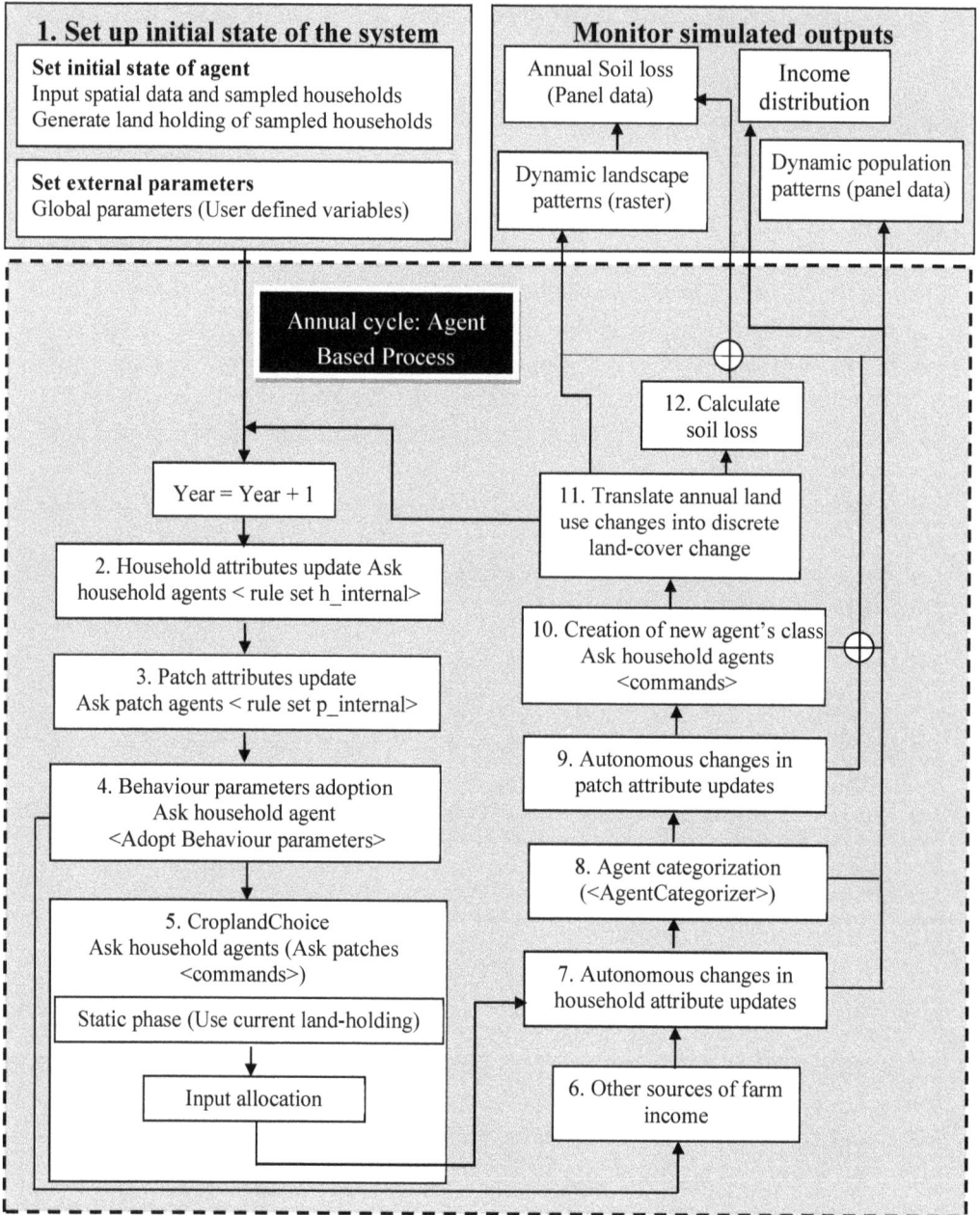

Figure A1. Main simulation steps for Vea-LUDAS (Note: Under the maize credit scenario simulation, step 5 is replaced with Maize credit adoption sub-model; dashes indicate the annual cycle of the agent-based process).

Agents and their state variables and scales: Human (household) agent and environmental (landscape) agent are the two agents in Vea-LUDAS, and each agent has numerous state variables. The human agent (*i.e.*, household agent) is represented in the model as farm household. Each household has its spatial location and can be identified with respect to its position. The state variables of human agent are household characteristics (e.g., age of household head, household size, household labor, household dependency ratio), human-plot characteristics (e.g., land holding per capita, rain-fed land holding, land area cultivated for different crops, household proximity to plots, and river and irrigation area), and household financial characteristics (e.g., income per capita, income from rain-fed crop). The landscape agent comprises the biophysical spatial raster layers and other variables in the form of GIS-raster layers. Landscape agent is also referred to as patch, and this includes biophysical features (e.g., land cover, elevation, upslope contributing area, wetness index, and soil texture components), and proximity layers (e.g., plot distance to river and plot distance to irrigation area).

Vea-LUDAS captures the whole Vea catchment (286 km^2) in the upper east region of Ghana, and is represented by grid or pixel layers (30 m × 30 m = 900 m^2). A 900-m^2 grid was used because of the form in which other spatial data were available and to avoid unnecessary delay in model computation. One year is equivalent to a time step; this is equivalent to one calendar cropping season in the study area where most of the crops cultivated are annual crops.

Process overview and scheduling: One simulation consists of 12 main steps (Figure A1). Each major time loop of the simulation program is referred to as an annual production cycle. Each cycle integrates agent-based and patch-based processes.

A2. Design Concepts (D)

The Vea-LUDAS model is designed to take into account variation of human behavior with respect to agricultural land-use change decision-making. The design of the model also considered the possible implications of policy scenarios on household agricultural land-use change decision and household lielihood.

Emergence: Land-use change is caused by household agents, as well as human agents' willingness to adopt maize cultivation credit. Annual change experienced in the total area cultivated is associated with increasing household number in the study area [18]. Crop yield is a result of household inputs' (e.g., seed, labor, and fertilizer) interaction with landscape features (e.g., upslope, wetness index). Farm income is estimated from crop yield generated by each household.

Adaptation/learning: A household agent chooses the best agricultural land-use with respect to preference coefficient. The behavior of the closest agent group is adopted by the household agent [62,63]. Furthermore, household status is updated

155

at the end of every time step and this influences their preference coefficient, hence their subsequent decision.

Prediction: A household agent is able to optimize spatial land-use choices only within his parcels.

Sensing: In evaluating land-use alternatives, it is assumed that household agents have absolute knowledge of the uniqueness of each landscape agent within their neighborhood spaces.

Interaction: Household agents interact directly when two or more households find their best land-use alternative in the same location. In this case, in a random manner one of them will have to leave that location and search for another plot [25]. Further, when a new household is created, information that will be useful for the new household agent is transferred from another household.

Stochasticity: Application of stochasticity in Vea-LUDAS occurs in four different processes, *i.e.*, (i) choosing plot locations for household agents (initialization), as well as the new household created at each time step; (ii) preference coefficients in the land-use choice function; (iii) ecological sub-models that produce variability in the process; and (iv) some status variables not affected by agent-based processes (all defined by even distribution and pre-defined bounds).

Observation: This includes annually successive charts that describe temporal patterns of land-use/cover coverage, landholdings, yield and components, income and income components, and soil loss.

A3. Details (D)

Initialization: Vea-LUDAS followed initialization steps similar to those of VN-LUDAS [62]. Simulation and analysis were based on the sample households (186). The data on the sampled household are imported first, followed by the spatial data (land cover, elevation, upslope contributing area, wetness index, soil texture components, plot distance to river, and plot distance to irrigation area). This is followed by the land holding generation of the household agents, and each patch is assigned to a household.

References

1. Darwin, R. A farmer's view of the Ricardian approach to measuring agricultural effects of climatic change. *Clim. Change* **1999**, *41*, 371–411.
2. Fischer, G.; Shah, M.; van Velthuizen, H. Climate change and agricultural vulnerability. A Special Report Prepared by the International Institute for Applied Systems Analysis (IIASA) under United Nations Institutional Contract Agreement No. 1113 on "Climate Change and Agricultural Vulnerability". In Proceedings of World Summit on Sustainable Development, Johannesburg, South Africa, 26 August–4 September 2002.

3. Vincent, K.; Joubert, A.; Cull, T.; Magrath, J.; Johnston, P. Overcoming the barriers: How to ensure future food production under climate change in Southern Africa. *Oxfam Policy Pract.: Agric. Food Land* **2011**, *11*, 183–242.

4. Gyasi, E.A.; Karikari, O.; Kranjac-Berisavljevic, G.; Vordzogbe, V.V. *Study of Climate Change Vulnerability and Adaptation Assessment Relative to Land Management in Ghana*; University of Ghana, Legon: Accra, Ghana, 2006.

5. Laube, W.; Awo, M.; Schraven, B. *Erratic Rains and Erratic Markets: Environmental Change, Economic Globalisation and the Expansion of Shallow Groundwater Irrigation in West Africa (No. 30)*; ZEF Working Paper Series; University of Bonn: Bonn, Germany, 2008.

6. Stanturf, J.A.; Warren, M.L.; Charnley, S., Jr.; Polasky, S.C.; Goodrick, S.L.; Armah, F.; Nyako, Y.A. *Ghana Climate Change Vulnerability and Adaptation Assessment*; United States Agency for International Development: Washington, DC, USA, 2011.

7. Ramos, M.C.; Martínez-Casasnovas, J.A. Erosion rates and nutrient losses affected by composted cattle manure application in vineyard soils of NE Spain. *Catena* **2006**, *68*, 177–185.

8. Gitas, I.Z.; Douros, K.; Minakou, C.; Silleos, G.N.; Karydas, C.G. Multi-temporal soil erosion risk assessment in N. Chalkidiki using a modified usle raster model. *EARSeL eProc.* **2009**, *8*, 40–52.

9. Reusing, M.; Schneider, T.; Ammer, U. Modelling soil loss rates in the Ethiopian Highlands by integration of high resolution MOMS-02/D2-stereo-data in a GIS. *Int. J. Remote Sens.* **2000**, *21*, 1885–1896.

10. Stocking, M. Erosion and crop yield. In *Encyclopedia of Soil Science*; Dekker: New York, NY, USA, 2003; pp. 1–4.

11. Nyakatawa, E.Z.; Reddy, K.C.; Lemunyon, J.L. Predicting soil erosion in conservation tillage cotton production systems using the revised universal soil loss equation (RUSLE). *Soil Tillage Res.* **2001**, *57*, 213–224.

12. Sthiannopkao, S.; Takizawa, S.; Homewong, J.; Wirojanagud, W. Soil erosion and its impacts on water treatment in the northeastern provinces of Thailand. *Environ. Int.* **2007**, *33*, 706–711.

13. Ontl, T.A.; Schulte, L.A. Soil carbon storage. *Nat. Educ. Knowl.* **2012**, *3*, 35.

14. Eswaran, H.; Lal, R.; Reich, P.F. Land degradation: An overview. In Responses to Land Degradation, Proceedings of the 2nd International Conference on Land Degradation and Desertification, Khon Kaen, Thailand, 25–29 January 1999.

15. Jebari, S.; Berndtsson, R.; Olsson, J.; Bahri, A. Soil erosion estimation based on rainfall disaggregation. *J. Hydrol.* **2012**, *436*, 102–110.

16. EPA. *National Action Programme to Combat Drought and Desertification*; Environmental Protection Agency: Accra, Ghana, 2003.

17. Evans, R. Mechanics of water erosion and their Spatial and Temporal controls: An Empirical view point. In *Soil Erosion*; Kirkby, M.J., Morgan, R.P.C., Eds.; John Wiley and Sons: Chichester, UK, 1980; pp. 88–91.

18. Ghana Statistical Services. *Population and Housing Census-Summary Report of Final Results*; GSS: Accra, Ghana, 2012.

19. Adu, S.V. Eroded savanna soils of the Navrongo-Bawku area, northern Ghana. *Ghana J. Agric. Sci.* **1972**, *5*, 3–12.

20. Ghana Statistical Services. *Ghana Living Standard Survey: Report of the Fourth Round (Ghana Living Standards Statistic 4)*; Ghana Statistical Services: Accra, Ghana, 2000.

21. IFAD. *Ghana: Upper East Region Land Conservation and Smallholder Rehabilitation Project (LACOSREP) Report 2007*; IFAD: Rome, Italy, 2007.

22. Mdemu, M.V. *Water Productivity in Medium and Small Reservoirs in the Upper East Region (UER) of Ghana*; Ecology and Development Series No. 59; University of Bonn: Bonn, Germany, 2008.

23. Van der Geest, K.; Dietz, T. A literature survey about risk and vulnerability in drylands with a focus on the Sahel. In *The Impact of Climate Change on Drylands, with a Focus on West Africa*; Dietz, A.J., Ruben, R., Verhagen, A., Eds.; Kluwer Academic Publishers: Dordrecht, The Netherlands; Boston, MA, USA; London, UK, 2004; pp. 117–146.

24. Badmos, B.K.; Villamor, G.B.; Agodzo, S.K.; Odai, S.N.; Guug, S.S. Examining agricultural land-use/cover change options in Rural Northern Ghana: A participatory scenario exploration exercise approach. *Int. J. Interdiscip. Environ. Stud.* **2014**, *8*, 15–35.

25. Le, Q.B.; Park, S.J.; Vlek, P.L.; Cremers, A.B. Land-Use Dynamic Simulator (LUDAS): A multi-agent system model for simulating spatio-temporal dynamics of coupled human–landscape system. I. Structure and theoretical specification. *Ecol. Inform.* **2008**, *3*, 135–153.

26. Schindler, J. *A Multi-Agent System for Simulating Land-Use and Land-Cover Change in the Atankwidi Catchment of Upper East Ghana*; Ecology and Development Series No. 68; Center for Development Research (ZEF): Bonn, Germany, 2009.

27. Kaplan, M. *Agent-Based Modelling of Land-Use Changes and Vulnerability Assessment in a Coupled Socio-Ecological System in the Coastal Zone of Sri Lanka*; Ecology and Development Series No. 77, 2011; Center for Development Research (ZEF): Bonn, Germany, 2011.

28. Villamor, G.B.; le, Q.B.; Djanibekov, U.; van Noordwijk, M.; Vlek, P.L. Biodiversity in rubber agroforests, carbon emissions, and rural livelihoods: An agent-based model of land-use dynamics in lowland Sumatra. *Environ. Model. Softw.* **2014**, *61*, 151–165.

29. Grimm, V.; Berger, U.; Bastiansen, F.; Eliassen, S.; Ginot, V.; Giske, J.; DeAngelis, D.L. A standard protocol for describing individual-based and agent-based models. *Ecol. Model.* **2006**, *198*, 115–126.

30. Grimm, V.; Berger, U.; DeAngelis, D.L.; Polhill, J.G.; Giske, J.; Railsback, S.F. The ODD protocol: A review and first update. *Ecol. Model.* **2010**, *221*, 2760–2768.

31. Wilensky, U. NetLogo 1999. Available online: http://ccl.northwestern.edu/netlogo/ (accessed on 24 June 2015).

32. Wischmeier, W.H.; Smith, D.D. *Predicting Rainfall Erosion Losses-A Guide to Conservation Planning*; Agriculture Handbook No. 537; USDA/Science and Education Administration, US. Govt. Printing Office: Washington, DC, USA, 1978.

33. Roose, E.J. Use of the universal soil loss equation to predict erosion in West Africa. In Soil Erosion: Prediction and Control, Proceedings of the National Conference on Soil Erosion; Soil Conservation Society of America: Ankeny, IA, USA, 1977; pp. 143–151.

34. Fournier, F. *Climate and Erosion*; University Press Paris: Paris, France, 1960.

35. Roose, E.J. Use of the universal soil loss equation to predict erosion in west Africa. In *Soil Erosion. Prediction and Control SCSA, Special Publication*; Soil Conservation Society of America: Ankeny, IA, USA, 1976; Volume 21, pp. 60–74.

36. Arnoldus, H.M.J. *Methodology Used to Determine the Maximum Potential Average Annual Soil Loss Due to Sheet and Rill Erosion in Morocco*; FAO Soils Bulletins; FAO: Rome, Italy, 1977.

37. Kassam, A.H. *Agro-Ecological Land Resources Assessment for Agricultural Development Planning: A Case Study of Kenya: Resources Data Base and Land Productivity*; Food Agriculture Organisation: Rome, Italy, 1991.

38. Singh, G.; Babu, R.; Narain, P.; Bhushan, L.S.; Abrol, I.P. Soil erosion rates in India. *J. Soil Water Conserv.* **1992**, *47*, 97–99.

39. Williams, J.R.; Jones, C.A.; Dyke, P. Modeling approach to determining the relationship between erosion and soil productivity. *Trans. Am. Soc. Agric. Eng.* **1984**, *27*, 129–144.

40. Shirazi, M.A.; Boersma, L. A unifying quantitative analysis of soil texture. *Soil Sci. Soc. Am. J.* **1984**, *48*, 142–147.

41. Zhang, K.L.; Shu, A.P.; Xu, X.L.; Yang, Q.K.; Yu, B. Soil erodibility and its estimation for agricultural soils in China. *J. Arid Environ.* **2008**, *72*, 1002–1011.

42. Wischmeier, W.H.; Johnson, C.B.; Cross, B.V. Soil erodibility nomograph for farmland and construction sites. *J. Soil Water Conserv.* **1971**, *26*, 189–193.

43. Geleta, H.I. Watershed Sediment Yield Modeling for Data Scarce Areas. Ph.D. Thesis, University of Stuttgart, Stuttgart, Germany, 2011.

44. FAO/Unesco/ISRIC. *Revised Legend of the Soil Map of the World*; World Soil Resources Report; FAO: Rome, Italy, 1990.

45. Borah, D.K.; Krug, E.C.; Yoder, D. Watershed sediment yield. *Sediment. Eng.: Process. Meas. Model. Pract. ASCE Man. Rep. Eng. Pract.* **2008**, *110*, 827–858.

46. Moore, I.D.; Burch, G.J. Physical basis of the length-slope factor in the Universal Soil Loss Equation. *Soil Sci. Soc. Am. J.* **1986**, *50*, 1294–1298.

47. Morgan, R.P.C.; Davidson, D.A. *Soil Erosion and Conservation*; Longman Group: Harlow, UK, 1991.

48. Moore, I.D.; Turner, A.K.; Wilson, J.P.; Jenson, S.K.; Band, L.E. GIS and land-surface-subsurface process modeling. In *Environmental Modeling with GIS*; Goodchild, M.F.R., Parks, B.O., Steyaert, L.T., Eds.; Oxford University Press, Inc.: Oxford, UK, 1993; pp. 196–230.

49. Van der Knijff, J.; Jones, R.J.A.; Montanarella, L. *Soil Erosion Risk Assessment in Italy*; European Soil Bureau, European Commission: Brussels, Belgium, 1999; p. 52.

50. Zhang, Y.; Liu, B.; Zhang, Q.; Xie, Y. Effect of different vegetation types on soil erosion by water. *Acta Bot. Sin.* **2002**, *45*, 1204–1209.

51. Hazarika, M.K.; Honda, K. Estimation of soil erosion using remote sensing and GIS: Its valuation and economic implications on agricultural production. In *Sustaining the Global Farm*; Scott, D.E., Mohtar, R.H., Steinhardt, G.C., Eds.; USDA-ARS national soil erosion research laboratory: Purdue, IN, USA, 2001; pp. 1090–1093.

52. Karaburun, A. Estimation of C factor for soil erosion modeling using NDVI in Buyukcekmece watershed. *Ozean J. Appl. Sci.* **2010**, *3*, 77–85.

53. Henao, J.; Baanante, C.A. *Estimating Rates of Nutrient Depletion in Soils of Agricultural Lands of Africa*; International Fertilizer Development Center: Muscle Shoals, AL, USA, 1999; p. 76.

54. Villamor, G.B.; Badmos, B.K. Grazing game: A learning tool for adaptive management in response to climate variability in semi-arid areas of Ghana. *Ecol. Soc.* **2015**. under review.

55. Angelucci, F. *Analysis of Incentives and Disincentives for Maize in Ghana*; Technical Notes Series; MAFAP, FAO: Rome, Italy, 2012.

56. Villamor, G.B. *Flexibility of Multi-agent System Models for Rubber Agroforest Landscapes and Social Response to Emerging Reward Mechanisms for Ecosystem Services in Sumatra, Indonesia*; Ecology and Development Series No. 88, 2012; Center for Development Research (ZEF): Bonn, Germany, 2012.

57. Villamor, G.B.; Troitzsch, K.G.; van Noordwijk, M. Validating human decision making in an agent-based land-use model. In Proceedings of the 20th International Congress on Modelling and Simulation (MODSIM): Adapting to Change: The Multiple Roles of Modelling, Adelaide, Australia, 1–6 December 2013.

58. Niazi, M.; Hussain, A. Agent-based computing from multi-agent systems to agent-based models: A visual survey. *Scientometrics* **2011**, *89*, 479–499.

59. Villamor, G.B.; van Noordwijk, M.; Troitzsch, K.G.; Vlek, P.L.G. Human decision making in empirical agent-based models: Pitfalls and caveats for land-use/change policies. In Proceedings of the 26th European Conference on Modelling and Simulation, Koblenz, 29 May–1 June 2012.

60. Villamor, G.B.; van Noordwijk, M.; Troitzsch, K.G.; Vlek, P.L.G. Human decision making for empirical agent-based models: construction and validation. In Proceedings of the 6th Biennial Meeting, International Environmental Modelling and Software, Managing Resources of a Limited Planet, Leipzig, Germany, 1–5 July 2012.

61. McSweeney, C.; Lizcano, G.; New, M.; Lu, X. The UNDP Climate change country profiles: Improving the accessibility of observed and projected climate information for studies of climate change in developing countries. *Bull. Am. Meteorol. Soc.* **2010**, *91*, 157–166.

62. Le, Q.B.; Park, S.J.; Vlek, P.L. Land Use Dynamic Simulator (LUDAS): A multi-agent system model for simulating spatio-temporal dynamics of coupled human–landscape system: 2. Scenario-based application for impact assessment of land-use policies. *Ecol. Inform.* **2010**, *5*, 203–221.

63. Le, Q.B.; Seidl, R.; Scholz, R.W. Feedback loops and types of adaptation in the modelling of land-use decisions in an agent-based simulation. *Environ. Model. Softw.* **2012**, *27*, 83–96.

How to Make a Barranco: Modeling Erosion and Land-Use in Mediterranean Landscapes

C. Michael Barton, Isaac Ullah and Arjun Heimsath

Abstract: We use the hybrid modeling laboratory of the Mediterranean Landscape Dynamics (MedLanD) Project to simulate *barranco* incision in eastern Spain under different scenarios of natural and human environmental change. We carry out a series of modeling experiments set in the Rio Penaguila valley of northern Alicante Province. The MedLanD Modeling Laboratory (MML) is able to realistically simulate gullying and incision in a multi-dimensional, spatially explicit virtual landscape. We first compare erosion modeled in wooded and denuded landscapes in the absence of human land-use. We then introduce simulated small-holder (e.g., prehistoric Neolithic) farmer/herders in six experiments, by varying community size (small, medium, large) and land management strategy (satisficing and maximizing). We compare the amount and location of erosion under natural and anthropogenic conditions. Natural (e.g., climatically induced) land-cover change produces a distinctly different signature of landscape evolution than does land-cover change produced by agropastoral land-use. Human land-use induces increased coupling between hillslopes and channels, resulting in increased downstream incision.

Reprinted from *Land*. Cite as: Barton, C.M.; Ullah, I.; Heimsath, A. How to Make a Barranco: Modeling Erosion and Land-Use in Mediterranean Landscapes. *Land* **2015**, *4*, 578–606.

1. Introduction

Characteristic features of many Mediterranean landscapes are deeply incised, intermittent watercourses, termed *barrancos* in Spanish. These can range from modest gullies a meter deep to extensive drainage systems that extend over tens of kilometers and tens of meters deep. While some deeper *barrancos* may intersect local water tables, most only carry water periodically during or after significant precipitation. Some *barrancos* are old and are related to characteristics of underlying lithologies, bedrock structure, long-term climate-driven changes in vegetation cover, and regional drainage networks [1]. However, many are clearly much younger, and are incised into unconsolidated sediments or soft calcareous bedrock [2–4]. Many older bedrock *barrancos* also show evidence of recent incision [1,5,6].

There is a widespread consensus that anthropogenic factors—especially agropastoral land-use—played a significant role in Holocene erosion and soil loss throughout the Mediterranean, although there remains considerable debate over the relative causal importance of human and natural processes at different temporal and

161

spatial scales [6–10]. Certainly, the Mediterranean region today and in the recent past is characterized by high sediment transport levels, as a result of both sheet erosion and incision [4,6]. There is also evidence of significant episodes of erosion, including incision, at various times in the historic and prehistoric past that seem coeval with changes in agropastoral land-use patterns [7,11–13].

While there has been considerable study of the impacts of land-use on and hillslope and rill erosion in the Mediterranean, the relationships between land-use and gullying and *barranco* formation are less well understood [4,6,14]. Moreover, quantitative studies of the effects of farming and herding practices on gully incision have been largely empirical and limited to short-term processes (from single events to several decades) that are observable (e.g., [2–4,6,15,16]). Some of the larger, and more areally extensive *barranco* systems have been forming over centuries or longer. Increased channel incision, along with increased sheet and rill erosion, is generally viewed as evidence of severe landscape degradation (*sensu* [17]). Yet, the creation or exacerbation of incision in *barrancos* is but one of many potential consequences of complex interactions between social and biophysical drivers of surface dynamics that have been shaping Mediterranean landscapes for millennia. The specific land-use histories of these coupled human and natural landscapes feedback into the earth-surface processes that shape them, in turn offering new constraints and opportunities for subsequent agropastoral and other land-uses [11]. It is therefore important to understand the potential long-term co-evolution between human land-use and *barranco* formation, however, the centuries-long time scales of these processes makes direct observation impossible. Proxy records of landscape change are widely and irregularly distributed in space and time, and often contain significant lacunae, allowing for multiple interpretations of the same evidence (e.g., [7,8]). Fortunately, advances in computational surface process modeling offer a way to investigate the complex, long-term interaction of anthropogenic and biophysical drivers of land-scape dynamics [18–20].

We describe the results of a series of modeling experiments, using a digital laboratory developed in the Mediterranean Landscape Dynamics (MedLanD) project, designed to explore the long-term consequences of small-holder agropastoral land-use for the evolution of *barrancos* in Mediterranean Spain. The MedLanD Modeling Laboratory (MML) is an open-source, integrated modeling environment that has the ability to couple spatially explicit (cellular automata) models of landscape evolution, agent-based and GIS-based models of human land-use, and regression or cellular models of vegetation and climate change to study the long-term dynamics of coupled human and natural landscapes [19,21–26]. In these experiments, we model the effects of increasing population, reducing fallowing intervals, and resource management strategies on *barranco* incision (Table 1). We situate these experiments in the real-world landscape of the Penaguila Valley in northern Alicante Province,

Spain, which is the location of one of the earliest farming settlements (*i.e.*, Neolithic) in the region [27,28]. Today, the valley is dissected by deeply entrenched *barrancos* containing incised sections that appear to postdate early Neolithic occupation of the valley.

Table 1. Table of modeling experiments conducted.

Experiment Number	Number of People	Land Tenure Type
1	30	Satisfice
2	30	Maximize
3	60	Satisfice
4	60	Maximize
5	120	Satisfice
6	120	Maximize

2. MedLanD Modeling Laboratory (MML)

2.1. Surface Process Model

Many of the details of the surface process model component of the MML—r.landscape.evol—are described in Mitasova and colleagues [19]. (See Table A1 for a description of the input parameters of the module). In brief, r.landscape.evol is written in Python to run within the open-source GRASS GIS environment, where it can take advantage of fast computational hydrology tools, a parallelized map calculator, and special Python library. It uses a 3D implementation of the Unit Stream Power Erosion/Deposition (USPED) equation [29–31] to estimate transport capacity on hillslopes and rills (Equation (1)), and the Stream Power equation [19] to estimate transport capacity in channels (Equation (2)):

$$T_{(hillslopes)} = R \ K \ C \ A^m \ (\sin \ \beta)^n \tag{1}$$

$$T_{(channels)} = K_t \ n^{-1} \ g_w \ h^m \ (\tan \ \beta)^n \tag{2}$$

where R, K, and C are the rain, soil erodibility, and land cover coefficients of the well-known RUSLE equation [32], A is the upslope accumulated area (per contour width), β is the local slope (in degrees), K_t is a coefficient of substrate erodibility in stream channels, n is Manning's coefficient, g_w is the gravitational power of flowing water, h is the depth of flow, and m and n are empirically derived transport coefficients (both 1 for sheetwash on hillslopes, and 1.5 and 1.6, respectively, for flow in channels). The implementation of r.landscape.evol used here does not use soil creep or shear stress equations mentioned in Mitasova and colleagues [19], but these are alternative modes that exist in the module, and which may be implemented if desired.

The model estimates meters of erosion or deposition (*ED*) in each cell of a DEM on the basis of 2D divergence in transport capacity across topography (Equation (3)):

$$ED = \frac{d\,(T \cdot \cos{(\alpha)})}{dx} + \frac{d\,(T \cdot \sin{(\alpha)})}{dy} \qquad (3)$$

where α is the topographical aspect (in degrees). A map of *ED* is converted to elevation changes by normalizing to sediment bulk density, and then added back to the DEM to create a new DEM to be the base layer for subsequent modeling. This process is iterated repeatedly to evolve the digital landscape. The model is transport capacity limited, but erosion is not unlimited. The model also requires a digital map of estimated bedrock elevations, and so tracks the amount of soil/regolith available for erosion. Erosion in excess of the available sediment in a given cell is not allowed. Although not implemented in the experiments presented here, bedrock erosion can be simulated by artificially deepening "soils" in places likely to experience significant erosion, and then "indurating" these soils with smaller values of K or K_t.

Sediment transport capacity can be altered by land cover, surface characteristics, or the amount of water available for runoff through the process equation coefficients [32,33] (Figure 1, see also Table A1). In the modeling experiments described here, K and R are kept constant, and take values empirically calculated for Mediterranean *terra rossa* soils [34] and mid-Holocene precipitation [22] (see Section 3.3, below). Human land-use activities, described below, can alter land cover. Land cover affects the calculation of erosion or deposition in two ways. On hillslopes the protective effect of vegetation on a plot of land is accounted for by C, affecting localized changes in transport capacity as estimated by the USPED equation. Additionally, the vegetation traps water, reducing runoff from a cell proportionally to the type of vegetation cover present (currently, runoff percentage is estimated from a linear regression of C *vs.* runoff water infiltration.). This changes the amount of water flowing though the downstream portions of the drainage, changing the estimated value of stream power in downstream reaches, and thereby affecting the calculation of erosion and deposition in those portions of the drainage. The input parameters of the surface process model are summarized in Table A1.

Finally, we parameterize the model to operate at a yearly interval to match better with the human agricultural cycle. This means that input parameters such as rainfall, storm frequency, vegetation growth, and land-use are all annualized, and that we do not explicitly model the occasional, very short term, extreme events that can have significant impacts on sediment transport (e.g., [14]). If these events were of significant effect in the formation of *barrancos*, then we may be underestimating the impact of human activity on *barranco* formation.

2.2. Human Land-Use Model

Although the MML contains a sophisticated Agent-Based land-use simulation engine, in this research we have chosen to use a more generalized model of Neolithic farming that simplifies land-use decisions with stochastic modeling techniques. This approach is more appropriate for modeling questions that do not require sophisticated agents (e.g., [22]), or that, as we do here, focus primarily on non-social aspects of socio-natural systems (but see [21,25,35] for modeling problems that do benefit from the ABM approach). Our simplified land-use model nevertheless encompasses a large range of human behaviors, however, and allows for several different land-use strategies to be modeled within a simple over-arching modeling framework that allows for faster simulation times.

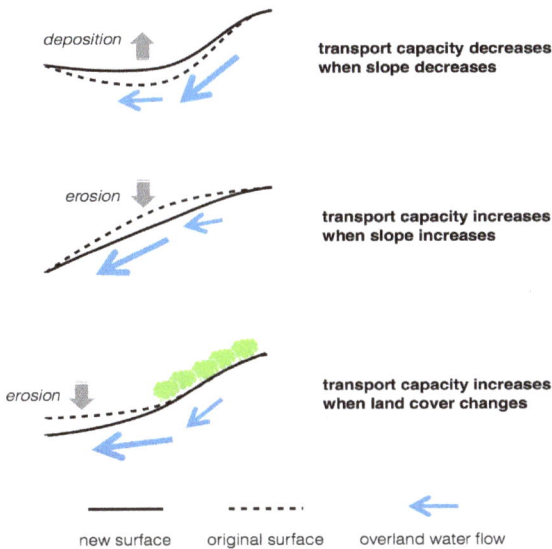

Figure 1. A schematic representation of the effect of topographic and vegetation change on the calculation of erosion and deposition in the MML surface process model. Our modeling approach assumes that the sediment load in flowing water is normally near transport capacity (a spatially and temporally dynamic equilibrium influenced by factors like the quantity of water, slope, and land-cover). In a three-dimensional landscape, changes in one of these factors will alter transport capacity. Sediment will then be entrained or deposited until a local equilibrium between sediment load and transport capacity is reached. Hence, increases in slope (or, convexity) will tend to result in erosion; decreases in slope (or, concavity) will tend to result in deposition. Reduction of vegetation (shown as green bushes) will tend to result in increased erosion or decreased deposition, depending on the localized change in slope. (Length of blue arrows schematically represents transport capacity of overland flow).

The simplified land-use model amalgamates the decisions of individuals (or households) at the coarser social level of a village, which we here take to mean a group of individuals that cohabit in a community, and that generally communicate or coordinate subsistence and land-use decisions. The model is divided into three sequential components: (1) land-use planning; (2) enactment of a land-use plan; and (3) calculation of direct land-use impacts. The land-use plan for the village is updated on annual basis, and balances rudimentary information about mean productivity in the village's catchments against a particular subsistence plan (ratio of agriculture to pastoralism), land choosing strategy (no land tenure with random allocation, satisficing subsistence needs, or maximizing agropastoral returns), and population, which are variables parameterized by the modeler at the start of a simulation run. Village labor availability and requirements, the general productivity and ecological characteristics of crops and herd animals, and the severity and spatial patterning of land-use impacts for herding or farming can all be parameterized to suit specific case-study conditions (see Table 2 for all land-use options of the model). Land-use occurs within predefined agricultural and pastoral catchments, which are created by the method described in [36], and which allows land above a slope threshold to be ignored (in our case, >10°). Each year, the village acquires new knowledge of the productivity of the landscape from the previous year's returns. These data are added to a "memory bank" of variable length, and the subsistence plan for the upcoming year is created using the averaged values over a predetermined number of previous years. This information is subjected to a Gaussian random perturbation to prevent perfect knowledge of past events, which more realistically simulates the vagaries of human memory and human ability to estimate averages [37,38].

The village agent then uses this information to derive the number of farm plots and grazing patches that it believes are required to sustain the target population level (which does not change during the simulation). The land-use plan is enacted via spatially-explicit stochastic farming and grazing land-use choice procedures.

Three different procedures may be employed for farming land-use behavior, depending on the style of land-tenuring strategy desired for a particular model run. In the case of a non-tenuring farming strategy, new plots are randomly chosen from within the farming catchment every year. For a satisficing strategy, plots are randomly chosen the first year, and are never relinquished. If more plots are needed, they are randomly selected from the unused portion of the catchment as needed, and remain under cultivation for the remainder of the simulation. For a maximizing strategy, plots are also randomly chosen the first year, but may be dropped in subsequent years if their productivity falls below a predetermined proportion of the average yield. Randomly chosen new fields are then added as necessary in order to meet farming goals. Previously used-and-relinquished fields are available for reuse in the future. Farming returns are calculated per-plot by the average of three regression formulas, each calculated from empirical data about the effect of soil depth, soil fertility, and rainfall on wheat and barley yields in Mediterranean climate regimes, and calibrated to typical yields of ancient wheat and barley varieties [39–48].

Grazing occurs within a subset of the grazing catchment that is determined by the current number of grazing patches required and a least-cost routing algorithm. The number of grazing patches is calculated by the village agent each year based on the current village population, the parameterized herding ratio, the fodder requirements of herd animals, and the agent's current knowledge of average grazing patch fodder yields. Optionally, grazing can be allowed on the fallowed portions of the farming catchment as well. Thus, non-denuded grazing patches near the village are always grazed, but farther patches may be added as necessary. Although all patches within the delimited subset of the grazing catchment will be grazed in a given year, some patches are grazed more intensively than others. The spatial patterning of this grazing intensity is determined by a Gaussian random function with tunable spatial autocorrelation that creates a patchy impact pattern. This more realistically simulates actual grazing patterns for site-tethered pastoralism [49–51]. Grazing returns are calculated from the known amount of digestible matter for given vegetation types, and the intensity of grazing [52–59]. People then gain calories from the milk and meat produced by the herds that were supported by the current grazing plan [60–64].

Once the amount and location of land-use activities are decided by the village agent for a given simulation-year, the model enacts the plan, assesses the amount of agricultural and pastoral returns that were gained, calculates the impacts of the land-use to vegetation and soil fertility, updates the landscape with the effect of these impacts, and then passes those new parameters to the surface process model. Impacts of farming a new plot include reduction of its vegetation cover to a value appropriate for cereals (*i.e.*, dense grassland), and all farmed plots also experience reduction in soil fertility. Soil fertility is reduced according to a Gaussian random function with tunable spatial autocorrelation, with mean and standard deviation set at the start of the modeling run. Grazing reduces land-cover by the amount determined by the grazing intensity map discussed in the previous paragraph. Grazing also affects soil fertility in that manure from grazing animals is added to the soil at a rate commensurate to the intensity of grazing on that patch [65].

The map of land cover (converted to values of C) for a given year is passed as input into the r.landscape.evol surface process model as described above. The calculated erosion or deposition changes the surface topography, affecting the depth of soils and local slopes, which feeds back into the farming and grazing planning process for the next year.

3. Modeling Experiment Design

3.1. Creating the Digital Landscape

We focus the experiments for studying the long-term impacts of land-use practices on *barranco* formation in a digital representation of the real-world landscape of the Rio Penaguila Valley, near the town of Benilloba, Alicante (Figure 2). We use the early Neolithic archaeological site of Mas d'Is [27,66] as the location of a small-holder farmer/herder village. Today, this area is dissected by five large *barrancos* that probably did not exist, or were much less deeply incised, in the mid-Holocene. Hence we need to initiate our surface process models on a landscape that lacked these erosional features. To do so, we conducted geomorphological and geoarchaeological fieldwork in the project area, compiling a variety of temporal data for the different landforms to create a landscape chronology. These data included the date of the earliest archaeological material recovered during pedestrian surveys, stratigraphic information, and OSL dates where possible. We were able to delineate terraces and other areas that were likely present during the Neolithic period ("Terrace A" in Figure 3), as well as those areas that are more recent. In general, post-Neolithic surfaces were located in the bottoms of the *barrancos* and on low alluvial terraces in *barranco*-bottoms. We masked these more recent surfaces in a GIS to remove them from the digital landscape (*i.e.*, DEM), and interpolated new terrain into the masked areas using the topographic information from the adjacent, older remnant landforms (e.g., "Terrace A"). The interpolation routine was tuned to also reduce the sharpness of slope curvatures (terrace edges), making slope breaks more natural. Any artificial internally-draining basins that were created in the valley floors by the interpolation routine were then filled so that the valley-bottoms were hydrologically continuous. This produced a landscape of broad U-shaped valleys, very different from the highly incised modern landscape (Figure 4).

3.2. Modeling Past Climate

We retrodicted climate variables for the late Neolithic I period of southern Spain (7550-6450 cal. BP) with a regional downscaling method that localizes paleoclimate retrodictions from large-scale Global Climate Models (GCM) based on a regression relationship between the GCM output for the 30-year climate norm, and the actual observed climate information observed at a weather station [67–69]. We have used this method in several other research projects with the MML (e.g., [21,25,34,70]). We used the average climate values derived across the entire Neolithic I period, downscaled at a weather station in Benifallim, a town close to the Penaguila valley project area. The specific climate values that we used are shown in Table A1.

168

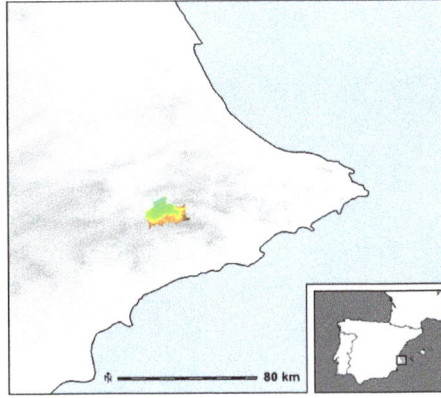

Figure 2. Map of Southeast Spain, showing the location of the Rio Penaguila valley project area.

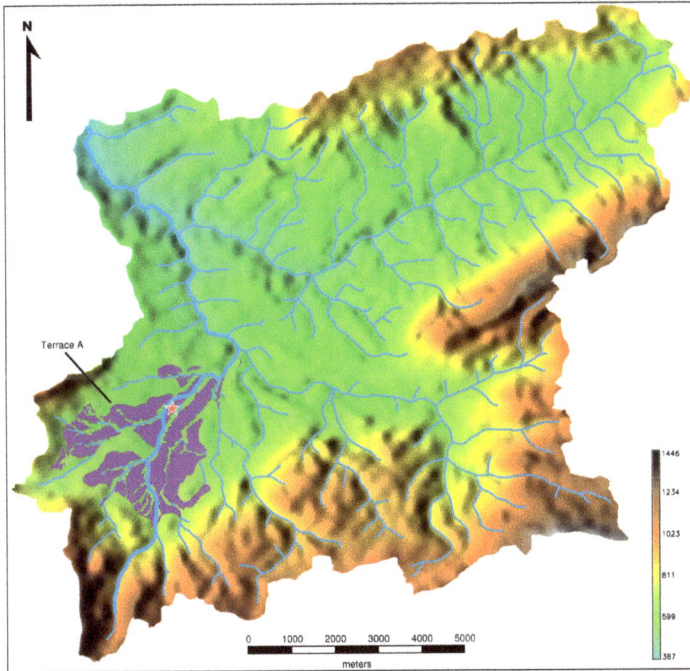

Figure 3. Reconstructed mid-Holocene topography in the Penaguila valley project area. The Neolithic site of Mas d'Is is shown, as are terraces near the site that are of Neolithic, or earlier, age (all labeled as "Terrace A"). The main watercourse passing by Mas d'Is is shown as a thick blue line, and the other portions of the extracted stream network are shown as thin blue lines. High elevations are shown in reds and brown, and lower elevations are shown in greens and yellows.

169

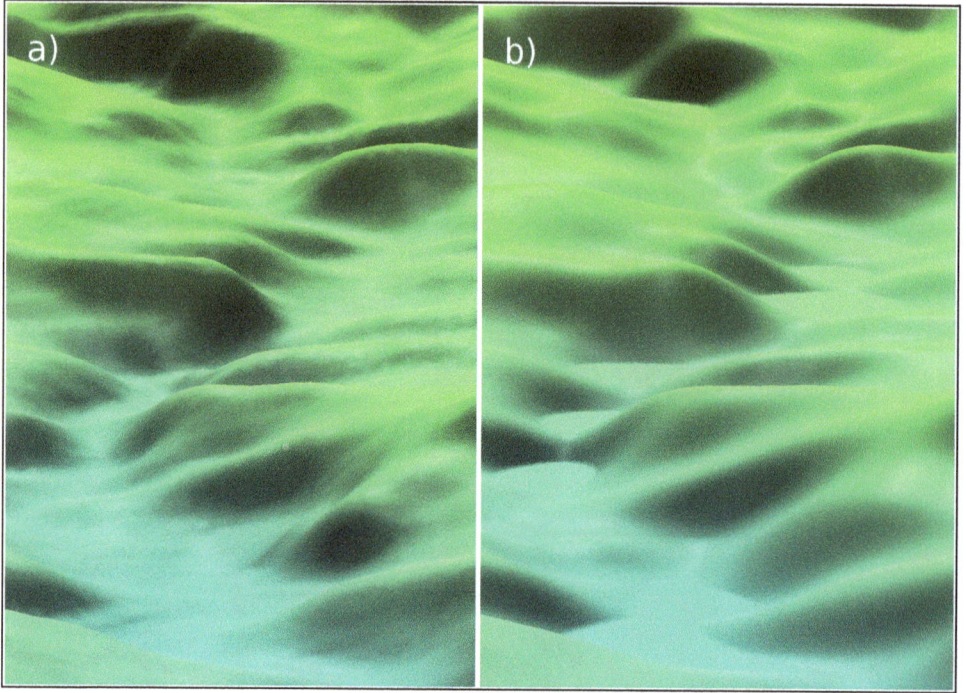

Figure 4. A comparison of the modern topography of a section of (**a**) the upper Rio Penaguila; and (**b**) the reconstructed Neolithic topography in the same area. The paleotopographic reconstruction resulted in a landscape of broad, U-shaped valleys, markedly different from the modern terrain.

3.3. Modeling Land-Use

We sited a small farming village at the location of the Mas d'Is archaeological site on the reconstructed mid-Holocene digital landscape, and performed a set of six experiments with the village population initialized at 30, 60, or 120 individuals using two different land-use strategies (Table 1). We constrained the area a village can use for cultivating cereal crops so that the villagers must reduce the fallow interval to produce sufficient food as the population grows. We also invoke two different strategies to guide agropastoral land use: satisficing and maximizing. With both strategies, a model begins by identifying all land that fits a set of basic criteria of suitability for cultivation and ovicaprine herding. In the experiments reported here, we parameterized the village based on ethnographic and archaeological data about small-scale, subsistence-focused, village-based Mediterranean agropastoralism. Neolithic subsistence in southern Europe was largely based on cereals [71]; therefore, we set 80% of a villages caloric needs to be met by consuming plants and 20% to be met by consuming animal products. Based on information about Neolithic agricultural practices [71–73], we parameterized 25% of the plant portion of the diet to be met by barley, and the remainder by wheat. Herds were parameterized as a 2:1 mix of goats and sheep, with the average herd animal requiring about 680 kg of fodder per year [36].

All together, this corresponds to a need of about six herd animals and about 230 kg of cereals per person per year. Herd animals were allowed to graze on fallowed portions of the agricultural catchment, as well as on the stubbles left after harvest of the farmed fields. Their manure contributed fertility gains of about 0.2% per year to the areas they grazed (including agricultural fields grazed for stubbles). Farm plots are modeled as discrete rectangular 20 m by 50 m regions, which are on par with the size of traditional, hand-worked historic peasant farm-plots in the southern Mediterranean region [74]. Farming reduced fertility at an average rate of 3% per year, and herd animals reduced land cover by an amount that would require about three years to regrow. Herds were parameterized as a 2:1 mix of goats and sheep, with the average herd animal requiring about 680 kg of fodder per year [36]. We kept the size and configuration of the farming and grazing catchments constant between all six experiments. This is not unreasonable, since real-world farming settlements are geographically constrained by the presence of other farmers in a region. Catchment modeling was conducted using a least-cost surfaces method described by Ullah [36]. For cereal cultivation, this included all land with slopes $\leq 10°$ within a 30 min walking distance of the village. Land suitable for herding was not limited by slope and extended to include all land within the Penaguila drainage (up to about an 8-h walk from the village).

Oversizing the farm catchment allows for the possibility of fallowing (particularly in maximizing strategies) on a time scale determined dynamically by the ratio of in-use fields to the catchment size. Initially, landscape cells are allocated to a village "agent" by randomly selecting enough suitable cells for cereal cultivation and enough suitable cells for animal pasturing to meet the caloric needs of the village (*i.e.*, depending on the initial village population). The total number of fields under cultivation can change over time, depending on the productivity of previously used fields within the time span of the agent memory (five years). For example, with a village population of 120 and a 313 ha farming catchment (used in the experiments reported here), about 3.5% of the possible farming catchment, would be cultivated initially. But as farming returns from the initially-used fields declines, the same village will need to increase the farmed area to 6%–7% of the total to meet their needs. In the same way, the percentage of the grazing catchment under use can change over time. The same 120 people would use between 10% and 17% of the grazing catchment, depending upon their perception of the current grazing productivity of vegetation within the catchment.

With a "satisficing" strategy, a village will compare land in use with caloric needs each subsequent annual time-step of the model. If caloric needs are exceeded, some land (with the lowest returns from cereal cultivation and animal herding) will be fallowed. If caloric needs are not met, then additional suitable landscape cells will be selected randomly to be put into use. With a "maximizing" strategy, at each time-step, the village will evaluate the performance of all land with regard to its caloric returns. Underperforming land (that is, fields that produce less than 80% of the current average yield) will be relinquished to fallow and will be replaced with new land chosen randomly from among the suitable and currently unused landscape cells. Table 2 summarizes the range of values used to parameterized the social portions of our simulation experiment.

3.4. Sensitivity Testing and Experiment Repetition

The stochasticity embedded in modeling procedure, and the complex interactions between land-use and landscape change means that the results differ in each model run, even if the initial parameters are the same. To more accurately assess the impacts of agropastoral land-use on *barranco* erosion, it is necessary to repeat every experiment (*i.e.*, model scenario with a particular set of initial parameters) multiple times, and calculate central tendencies and variance measures across multiple dimensions. While this does mean that the current research cannot focus on infrequent, "extreme" results (which may be important drivers of *barranco* evolution in some cases) the use of central tendencies allows us to understand the base-line (averaged) affects of human-land use on *barranco* formation under different land-use scenarios. The number of repetitions needed to adequately encompass the range of variation in land-use and landscape interactions can vary greatly with different model algorithms and model designs. Hence, we carried out a series of tests to assess model sensitivity to variation in initial parameters and estimate an optimum number of repetitions needed.

We carried out an initial run of the 60-person satisficing model, with 100 repetitions of 500 years each. We extracted several types of statistics from each model run (e.g., average amount of erosion/deposition per year, village population, proportion of different vegetation classes over time), and calculated the standard error of these statistics for all permutations of repetitions (1 through 100 repetitions). We repeated this process 100 times, shuffling and re-sampling the set of runs between each repeat using Monte Carlo methods. Example results of this sensitivity analysis show how the standard error for mean erosion/deposition in the stream channels changes as the number of run repetitions (n) increases (Figure 5). These tests indicated that error ceases to significantly improve after about 40 repetitions, and so we limited our other experiments to this number of repetitions. We also observed that the initial runs of 500 model years did not yield significantly different results from shorter runs. Additional experiments, therefore, ran for a total of 300 model years, significantly reducing processing-time and storage requirements. Running the model for 300 years still allowed for a period of about 50 years in which village land-use stabilized, followed by 250 years in which the land-use pattern of each experiment impacted erosion and deposition in the watershed. The combination of six experiments, repeated 40 times resulted in a total of 240 individual model runs of 300 years each (Table 1). These produced 40 time series of 300 maps for annual erosion/deposition, soil depth, land-cover, and topography for each experiment.

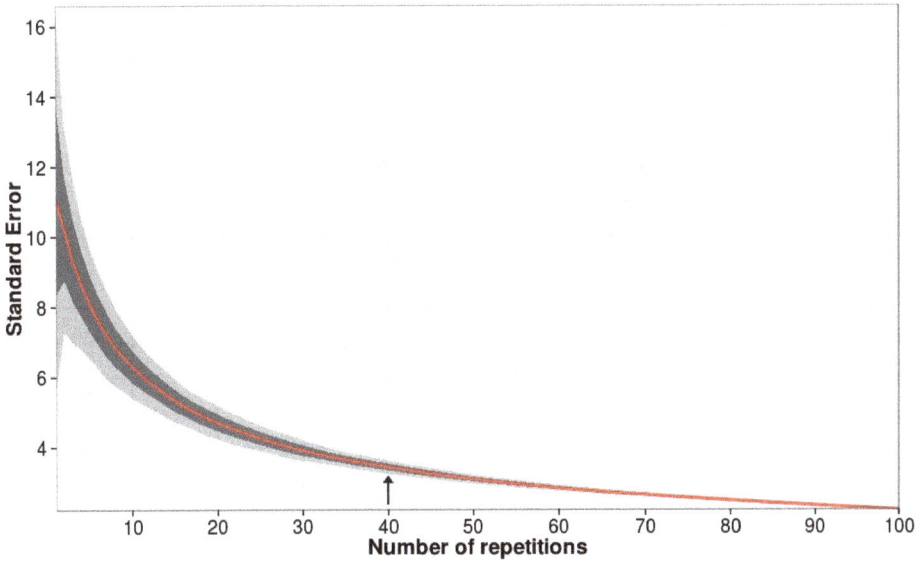

Figure 5. Results of sensitivity analysis about the effect of the number of experiment repetitions on the error surrounding mean estimates of erosion in the main *Barranco* near Mas d'Is. Red line is the mean of all recombinations at the specified number of repetition. The dark grey shaded region denotes 1 SD, and the lighter grey shaded region denotes 2 SD, based on the range of values from all 100 recombinations at each number of repetitions. The rate at which the standard error decreases with sample size (*n*) greatly tapers off after about 40 repetitions (indicated by the arrow).

Because landscapes are also dynamic in the absence of any human presence, we conducted two additional experiments without any human land-use. In one experiment, we covered the landscape with "typical" Mediterranean woodland as would be present under mild climatic conditions with no farming or grazing; in the other one, we mantled it with sparse grass/herb cover as might be present under harsher climatic conditions. These provide points of comparison of human impacts with extremes of non-anthropogenic climate-induced landscape changes in this region.

For all experiments, we calculated total, cumulative maps of landscape change, as well as statistical maps of central tendencies and variance. The results of all repeated runs were statistically combined for each of the six experiments as maps and summary values, to allow comparison of the landscape consequences of different land-use configurations. We refer to these syntheses of landscape change in the discussions below. The results of these computational experiments in socio-ecological system dynamics are summarized below.

4. Simulation Results

4.1. Erosion and Deposition in Barrancos

In order to use quantitative modeling to investigate the potential impacts of rural land-use on gullying and *barranco* incision it is necessary to have a surface process model that can accurately represent incision. While we have been developing the surface process model of the MML for nearly a decade, our focus was more on hillslope processes than incision [21,34,35]. Recently, we enhanced this model to better represent erosion and deposition in channels, combining well-known formulas to represent surface processes across the different landforms of small watersheds [19].

The results are encouraging. An oblique detailed view of the final topography for one run of one experiment, colored by the 300-year sum-total erosion/deposition for each map pixel, shows that the surface process model is capable of creating significant erosional dynamics in stream channels (Figure 6). Streams incise at the outer edge of channel bends, and at places where the channel gradient rapidly increases. Material is deposited where the gradient decreases and at the inside of channel bends. Flat areas are characterized by meanders and channel levies, and there are several remnants of abandoned channels throughout the stream course.

We sampled one of these digital landscapes through time at four cross sectional transects of the middle reaches of the major *barranco* that passes near Mas D'is, from its upstream to downstream reaches, indicated by the labeled horizontal lines in Figure 6. We recorded elevation cross-sections at these sampling transects at 50-year intervals in the 300 model-year time series at each of these transects (Figure 7). These cross-sections show the temporal dynamics of downcutting in different places along the drainage network. *Barranco* incision increases from upstream to downstream in the middle reaches of the *barranco*. This is to be expected as a result of increased water flow through the channel. Lateral incision is more prominent than absolute downcutting, but distinct channelization and incision did occur in all experiments. This is similar to empirical observations of the process of erosion in real *barrancos* [5]. Channel bar and levy development may offset some of the total erosion over time,

174

however, particularly as the main channel of the *barranco* migrates within the valley-bottom. This indicates that incision and erosion of these *barrancos* may be a temporally complex phenomenon, with time-lag in sediment transport obscuring the observability of the total incision that is occurring in the system. This would mean that *barranco* downcutting would not have been immediately obvious to local farming peoples.

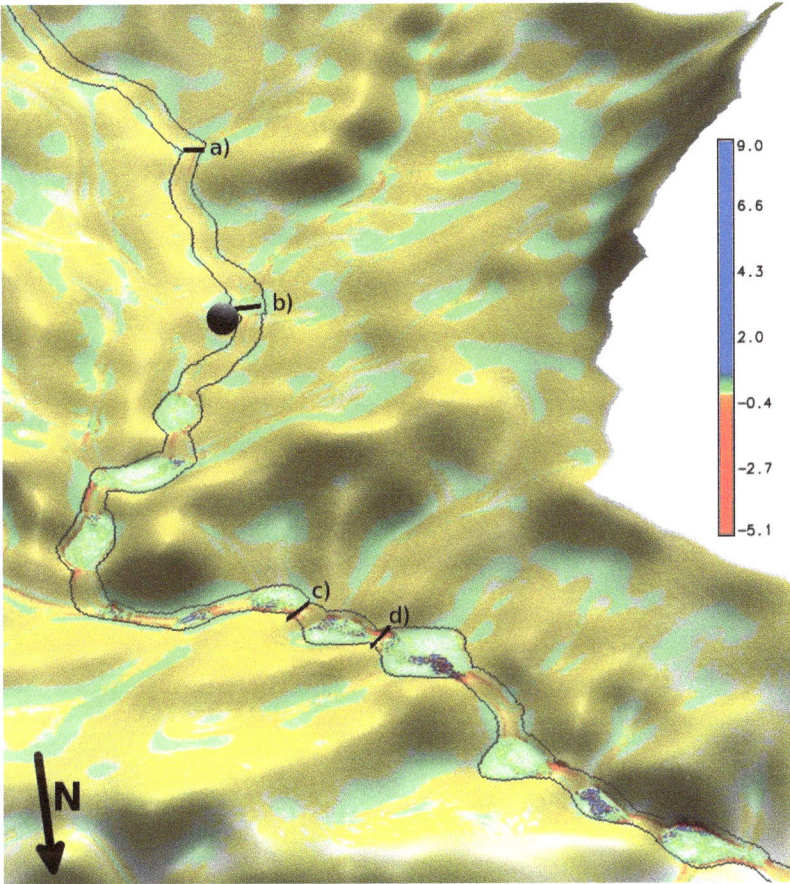

Figure 6. An oblique, three-dimensional view of a portion of the region shown in Figure 4. Warm colors (yellow through red) indicate cumulative erosion, and cool colors (green through blue) indicate cumulative deposition. The locations of the four cross-sections are indicated, as is the position of the Neolithic site of Mas D'is. Note that channel erosion occurs most intensively where narrowing of the channel bottom, coincides with a rapid convex change in slope in the channel gradient (e.g., points "c" and "d"). These are similar to the process of "head cuts" in real channels (e.g., [5]).

175

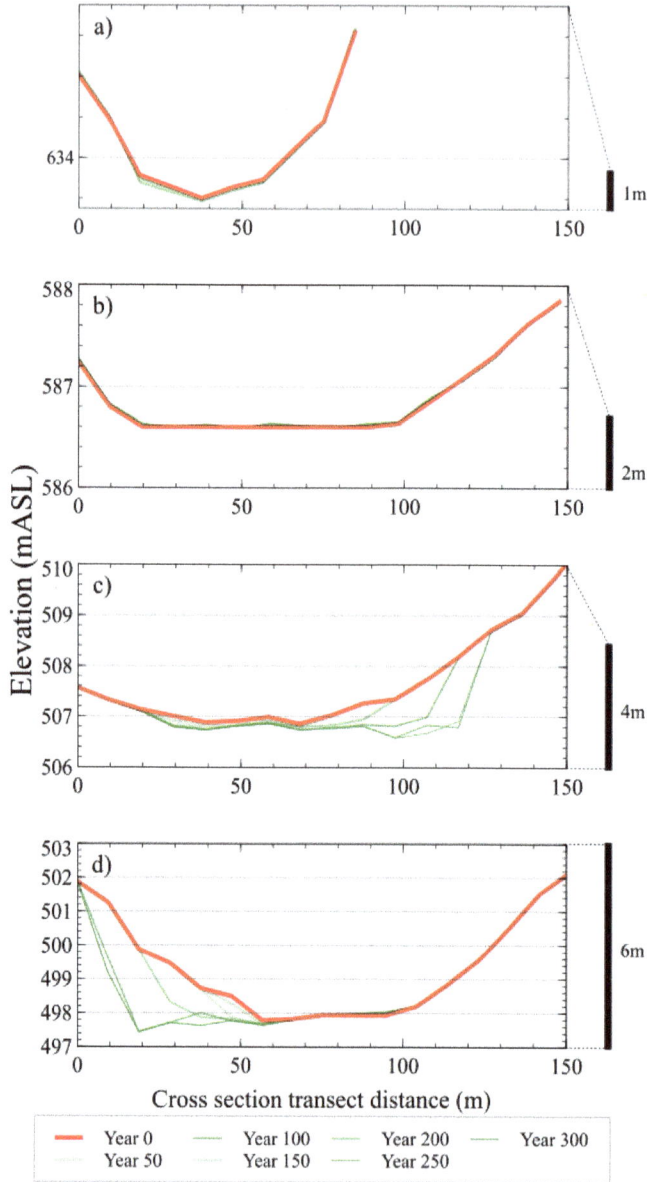

Figure 7. (**a–d**) Four cross-sections of the main *Barranco*: two profiles upstream from Mas d'Is, near Mas d'Is, and downstream from Mas d'Is. The red line shows the starting reconstructed mid-Holocene cross-section, and the green lines show the cross-section at 50-year intervals during one run of the maximizing large-population experiment. Incison and lateral erosion are apparent at all cross sections, although the absolute amount of incision is greater in the downstream cross sections. The locations of the cross-sections are shown in Figures 5 and 8.

Over the total length of a *barranco* channel, however, erosion should decrease in the furthest downstream section, as local base-level is approached, to create a graded profile. We kept track of the longitudinal profile of the main Mas D'is *barranco* (indicated by the thick blue line in Figure 3) as it changed over the 300 model-years of the simulation experiments (Figure 8). The *barranco* profile elongates, and does become increasingly graded over time. The elongation is likely due to increased channel sinuosity as meanders form in the flats, and as lateral incision cuts into the valley walls. Overall, erosion tends to occur in punctuated stretches of the channel, producing a stepped longitudinal profile similar to head cuts in real-world gullies (Figures 5 and 8).

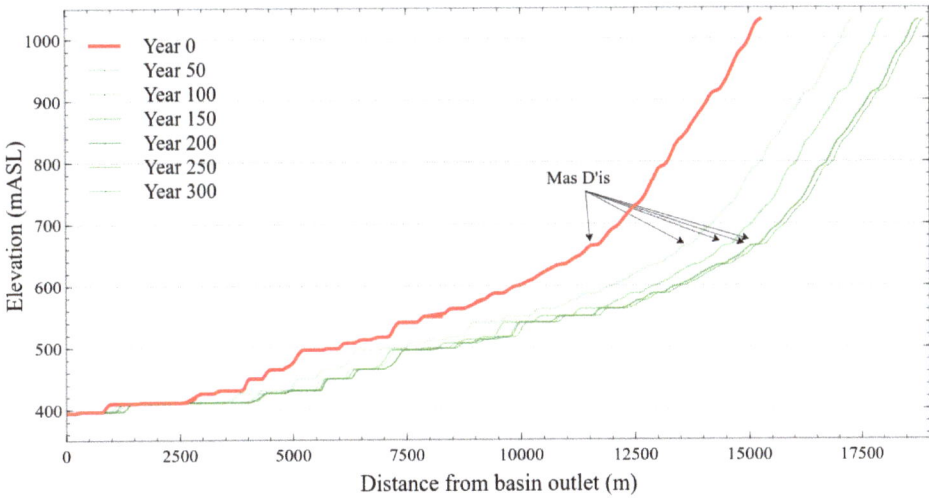

Figure 8. The longitudinal profile of the main watercourse near Mas d'Is (see Figure 2). The red line indicates the profile as extracted from the starting reconstructed mid-Holocene topography. The green lines show the profile in 50-year increments during one run of the maximizing large-population experiment. The overall length of the profile increases over time due to changes in channel morphology (such as the growth of meanders and lateral incision). The profile also becomes more graded over time, although erosion seems to occur mainly in punctuated stretches of the channel bottom.

The range of values for average channel incision after 300 model-years varies from 0.062 to 0.083 m (or, 0.21 to 0.28 mm/yr), with a mean value for average channel incision of 0.07 m (0.23 mm/yr) for all experiments combined. These rates are similar to those observed for recent erosion in *barrancos* [3,15,16]. In each of the experiments, there were locations in the channel that were eroded to depths of 5 to 6 m at the end of the simulation period, again similar to rates of head cutting in modern *barrancos* [3,15,16]. To be clear, although we initialized the MML surface process model with realistic values, we have not yet attempted to fine-tune the internal model dynamics to match those of real-world landscapes. In part, this is a function of the lack of detailed, quantitative data on *barranco* erosion rates at century to millennial scales. What we have tested here is whether our modeling environment can simulate real-world erosional processes, even if the simulations of specific quantities of sediment moved may not match those of Mediterranean *barrancos* in the Holocene. In fact, the tests summarized here indicate that the surface process component of the MML creates realistic patterns of channel incision in both the longitudinal and lateral dimensions of *barrancos*. This improves our confidence in the results when we use this environment to examine interactions between human land-use and landscape change. We review these socio-ecological interactions below.

4.2. Effects of Land-Use on Barranco *Incision*

As noted above, we evaluated the consequences for landscape change of six combinations of population size and decision strategies (Table 1). We parsed the landscape into three socio-ecological meaningful units: the Mas D'is *barranco* valley-bottom (Figure 9, transparent outlined area), the portions of the landscape outside of this *barranco* (mostly hillslopes and smaller channels), and the modeled Mas D'is farming catchment (Figure 9, grey outlined area). Partitioning the landscape in this way helps us to better understand the connection between human land use and the spatial patterning of erosive impacts. We do this by separately calculating the cumulative amount of sediment eroded and deposited in each of these three regions after 300 simulated years in each experiment.

At the scale of the watershed (Figure 10), 300 years of human farming and animal herding produces erosion rates on hillslopes and small channels that slightly exceed those for the natural wooded landscape (transparent yellow line), but are considerably less than erosion in natural sparse, open vegetation (solid yellow line). The difference between these two control runs simulates large-scale shifts in vegetation cover that would accompany climate change, and provide interesting benchmarks to compare to the simulations that included agropastoral activity. Sediment deposited on hillslopes in human-occupied landscapes is roughly the same as in an unoccupied wooded landscape. Overall, small-holder agropastoralism has limited effects on hillslope dynamics, at least initially. Unsurprisingly, the simulated community with the largest population and following a returns-maximizing strategy had the strongest impact, but only by a comparatively small amount. Population makes a slightly larger difference than satisficing *vs.* maximizing strategy.

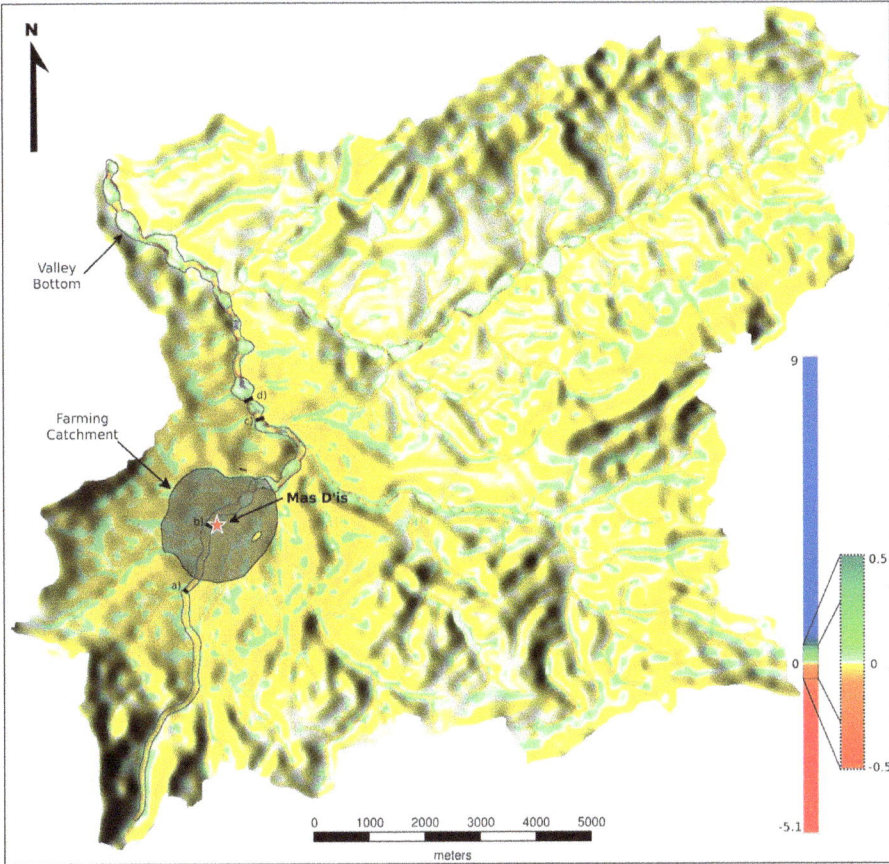

Figure 9. Map of cumulative erosion and deposition after 300 simulated years for one run of the maximizing large population scenario (see Table 1). The farming catchment is shown as a shaded region surrounding the site of Mas d'Is (indicated by the red star). The outline of the main *Barranco* that passes near Mas d'Is is also shown, as are the locations of the four cross-sections shown in Figure 6. Warm colors (yellow through red) indicate cumulative erosion, and cool colors (green through blue) indicate cumulative deposition. Scale is in meters.

Within the farming catchment (Figure 11), anthropogenic erosion from all scenarios exceeds that of natural vegetation change, while anthropogenically-driven deposition falls between the wooded and open landscapes. Interestingly, a maximizing strategy and medium sized population (solid blue line) produces considerably more erosion than other agropastoral land-use scenarios, including those with higher populations. In other words, small-holder agropastoral land-use has significant impact at local scales, but is less important than large-scale vegetation change (e.g., due to climate change). Additionally, while the degree of impact seems to scale with the number of people and their subsistence activities at the watershed scale, at the local level, the strategy used to make land-use decisions can exceed the landscape impacts of increased population density alone.

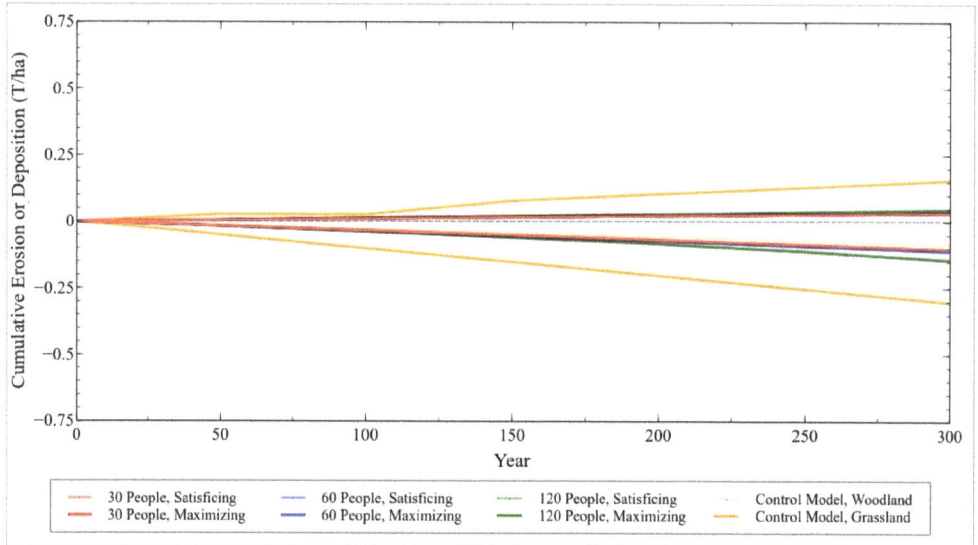

Figure 10. Cumulative amount of sediment eroded or deposited (tons/ha) on hillslopes for each of the six experiments and two control runs during the 300-year modeling interval. Values are averaged across the 40 repetitions of each experiment. The cumulative tons/ha of eroded sediment are shown below the center lines of each graph, and cumulative tons/ha of deposited sediment are above the center lines for the 300 year modeling interval. We show erosion and deposition separately because sediment removal and accumulations often happen in different landscape contexts, with different consequences for agropastoral and other land-use.

The natural and human effects on *barranco* incision are much different from those on the rest of the watershed. Whereas a large-scale shift from natural woodland to open vegetation on the hillslopes and small channels increases both erosion and deposition by roughly equal amounts within the watershed (Figure 10), within the *barranco* valley, non-anthropogenic vegetation change has disproportionate effects on erosion and deposition. That is, at the temporal scale of 300 model years, the effect of large-scale denudation is to increase the overall sedimentary dynamics of hillslopes, but not in the main channel. Erosion does increase slightly in the main *barranco* valley (at a level more or less equivalent to that in the rest of the watershed) but deposition increases by several orders of magnitude. The major effects of climate-induced large-scale devegetation, then, would be hillslope erosion and valley alluviation, agreeing with observations of field geomorphologists (e.g., [75]).

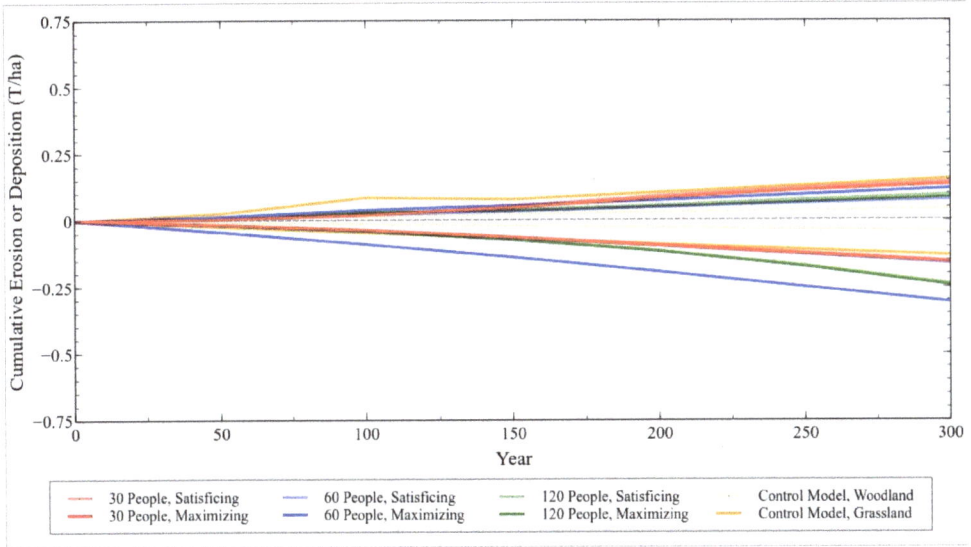

Figure 11. Cumulative amount of sediment eroded or deposited (tons/ha) in the delineated Mas d'Is farming catchment (see Figure 4) for each of the six experiments and two control runs during the 300-year modeling interval. Values are averaged across the 40 repetitions of each experiment. The cumulative tons/ha of eroded sediment are shown below the center lines of each graph, and cumulative tons/ha of deposited sediment are above the center lines for the 300 year modeling interval. We show erosion and deposition separately because sediment removal and accumulations often happen in different landscape contexts, with different consequences for agropastoral and other land-use.

Anthropogenic impacts on the main *barranco* differ significantly from non-human vegetation changes (Figure 12). Starting with a wooded landscape, both erosion and deposition are greatly increased by small-holder agropastoral land-use of all kinds modeled here. Anthropogenic deposition rates exceed those of non-human open woodland, and anthropogenic erosion rates in *barrancos* exceed by order of magnitude those driven by non-human vegetation change, as well as all erosion rates in the rest of the watershed. So, human activity contributes little in the way of sediment to these watercourses, but greatly increases the dynamics of sediments already in the floors and banks of *barrancos*. This effect seems more strongly affected by the number of people engaged in agropastoral activities than the strategy they use, at least for erosion. However, the overall rate of erosion slows over time in the anthropogenic cases, while the rate of deposition does not. This may indicate that incision in *barranco* systems begins to occur relatively quickly after initial anthropogenic land cover changes, but, without further changes in land-use or climate, may eventually achieve a metastable equilibrium (balance of erosion and deposition).

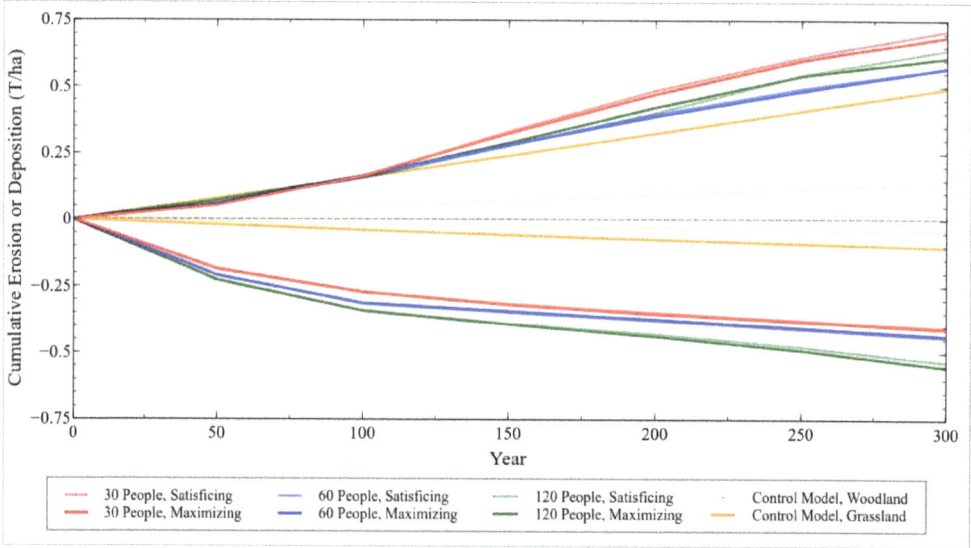

Figure 12. Cumulative amount of sediment eroded or deposited (tons/ha) in the Mas d'Is *Barranco* for each of the six experiments and two control runs during the 300-year modeling interval. Values are averaged across the 40 repetitions of each experiment. The cumulative tons/ha of eroded sediment are shown below the center lines of each graph, and cumulative tons/ha of deposited sediment are above the center lines for the 300 year modeling interval. We show erosion and deposition separately because sediment removal and accumulations often happen in different landscape contexts, with different consequences for agropastoral and other land-use.

5. Conclusions

The MedLand Modeling Laboratory provides a useful tool to examine surface dynamics and the impacts of rural land-use on landscapes [19,21,34]. The inclusion of separate hillslope- and stream-process models allow it to achieve reasonable results on slopes, while also more realistically simulating erosion and deposition in intermittent watercourses like *barrancos* that dissect landscapes of the western Mediterranean. The results presented here show that the MML can simulate simultaneous vertical and lateral erosion, generate head-cuts, and produce graded profiles over time (Figures 6–8). This allows us to investigate the impacts of the spread of agropastoral communities and associated land-use practices in the western Mediterranean, and clarifies the potential role of humans in *barranco* formation. Whereas our simulation experiments show that large-scale (climate-driven) vegetation change does greatly increase incision in these watercourses, they also show that smaller-scale, localized vegetation changes brought about by human land-use can have similar effects on *barranco* erosion rates, *without* climate change (Figures 10–12).

McClure and colleagues [76,77] suggest on the basis of archaeological evidence that the earliest Neolithic agropastoralists in Mediterranean Spain preferentially cultivated the alluvial soils of valley bottoms, but subsequently shifted to upland localities. They hypothesize that a combination of expanding agropastoral land-use and changing climate made these valley-bottom locales vulnerable to severe erosion, forcing a reorganization of settlement and agricultural practices that appear archaeologically as the Late Neolithic (Neolithic II-b locally). Using an earlier version of the MML, Barton and colleagues [34,35] show how modest increases in the size of small-holder agropastoral communities in northern Jordan, engaged in wadi-bottom farming, could have triggered a phase shift in erosion/deposition ratios, leading to declining agricultural productivity.

The modeling results presented here appear to confirm these earlier studies and offer a more nuanced view of the relationships between human activity, climate change, and the location and intensity of landscape change. Climate-driven decrease in vegetation cover can increase the surface dynamics and sediment transport rates on hillslopes and within watersheds (Figure 10), with increased alluviation in intermittent watercourses like *barrancos* and *wadis* (Figure 12). While small-holder agropastoralism can have significant local effects on hillslopes within the catchment actively cultivated (Figure 11), its impacts on the broader landscape are minimal (Figure 10)—at least until farms and pasturage cover significant proportions of a region. However, even very limited settlement and agropastoral land-use appears to have impacts on the local and downstream dynamics of intermittent watercourses that are orders of magnitude greater than non-anthropogenic vegetation change at regional scales (Figure 12). The greatly increased sedimentary dynamics in the farming catchment erodes farmed fields in some locales and buries them in others, with the potential for significant loss of productivity and greatly increased risk for subsistence farmer/herders (both in terms of uncertainty and potential for loss). The easily-cultivated and fertile soils of these watercourses, which are so common in Mediterranean landscapes, would have made them especially attractive to the earliest farmers of the region, who relied on hoes and digging sticks instead of plows. Cultivating these valley-bottoms, however, could lead to their increasing unsuitability for agriculture within a few generations—potentially with correlated impacts spanning areas much larger than just that of the farming communities or their catchments of active use. The increased downstream incision that results from valley-bottom farming would decrease the number of viable fields in the alluvial zone, limiting future options for valley-bottom cultivation when the fertility of the local catchment is depleted, or when local agricultural soils have themselves been severely eroded. McClure and colleagues suggest that the loss of these initially highly productive patches of the Mediterranean landscape created strong socioeconomic incentives for significant reorganization of both subsistence practices and society toward greater complexity [76,77].

The kinds of processes that we have presented here also exemplify the kinds of non-linear dynamics that typify human-environmental interactions in complex, coupled socio-natural systems. The difficulty of predicting these non-linear dynamics through normal reductionist science or empirical statistical analyses underscores the importance of the kinds of modeling illustrated in this Special Issue of *Land*. Computational modeling allows investigation of feedbacks between the human and natural components of earth-systems processes with reproducible and controlled experiments that have tangible implications for the way we understand the empirical record of landscape change. We wish to stress that computational and empirical research techniques complement, rather than compete with, each other, and integrating the two approaches is essential for a holistic approach to understanding earth-surface change. For example, the simulations discussed in this paper were designed to extend information gained from empirical observation of human land-use and *barranco* erosion over short periods to gain insight into socio-ecological processes that play out over centuries. Ultimately, we learn not only that relatively moderate changes to vegetation by small populations of agropastoral villagers can greatly affect the capacity for incision in *barrancos*, but that very deep *barranco* incision may require continual changes to land-use (or climate) over long periods of time. It may also be that self-amplifying feedback processes between land-use and erosion of easily cultivated land in *barranco* bottoms could have driven the kinds of increases in social complexity and land-management that would have continually reinforced erosive regimes, producing the deeply incised landscapes that exist throughout southern Europe today. These insights were made possible by the combination of traditional and simulation research approaches.

Acknowledgments: This work was supported by a grant from the US National Science Foundation, Coupled Natural and Human Systems Program, Grant #DEB-1313727.

Author Contributions: C. Michael Barton directs the Mediterranean Landscape Dynamics project, planned and directed this research component of the larger project, and wrote the majority of the paper. Isaac Ullah carried out the modeling experiments and wrote a significant portion of the paper. Arjun Heimsath advised Barton and Ullah on geomorphic dyamics modeled in this research and wrote sections of the paper.

Appendix A.

Table A1. Table of landscape evolution parameters used in the surface process modeling component of the MML.

Landscape Evolution Parameter	Typical Range of Values	Units	Value(s) Used in This Paper	Explanation
USPED K-factor	0.01–0.75	$(T \cdot ha \cdot hr)/(ha \cdot MJ \cdot mm))$	0.42	Soil erodibility index from RUSLE.
USPED R-factor	0–50	$((MJ \cdot mm)/(ha \cdot hr \cdot yr))$	4.54	Rainfall factor from RUSLE.
USPED C-factor	0.005–0.5	unitless	0.005–0.5	Vegetation cover factor from RUSLE.
Kt	0.001–0.000001	unitless	0.0001	Stream transport efficiency variable (erodibility of stream substrate).
Sediment load exponent	1.5,2.5	unitless	1.5	Stream transport type variable (1.5 for mainly bedload transport, 2.5 for mainly suspended load transport).
Manning's N	0.01–0.16	$s/m^{1/3}$	0.05	Average value of Manning's surface roughness coefficient value for channelized flow in the drainage.
Flow speed	0–3	m/s	1.4	Average velocity of flowing water in the drainage.
Soil density	0–3	T/m^3	1.2184	Soil density map or constant for conversion from mass to volume.
Transition point	10–500	number of raster cells	100	Flow accumulation breakpoint value for shift from hillslopes to stream flow.
Per-storm precipitation totals	0–10,000	mm	20.61	Precipitation totals for the average storm.
Number of storms	0–300	storms/yr	25	Average number of storms per year.
Storm length	0–72	hr	24	Length of the average storm.

Table 2. Table of agropastoral parameters used in the stochastic agropastoral simulation component of the MML.

Village Subsistence Characteristic	Typical Range of Values	Units	Value(s) Used in This Paper	Explanation
General Village Characteristics				
Number of people in the village	0–1000	number of individuals	30, 60, 120	The "target" number of people to be fed every year. Stays constant throughout the simulation.
Length of village "memory"	0–59	yr	5	Length of the "memory" of the agent in years. The agent will use the mean surplus/deficit information from this many of the most recent previous years when making a subsistence plan for the current year.
Amount of agricultural labor available	0–365	person-days	300	The amount of agricultural labor an average person of the village can do in a year.
Required amount of cereals	300–500	kg	370	Amount of cereals that would be required per person per year if cereals were the only food item being consumed.
Required number of animals	40–100	number of individuals	60	Number of herd animals that would be needed per person per year if pastoral products were the only food item being consumed.
Required amount of animal fodder	500–1000	kg	680	Amount of fodder required per herd animal per year.
Agropastoral Ratio	0–1	unitless ratio	0.2	Actual ratio of agricultural to pastoral foods in the diet, where 0 = 100% agricultural and 1 = 100% pastoral.

Table 2. *Cont.*

Village Subsistence Characteristic	Typical Range of Values	Units	Value(s) Used in This Paper	Explanation
Village Farming Characteristics				
Agricultural mix	0–1	unitless ratio	0.25	The wheat/barley ratio (e.g., 0.0 for all wheat, 1.0 for all barley, 0.5 for an equal mix).
Field dimensions	5–100	m	20, 50	North-South and East-West dimensions of agricultural fields.
Labor per field	5–100	person-days	50	Number of person-days required to till, sow, weed, and harvest one farm field in a year.
Field landcover value	0–50	succession stage	5	The landcover value for farmed fields (corresponds to an appropriate value from the landcover regrowth scheme).
Farming impact	0–10 (0–5)	% of maximum fertility	3 (2)	The mean and standard deviation of the amount to which farming a patch decreases its fertility (in percentage points of maximum fertility). Fertility impact values of individual farm plots is randomly chosen from a gaussian distribution that has this mean and standard deviation.
Maximum wheat	3000–4000	kg/ha	3500	Maximum amount of wheat that can be grown.
Maximum barley	2000–3000	kg/ha	2500	Maximum amount of barley that can be grown.
Satisficing farming strategy	Y/N	boolean	Both Y and N	Land is never dropped, only added if needed.

Table 2. *Cont.*

Village Subsistence Characteristic	Typical Range of Values	Units	Value(s) Used in This Paper	Explanation
Maximizing Farming strategy	Y/N	boolean	Both Y and N	Land is dropped if below a previously defined threshold in productivity.
Productivity threshold	0–1	unitless ratio	0.2	Threshold for dropping land out of tenure with a maximizing strategy, interpreted as a percentage below the yearly average yield of all farm cells.
Fertility regain rate	0–100 (0–100)	% of maximum fertility	2 (0.5)	The mean and standard deviation of the natural fertility recovery rate (percentage by which soil fertility increases per year if not farmed). Fertility recovery values of individual landscape patches will be randomly chosen from a gaussian distribution that has this mean and standard deviation.
Village Grazing Characteristics				
Minimum grazability	0–50	succession stage	2	Minimum amount of vegetation on a cell for it to be considered grazable by ovicaprines (corresponds to an appropriate value from the landcover regrowth scheme).

188

Table 2. *Cont.*

Village Subsistence Characteristic	Typical Range of Values	Units	Value(s) Used in This Paper	Explanation
Grazing spatiality coefficient	0–200	m	50	Spatial dependency of the grazing pattern in map units. This value determines how "clumped" grazing patches will be. A value close to 0 will produce a perfectly randomized grazing pattern with patch size equal to raster cell resolution, and larger values will produce increasingly clumped grazing patterns, with the size of the patches corresponding to the value given.
Grazing patchiness coefficient	0–1	unitless	1	Coefficient that, along with the spatiality coefficient, determines the patchiness of the grazing pattern. Value must be non-zero, and usually will be ≤1.0. Values close to 0 will create a patchy grazing pattern, values close to 1 will create a "smooth" grazing pattern. Actual grazing patches will be sized to the resolution of the input landcover map.
Maximum grazing impact	0–50	succession stage	3	Maximum impact of grazing in units of "landcover succession" per annual grazing event. Grazing impact values of individual patches will be chosen from a gaussian distribution between 1 and this maximum value (*i.e.*, most values will be between 1 and this value). Value must be ≥1.

Table 2. *Cont.*

Village Subsistence Characteristic	Typical Range of Values	Units	Value(s) Used in This Paper	Explanation
Manuring rate	0–100	% of maximum fertility	0.2	Base rate that animal dung contributes to fertility increase on a grazed patch in units of percentage of maximum fertility regained per increment of grazing impact. Actual fertility regain values are thus calculated as "manuring rate x grazing impact", so this variable interacts with the grazing impact settings.
Avoid grazing in agricultural catchment	Y/N	boolean	N	If turned on, ovicaprines will not graze in unused portions of the agricultural catchment (*i.e.*, do not graze on "fallowed" fields, and thus no "manuring" of those fields will occur).
Avoid grazing on field stubbles	Y/N	boolean	N	If turned on, ovicaprines will not do any "stubble grazing" on harvested fields (and thus no "manuring" of fields).

Conflicts of Interest: The authors declare no conflict of interest.

References

1. Nogueras, P.; Burjachs, F.; Gallart, F.; Puigdefàbregas, J. Recent gully erosion in the El Cautivo badlands (Tabernas, SE Spain). *Catena* **2000**, *40*, 203–215.
2. Casalí, J.; López, J.J.; Giráldez, J.V. Ephemeral gully erosion in southern Navarra (Spain). *Catena* **1999**, *36*, 65–84.
3. Valcárcel, M.; Taboada, M.T.; Paz, A.; Dafonte, J. Ephemeral gully erosion in northwestern Spain. *Catena* **2003**, *50*, 199–216.
4. Gutiérrez, Á.G.; Schnabel, S.; Contador, F.L. Gully erosion, land use and topographical thresholds during the last 60 years in a small rangeland catchment in SW Spain. *Land Degrad. Dev.* **2009**, *20*, 535–550.
5. Marzolff, I.; Ries, J.B.; Poesen, J. Short-term *versus* medium-term monitoring for detecting gully-erosion variability in a Mediterranean environment. *Earth Surf. Process. Landf.* **2011**, *36*, 1604–1623.
6. Hooke, J.M. Human impacts on fluvial systems in the Mediterranean region. *Geomorphology* **2006**, *79*, 311–335.
7. Van Andel, T.H.; Zanagger, E. Landscape stability and destabilization in the prehistory of Greece. In *Man's Role in the Shaping of the Eastern Mediterranean Landscape*; Bottema, S., Entjes-Nieborg, G., van Zeist, W., Eds.; A.A. Balkema: Rotterdam, The Netherlands, 1990; pp. 139–157.
8. Bintliff, J. Time, process and catastrophism in the study of Mediterranean alluvial history: A review. *World Archaeol.* **2002**, *33*, 417–435.
9. Redman, C.L.; Fish, P.R.; James, S.R.; Rogers, J.D. *The Archaeology of Global Change: The Impact of Humans on Their Environment*; Smithsonian Books: Washington, DC, USA, 2004.
10. Perevolotsky, A.; Seligman, N.G. Role of grazing in Mediterranean rangeland ecosystems. *BioScience* **1998**, *48*, 1007–1017.
11. Barton, C.M.; Bernabeu Auban, J.; Garcia Puchol, O.; Schmich, S.; Molina Balaguer, L. Long-term socioecology and contingent landscapes. *J. Archaeol. Method Theory* **2004**, *11*, 253–295.
12. Fumanal Garcia, M.P. Dinámica sedimentaria Holocena en valles de cabecera del País Valenciano. *Cuatern. Geomorfol.* **1990**, *4*, 93–106.
13. Hill, J.B. Land use and an archaeological perspective on socio-natural studies in the Wadi Al-Hasa, West-Central Jordan. *Am. Antiq.* **2004**, *69*, 389–412.
14. García-Ruiz, J.M. The effects of land uses on soil erosion in Spain: A review. *Catena* **2010**, *81*, 1–11.
15. Martínez-Casasnovas, J.A. A spatial information technology approach for the mapping and quantification of gully erosion. *Catena* **2003**, *50*, 293–308.
16. Vandekerckhove, L.; Poesen, J.; Govers, G. Medium-term gully headcut retreat rates in Southeast Spain determined from aerial photographs and ground measurements. *Catena* **2003**, *50*, 329–352.

17. Hill, J.B. What difference does environmental degradation make? In *The Archaeology of Environmental Change*; Fisher, C.T., Hill, J.B., Feinman, G.M., Eds.; The University of Arizona Press: Tucson, AZ, USA, 2009; pp. 160–173.

18. Clevis, Q.; Tucker, G.E.; Lock, G.; Lancaster, S.T.; Gasparini, N.; Desitter, A.; Bras, R.L. Geoarchaeological simulation of meandering river deposits and settlement distributions: A three-dimensional approach. *Geoarchaeology* **2006**, *21*, 843–874.

19. Mitasova, H.; Barton, C.M.; Ullah, I.I.T.; Hofierka, J.; Harmon, R.S. GIS-based soil erosion modeling. In *Treatise in Geomorphology: Vol. 3 Remote Sensing and GI Science in Geomorphology*; Shroder, J., Bishop, M., Eds.; Academic Press: San Diego, CA, USA, 2013; pp. 228–258.

20. Wainwright, J. Can modelling enable us to understand the rôle of humans in landscape evolution? *Geoforum* **2008**, *39*, 659–674.

21. Barton, C.M.; Ullah, I.I.T.; Bergin, S.M.; Mitasova, H.; Sarjoughian, H. Looking for the future in the past: Long-term change in socioecological systems. *Ecol. Model.* **2012**, *241*, 42–53.

22. Barton, C.M.; Ullah, I.I.T.; Mitasova, H. Computational modeling and Neolithic socioecological dynamics: A case study from Southwest Asia. *Am. Antiq.* **2010**, *75*, 364–386.

23. Mayer, G.R.; Sarjoughian, H.S. Composable cellular automata. *Simulation* **2009**, *85*, 735–749.

24. Mayer, G.R.; Sarjoughian, H.S.; Allen, E.K.; Falconer, S.E.; Barton, C.M. Simulation modeling for human community and agricultural landuse. In Agent-Directed Simulation, Proceedings of the Agent-Directed Simulation Multi-Conference, Huntsville, AL, USA, 2–6 April 2005; Society for Computer Simulation International: San Diego, CA, USA, 2006; pp. 65–72.

25. Ullah, I.I. T.; Bergin, S. Modeling the consequences of village site location: Least cost path modeling in a coupled GIS and agent-based model of village agropastoralism in eastern Spain. In *Least Cost Analysis of Social Landscapes: Archaeological Case Studies*; White, D.A., Surface-Evans, S.L., Eds.; University of Utah Press: Salt Lake City, UT, USA, 2012; pp. 155–173.

26. Barton, C.M.; Ullah, I.I.; Mayer, G.R.; Bergin, S.M.; Sarjoughian, H.S.; Mitasova, H. *MedLanD Modeling Laboratory v.1*. CoMSES Computational Model Library. Available online: https://www.openabm.org/model/4609/version/1 (accessed on 23 April 2015).

27. Bernabeu Auban, J.; Orozco Köhler, T.; Diez Castillo, A.; Gomez Puche, M. Mas d'Is (Penàguila, Alicante): Aldeas y recintos monumentales del Neolítico Antiguo en el Valle del Serpis. *Trab. Prehist.* **2003**, *60*, 39–59.

28. Bernabeu, J.; García Puchol, O.; Pardo, S.; Barton, M.; McClure, S.B. AEA 2012 Conference reading: Socioecological dynamics at the time of Neolithic transition in Iberia. *Environ. Archaeol.* **2014**, *19*, 214–225.

29. Mitas, L.; Mitasova, H. Distributed soil erosion simulation for effective erosion prevention. *Water Resour. Res.* **1998**, *34*, 505–516.

30. Mitasova, H.; Hofierka, J.; Zlocha, M.; Iverson, R. Modeling topographic potential for erosion and deposition using GIS. *Int J. Geogr. Inf. Syst.* **1996**, *10*, 629–641.

31. Moore, I.D.; Burch, G.J. Physical basis of the length-slope factor in the Universal Soil Loss Equation. *Soil Sci. Soc. Am. J.* **1986**, *50*, 1294–1298.

32. Renard, K.G.; Foster, G.R.; Weesies, G.A.; Porter, J.P. RUSLE: Revised Universal Soil Loss Equation. *J. Soil Water Conserv.* **1991**, *46*, 30–33.

33. Renard, K.G.; Foster, G.R.; Weesies, G.A.; McCool, D.K.; Yoder, D.C. Predicting soil erosion by water: A guide to conservation planning with the Revised Universal Soil Loss Equation (RUSLE). In *Agriculture Handbook*; US Department of Agriculture: Washington, DC, USA, 1997; Volume 703, pp. 1–251.

34. Onori, F.; de Bonis, P.; Grauso, S. Soil erosion prediction at the basin scale using the Revised Universal Soil Loss Equation (RUSLE) in a catchment of Sicily (southern Italy). *Environ. Geol.* **2006**, *50*, 1129–1140.

35. Barton, C.M.; Ullah, I.I.; Bergin, S. Land use, water and Mediterranean landscapes: Modelling long-term dynamics of complex socio-ecological systems. *Philos. Trans. R. Soc. A: Math. Phys. Eng. Sci.* **2010**, *368*, 5275–5297.

36. Ullah, I.I.T. A GIS method for assessing the zone of human-environmental impact around archaeological sites: A test case from the Late Neolithic of Wadi Ziqlâb, Jordan. *J. Archaeol. Sci.* **2011**, *38*, 623–632.

37. Koriat, A.; Goldsmith, M.; Pansky, A. Toward a psychology of memory accuracy. *Annu. Rev. Psychol.* **2000**, *51*, 481–537.

38. Schacter, D.L. The seven sins of memory: Insights from psychology and cognitive neuroscience. *Am. Psychol.* **1999**, *54*, 182–203.

39. Quiroga, A.; Funaro, D.; Noellemeyer, E.; Peinemann, N. Barley yield response to soil organic matter and texture in the Pampas of Argentina. *Soil Tillage Res.* **2006**, *90*, 63–68.

40. Araus, J.L.; Febrero, A.; Buxó, R.; Camalich, M.D.; Martin, D.; Molina, F.; Rodriguez-Ariza, M.; Romagosa, I. Changes in carbon isotope discrimination in grain cereals from different regions of the western Mediterranean Basin during the past seven millennia. Palaeoenvironmental evidence of a differential change in aridity during the late Holocene. *Glob. Change Biol.* **1997**, *3*, 107–118.

41. Araus, J.L.; Amaro, T.; Zuhair, Y.; Nachit, M.M. Effect of leaf structure and water status on carbon isotope discrimination in field-grown durum wheat. *Plant Cell Environ.* **1997**, *20*, 1484–1494.

42. Barzegar, A.R.; Yousefi, A.; Daryashenas, A. The effect of addition of different amounts and types of organic materials on soil physical properties and yield of wheat. *Plant Soil* **2002**, *247*, 295–301.

43. Carter, D.L.; Berg, R.D.; Sanders, B.J. The effect of furrow irrigation erosion on crop productivity. *Soil Sci. Soc. Am. J.* **1985**, *49*, 207–211.

44. Pswarayi, A.; van Eeuwijk, F.A.; Ceccarelli, S.; Grando, S.; Comadran, J.; Russell, J.R.; Francia, E.; Pecchioni, N.; Li Destri, O.; Akar, T.; *et al.* Barley adaptation and improvement in the Mediterranean basin. *Plant Breed.* **2008**, *127*, 554–560.

45. Sadras, V.O.; Calvino, P.A. Quantification of grain yield response to soil depth in soybean, maize, sunflower, and wheat. *Agron. J.* **2001**, *93*, 577.

46. Araus, J.L.; Slafer, G.A.; Romagosa, I.; Molist, M. FOCUS: Estimated wheat yields during the emergence of agriculture based on the carbon isotope discrimination of grains: Evidence from a 10th millennium BP site on the Euphrates. *J. Archaeol. Sci.* **2001**, *28*, 341–350.

47. Wong, M.T.F.; Asseng, S. Yield and environmental benefits of ameliorating subsoil constraints under variable rainfall in a Mediterranean environment. *Plant Soil* **2007**, *297*, 29–42.

48. Slafer, G.A.; Romagosa, I.; Araus, J.L. Durum wheat and barley yields in antiquity estimated from 13C discrimination of archaeological grains: A case study from the western Mediterranean Basin. *Funct. Plant Biol.* **1999**, *26*, 345–352.

49. Adler, P.; Raff, D.; Lauenroth, W. The effect of grazing on the spatial heterogeneity of vegetation. *Oecologia* **2001**, *128*, 465–479.

50. Alados, C.L.; ElAich, A.; Papanastasis, V.P.; Ozbek, H.; Navarro, T.; Freitas, H.; Vrahnakis, M.; Larrosi, D.; Cabezudo, B. Change in plant spatial patterns and diversity along the successional gradient of Mediterranean grazing ecosystems. *Ecol. Model.* **2004**, *180*, 523–535.

51. Alados, C.L.; Pueyo, Y.; Barrantes, O.; Escós, J.; Giner, L.; Robles, A.B. Variations in landscape patterns and vegetation cover between 1957 and 1994 in a semiarid Mediterranean ecosystem. *Landsc. Ecol.* **2004**, *19*, 543–559.

52. Al-Jaloudy, M.A. *Country Pasture/Forage Resource Profiles: JORDAN*; Food and Agriculture Organization of the United Nations: Rome, Italy, 2006.

53. Gulelat, W. Household Herd Size among Pastoralists in Relation to Overstocking and Rangeland Degradation (Sesfontein, Namibia). Master's Thesis, International Institute for Geo-Information Science and Earth Observations, Enschede, The Netherlands, 2002.

54. Lubbering, J.M.; Stuth, J.W.; Mungall, E.C.; Sheffield, W.J. An approach for strategic planning of stocking rates for exotic and native ungulates. *Appl. Anim. Behav. Sci.* **1991**, *29*, 483–488.

55. Nablusi, H.; Ali, J.M.; Abu Nahleh, J. *Sheep and Goat Management Systems in Jordan: Traditional and Feedlot—A Case Study*; Task Force Documents; Amman, Jordan, 1993.

56. Ngwa, A.T.; Pone, D.K.; Mafeni, J.M. Feed selection and dietary preferences of forage by small ruminants grazing natural pastures in the Sahelian zone of Cameroon. *Anim. Feed Sci. Technol.* **2000**, *88*, 253–266.

57. Hocking, D.; Mattick, A. *Dynamic Carrying Capacity Analysis as Tool for Conceptualising and Planning Range Management Improvements, with a Case Study from India*; Overseas Development Institute, Pastoral Development Network: London, UK, 1993.

58. Stuth, J.W.; Sheffield, W.J. Determining carrying capacity for combinations of livestock, white-tailed deer and exotic ungulates. In *Wildlife Managagement Handbook*; Texas A&M University: College Station, TX, USA, 2001; pp. 5–12.

59. Stuth, J.W.; Kamau, P.N. Influence of woody plant cover on dietary selection by goats in an *Acacia senegal* savanna of East Africa. *Small Rumin. Res.* **1990**, *3*, 211–225.

60. Degen, A.A. Sheep and goat milk in pastoral societies. *Small Rumin. Res.* **2007**, *68*, 7–19.

61. Haenlein, G.F. W. The Nutritional Value of Sheep Milk. Available online: http://www.smallstock.info/issues/sheepmilk.htm (accessed on 21 October 2010).

62. Maltz, E.; Shkolnik, A. Milk production in the desert: Lactation and water economy in the black Bedouin goat. *Physiol. Zool.* **1980**, *53*, 12–18.

63. Meged', S.S.; Torkaev, A.N.; Egorov, S.V.; Storozhuk, S.I. Milk productivity of breeding sheep of the Altai finewool breed. *Russ. Agric. Sci.* **2008**, *34*, 52–54.

64. Thomson, E.F.; Bahhady, F.; Termanini, A.; Mokbel, M. Availability of home-produced wheat, milk products and meat to sheep-owning families at the cultivated margin of the NW Syrian steppe. *Ecol. Food Nutr.* **1986**, *19*, 113–121.

65. Harris, F. Management of manure in farming systems in semi-arid West Africa. *Exp. Agric.* **2002**, *38*, 131–148.

66. Bernabeu Auban, J.; Molina Balaguer, L.; Orozco Köhler, T.; Díaz Castillo, A.; Barton, C.M. Early Neolithic at the Serpis Valley, Alicante, Spain. In The Early Neolithic in the Iberian Peninsula. Regional and Transregional Components, Proceedings of the XV World Congress, Lisbon, Portugal, 4–9 September 2006; Diniz, M., Ed.; BAR International Series: Oxford, UK, 2008; pp. 53–59.

67. Bryson, R.A.; Bryson, R.U. High resolution simulations of regional Holocene climate: North Africa and the Near East. In *Third Millennium BC Climate Change and Old World Collapse*; Nato ASI Series I: Global Environmental Change; Nüzhet Dalfes, H., Kukla, G., Eds.; Springer Verlag: Berlin, Germany, 1996; Volume 49, pp. 565–593.

68. Ruter, A.; Arzt, J.; Vavrus, S.; Bryson, R.A.; Kutzbach, J.E. Climate and environment of the subtropical and tropical Americas (NH) in the mid-Holocene: Comparison of observations with climate model simulations. *Quat. Sci. Rev.* **2004**, *23*, 663–679.

69. Arıkan, B. Macrophysical climate modeling, economy, and social organization in Early Bronze Age Anatolia. *J. Archaeol. Sci.* **2014**, *43*, 38–54.

70. Ullah, I. The Consequences of Human Land-Use Strategies during the PPNB-LN Transition: A Simulation Modeling Approach. Ph.D. Thesis, Arizona State University, Tempe, AZ, USA, 2013.

71. Bogaard, A. "Garden agriculture" and the nature of early farming in Europe and the Near East. *World Archaeol.* **2005**, *37*, 177–196.

72. Conolly, J.; Colledge, S.; Shennan, S. Founder effect, drift, and adaptive change in domestic crop use in early Neolithic Europe. *J. Archaeol. Sci.* **2008**, *35*, 2797–2804.

73. Colledge, S.; Conolly, J.; Shennan, S. The evolution of Neolithic farming from SW Asian origins to NW European limits. *Eur. J. Archaeol.* **2005**, *8*, 137–156.

74. Halstead, P. Traditional and ancient rural economy in Mediterranean Europe: Plus ça change? *J. Hell. Stud.* **1987**, *107*, 77–87.

75. Butzer, K.W. *Archaeology as Human Ecology*; Cambridge University Press: Cambridge, UK, 1982.

76. McClure, S.B.; Barton, C.M.; Jochim, M.A. Human behavioral ecology and climate change during the transition to agriculture in Valencia, eastern Spain. *J. Anthropol. Res.* **2009**, *65*, 253–269.

77. McClure, S.; Jochim, M.A.; Barton, C.M. Behavioral ecology, domestic animals, and land use during the transition to agriculture in Valencia, eastern Spain. In *Foraging Theory and the Transition to Agriculture*; Kennett, D., Winterhalder, B., Eds.; Smithsonian Institution Press: Washington, DC, USA, 2006; pp. 197–216.

Examining Social Adaptations in a Volatile Landscape in Northern Mongolia via the Agent-Based Model *Ger Grouper*

Julia K. Clark and Stefani A. Crabtree

Abstract: The environment of the mountain-steppe-taiga of northern Mongolia is often characterized as marginal because of the high altitude, highly variable precipitation levels, low winter temperatures, and periodic droughts coupled with severe winter storms (known as *dzuds*). Despite these conditions, herders have inhabited this landscape for thousands of years, and hunter-gatherer-fishers before that. One way in which the risks associated with such a challenging and variable landscape are mitigated is through social networks and inter-family cooperation. We present an agent-based simulation, Ger Grouper, to examine how households have mitigated these risks through cooperation. The Ger Grouper simulation takes into account locational decisions of households, looks at fission/fusion dynamics of households and how those relate to environmental pressures, and assesses how degrees of relatedness can influence sharing of resources during harsh winters. This model, coupled with the traditional archaeological and ethnographic methods, helps shed light on the links between early Mongolian pastoralist adaptations and the environment. While preliminary results are promising, it is hoped that further development of this model will be able to characterize changing land-use patterns as social and political networks developed.

Reprinted from *Land*. Cite as: Clark, J.K.; Crabtree, S.A. Examining Social Adaptations in a Volatile Landscape in Northern Mongolia via the Agent-Based Model *Ger Grouper*. *Land* **2015**, *4*, 157–181.

1. Introduction

Sharing and cooperation between individuals and among groups can increase carrying capacity and survivability [1,2]. However, sharing and cooperation can take many forms [1,3–5], some more beneficial to the group, or individuals, than others. Here we ask "How do different sharing strategies impact survivability in a mobile pastoralist case?"

This work is built on theory developed in the U.S. Southwest among sedentary farming populations, which we adapt and apply to mobile pastoralists of Mongolia. Specifically, we use theory developed by Hegmon [6] who simulated the rationale for exchange among Hopi based on three forms of logic: pooling of resources, independence (or hoarding of resources), and restricted sharing [6]. Her research showed that in general restricted sharing is the best strategy, often working better than the other two strategies for both low- and high-production years. By creating rules for whom to share with and when, the Hopi are able to take control of their own needs first before assessing the needs of the community [6,7]. We hypothesize that similar mechanisms were at play with Mongolian pastoralists in prehistory and that rules for whom to share with and when structure modern household configurations.

Seasonal mobility is a common strategy employed primarily by hunter-gatherers and pastoralists living in highly variable, low productivity environments. These environments are characterized by little precipitation, high altitude/latitude, and/or extreme temperature (cold or hot). In these environments, people migrate within the landscape to take advantage of spatially dispersed, seasonally available resources. These patterns are not random, but rather the culmination of generations of accumulated traditional ecological knowledge [8,9]. Mobility can be a wise economic adaptation with many variant forms (*i.e.*, degree, frequency) [10,11], allowing mobile groups to inhabit regions that are not easily occupied by settled groups. Since the individual household units of a group are willing and able to move easily, the group by default is flexible, able to adapt or react to changing environmental, political and social challenges on short notice. In moments of crisis (*i.e.*, high risk), adaptive solutions can be immediately implemented that will carry the household units through until the previously established habitation pattern can be resumed or a new pattern developed.

In central and northern Mongolia, it has been noted [12] that following years of environmental catastrophe (usually resulting in great losses of livestock) household units, which usually numbered from two to four households, clustered into larger groups of five to seven—a cluster similar in size to Hegmon's ideal restricted sharing group [7]. Over time, after households had recovered from herd losses, the units once again dispersed. This temporary fission-fusion cycle is an adaptation to the inherent risk of the low-productivity, highly variable environment in which these populations live. Because these households move every few months anyway, this fission-fusion cycle can occur rather rapidly. However, cooperation was not random, though the rules about who would help whom and under what circumstances were not immediately apparent. While this has been observed anecdotally, ethnographic data continues to be compiled to more rigorously characterize these cycles [12].

In patchy environments (*i.e.*, environments where productivity is spatially and/or temporally variable), the ability to count on kin and neighbors during years of low productivity is essential for survival. Sahlins [5] demonstrated that cross-culturally there are distinct rules for the sharing of resources, and that small-scale societies worldwide have tactics for surviving bad years. Hegmon [6,7] has shown that restricted-sharing tactics are reliable for most years when both the pooling of resources and hoarding of resources are not optimal. Such strategies appear to be employed by mobile pastoral groups of modern Mongolia. The decision to aggregate with some groups as a form of risk management, while still excluding other groups from aggregation, exemplifies a strategy of counting on trusted kin or neighbors when times are difficult.

For this research, developing agent-based models that imbue agents with decisions on where to locate and how to form cohesive groups will enable the examination of individual-level processes as reactions to environmental pressures. Costopoulos, Lake and Gupta tell us that *"simulations can surprise us. Whether the surprises are due to our faulty understanding of the reality we are modeling or to our faulty modeling of the reality we are seeking to understand, they can force us to reexamine our assumptions and to push beyond the intuitive models of the past for which we often settle too easily"* [13].

While decades of research have focused on cross-cultural studies of human systems, model building and theory testing provide a novel way to examine the world, helping to answer questions that would be unanswerable from traditional approaches [14]. Instead of seeing the panoply of human culture and searching for patterns, we create theory, build models based on theory, and then compare output to data. Simulation enables us to test theories developed by anthropologists and historians from years of cross-cultural research [14]. Lake estimates that works based on 54 different archaeological simulations were published between 2001 and 2010, showing the increasing value of agent-based modeling in archaeology [14], and the increasing ability for agent-based modeling to assess archaeological theories. Simulation does not more correctly address the archaeological record, but can address different questions than cross-cultural research can, and can easily help refine hypotheses of the archaeological record.

Our paper explores the extent to which sharing practices would have helped the survival of mobile pastoralists in Mongolia and the surrounding regions of northeast Asia, and how a patchy environment led to the profusion of fission/fusion dynamics in Mongolia. In this model we define sharing and cooperation very simply: the likelihood that one household will merge with another household in need of assistance for one timestep, dividing resources equally between households. Seasonal movements characteristic of the semi-nomadic inhabitants of the region provide ample opportunity to examine such fusion and fission events. Groups fuse together when it is beneficial to do so, and then part ways when this approach becomes more advantageous. The presented model will help us to understand when fusion, fission, and sharing may be sought as a risk management strategy.

Computer modeling is not a new approach for Mongolian case studies [15,16]. However, these models approach the question of the emergence of empires and other large political formations based on a number of environmental and historical parameters. The model presented here is of an entirely different scale and is based in ethnographic and historical data. While previous models are designed to investigate political processes on an inter-regional scale, the model we are presenting here approaches the economic sphere from the domestic (*i.e.*, household) viewpoint with the intention of creating results that are compatible with available ethnographic and archaeological data from the region.

This paper is structured in the following way. First, we present the necessary background for how sharing strategies structure populations in northern Mongolia. We discuss ethnographic and archaeological evidence for sharing both in our study area and in other small-scale societies worldwide. We then present how agent-based modeling can help to examine sharing strategies, exploring how four different sharing strategies create different population levels in a variable environment. In the conclusion, we discuss the significance of our findings from employing a simple agent-based model and suggest ways in which this model may be refined for further future use.

1.1. Background

Mongolia is located in northeast Asia and is home to a primarily pastoralist population. In this study we focus on the inhabitants of the steppe and forest steppe in the central and northern portions of the country. These individuals primarily keep sheep and goats, with horses, cows, yaks and camels making up lesser percentages of their stock. Mongolian pastoralists derive much of what they consume from their livestock, and spend considerable time and energy ensuring the survival of their flocks. They rely on extensive traditional ecological knowledge that has been passed from generation to generation in order to minimize herd deaths during the difficult winter months. This knowledge includes ways to navigate both environmental landscapes and social networks. These modern day herders provide a useful ethnographic analogy, when applied cautiously, for the semi-nomadic nature of the early herders of Mongolia [12,17,18].

Today, Mongolian pastoralists move seasonally between summer and winter pastures. During summer, grazing conditions are good and herds are fattened for the long winters when grazing conditions are poor because of extended cold periods, little forage, and snow cover. These movements vary from a few kilometers to over 100 km between camps, though in central and northern Mongolia, where the authors have collected data, the average is usually 10–20 km [12]. Typically households move two to five times annually following a similar mobility pattern year after year, returning to the same location at roughly the same time each season [12,19–22]. However, this pattern may shift from time to time in order to address a number of factors, including social conventions and environmental degradation or disaster.

Ethnographic observation has shown that group size is not consistent from season to season or year-to-year [19]. Each group of households, known as a *khot ail*, is made up of a number of nuclear families, each occupying their own dwelling called a *ger* (a round tent made of wood, felt and canvas or hides—also known by the Russian term *"yurt"*). The size of the *khot ail* may vary from a single *ger* to more than 20 [23], although most never exceed 10 households. Average camp size appears to increase following environmental disasters as individual *khot ails* band together utilizing kinship and social ties as a failsafe to help recover from the losses of herd animals following these events. *Gers* from the same valley may group together, but larger risk mitigating groups that extend beyond valleys are also normal [12]. If the individual *khot ails* are able to rebuild their herds, they may once again disband into smaller groups.

A number of environmental conditions might present risk to the herds of Mongolia's rural populations. These include drought, bad winter storms locally known as *dzud*s, and the outbreak of epizootic diseases [24,25]. *Dzud*s come in several varieties depending upon the particular environmental conditions. Types of *dzud*s include: deep snows, no snows, ice sheets, extended or extreme cold spells, and extreme overgrazing and trampling. These events occur periodically–every 5–10 years according to some studies [25]. *Dzud*s may not impact regions equally creating a "patchy" environment on the large scale. While much of the discussion about mitigating the effects of *dzud*s has focused on aid efforts and observed rural to urban migration, a few sources have attempted to document the local adaptations and coping methods used by herders [25,26]. Shelter may be improved including: alterations to structures, tunneling, insulating structures with dung, and bringing animals into the family *ger*. Of interest to this project are those strategies that rely upon social and kin networks to mitigate the impact of *dzud*s. Such adaptations include movement to other, less impacted areas (known as *Otor*, the movement from adjacent valleys up to hundreds of kilometers away), or joining forces with local family or friends in which mutual assistance may increase the chances of survival. Though these are short-lived events, they can be devastating. Cooperation is needed not to survive the *Dzud* itself, but to recover after great losses following the event. While there are clear advantages to the "movers", the "hosts" are willing participants in this coping method because of expected future reciprocity (much like insurance) and cultural expectations (e.g., an expectation to help out extended family members) [5].

It is clear that modern day Mongolia has a culturally dictated set of rules regarding sharing and cooperation. But how do these sharing strategies develop? A study by Fitzhugh *et al.* [27] helps inform us of the development of sharing strategies. They suggest that hunter-gatherer populations use exchange to build information networks that help establish relationships among different bands. These information networks connect households to an expanded pool of bands and/or tribes, allowing for group survival during catastrophic events. Additionally, they argue that high cost and low predictability/low productivity landscapes exhibit higher network connectivity than highly predictable landscapes. Furthermore, as populations become entrenched in an area they adapt to the environment and will rely on information networks only for highly unpredictable and catastrophic events, not for more predictable events. The high climatic variability of Mongolia combined with the potential for (and reality of) catastrophic failure would make the region more reliant on networks, according to this model [27].

Fitzhugh *et al.* [27] also state that groups should rely on more proximal bands for regularly occurring crises, such as low food production and droughts, while more irregular crises, such as earthquakes, would require a longer temporal memory of alliances with more distant allies. Therefore, since *dzuds* are unpredictable, but frequently recurring disasters, we can infer from Fitzhugh *et al.*'s model that Mongolian households would rely more on their neighbors for economic stability than on more distant allies.

1.2. The Model

A model is an idealized microcosm of a real system and is built on theory, or, as Clarke [28] states "models are pieces of machinery that relate observations to theoretical ideas." Using models built on simple rules can help eliminate poor hypotheses, and can help enable better understanding of a system. Even when a model is wrong (as *"all models are wrong, but some are useful"* [29] we can glean a better understanding of the system by slowly building the model up and studying simplified processes of complex systems.

The agent-based model detailed in this paper was generated in NetLogo, although could have easily been written for any other modeling platform. The agents in this model represent an economic production unit, in this case a household (*sensu* [30]). There are twenty agents randomly seeded on the landscape at the beginning of the simulation. Each agent represents one of four distinct sharing scenarios, discussed below. The landscape is 40-cells by 40-cells wide, making a total of 1600 cells for the simulation window; each of these cells correspond to a catchment area (the area within which *most* household activities will take place) of a typical household of two square kilometers.

The simulation window is divided into two sections—a summer landscape and a winter landscape. Each of these comprises 800 cells. This is admittedly reduced (modern herders may move several times in a single year) in order to preserve the simplicity of the model. The agents themselves migrate between the summer and winter landscapes each season (represented by one timestep, or *tick* in the model). In summer all land is productive. In winter, however, only half of the landscape (400 cells) has the possibility of being productive, with the other half of the landscape being composed of barren patches. These barren patches are populated in random locations at the beginning of the simulation. Additionally, 2/3 of the remaining winter cells (264 cells) begin as "brown" and regenerate according to the parameter "grass regrowth time", which was set at five timesteps for this simulation (five timesteps being the equivalent of five seasons, so if a patch dies during summer, it will regenerate five seasons later in winter. The decision for five timesteps is not based on any ethnographic fact, but was used for simplicity in this simulation. Future studies may test and alter this parameter.) While five timesteps may seem long, in northern Mongolia, at least, areas of intense utilization are still visible one or more years after a household has abandoned that area.

To summarize, green patches are productive, brown patches are currently unproductive and symbolize those areas that can regenerate with time, while barren patches are never productive and symbolize those areas that will always be dead in winter. Both summer and winter patches can become brown with use, while only some winter patches will be barren. Barren and brown patches are not only representative of the absence of grass, but by logical extension, any reduction in productivity. For example, a *dzud* may not have a long term impact on grass growth, but the impact on productivity is great due to herd loss.

When an agent lands on a cell, the agent automatically takes the resources that grow on that patch—in the simulation we call these resources "energy" and energy gained from patches is set by the parameter "ger gain from food". In this sweep energy was set to five. Here we have the logical proxy that a household is dependent on its herd, and herds depend on grass, so the quantity of energy (as measured by converting grass to stock) equals the quantity of sheep a household could have. While there may be more sophisticated ways of modeling energy as it moves through trophic levels, the correlation of herd size and grass was maintained in order to preserve the simplicity of the model. When a patch has all of its grass eaten, the patch turns brown and is unproductive; it will regrow the grass when agents move off of it according to the parameter "grass regrowth time".

There is one final parameter related to patch productivity: the parameter "energy loss from dead patches". If at the end of an agent's move but before the end of the timestep an agent lands on a brown patch, that agent is charged energy according to that parameter. In this sweep that parameter was also set at 5. For clarification, while an agent will, in the end, be on a brown patch (because it eats the grass there) the agent is only penalized if it lands on a patch where there was no grass to begin with (if the patch was brown or barren upon landing there). This penalty is meant to simulate the costs that herders who are unable to find suitable locations in patchy environments may have to endure, which may include camping in less than ideal locations.

Agents move each summer and each winter (mimicking Mongolian semi-nomadic seasonal shifts) by randomly choosing an unoccupied patch on the opposite side of their current simulation window (in summer they move to winter, and vice versa). If the agent lands on an unproductive patch, it checks its Moore neighborhood radius (each adjacent cell) and moves to a green patch in the radius; if there are no productive cells in the Moore radius the agent stays put until the next season. Agents are charged one energy unit to move, but are penalized five energy units if they stay on an unproductive patch. In the system we are simulating here, Mongolian pastoralists choose to move seasonally as the long term benefits of fresh pasture outweigh the relatively low, short term costs associated with moving.

Agents in this simulation are incredibly myopic and have limited memory. However, agents do track the productive patches they have visited in winter and will choose to move to a previously visited patch (as long as that patch is empty, as only one agent can be on a patch at a time). If a productive patch they have previously visited is not available, the agent will simply move to an empty winter patch. Since half of the winter landscape is composed of patches that cannot produce food, remembering (and moving to) a patch that previously was productive gives the agents the ability to avoid accidentally landing on a completely unproductive patch. In this sense the agents are reactive to their environmental conditions, and can only work to improve their quest for energy in two ways: moving, or asking a neighbor of a similar strategy for help.

Each winter, agents move from the summer cells to the winter cells. This migration is costless as long as a ger lands on a productive patch. If they land on an unproductive patch they are charged one energy unit to move in their Moore radius to a productive patch. Agents get five energy units each time they eat grass, and if they land on an unproductive patch they are charged five energy units at the end of the timestep. A lucky ger, landing regularly on good winter pasture, will be able to sustain and grow its energy stock.

In summer, if agents have stored more than 20 energy units they have a 5% chance of reproduction by fissioning. When agents reproduce, the daughter household is spawned one cell distant from the parent cell and the stored energy of the parent household is divided evenly between parent and daughter households.

Agents are initially created with four distinct sharing strategies. These strategies are related to the storage of resources and are tracked based on lineage. When agents are created they track their strategy as their lineage, and they never change strategies (agents do not learn). They pass these strategies on to their daughter households.

Strategy A—agents will always merge with another household when asked
Strategy B—agents have a 50% likelihood of accepting an offer of merger
Strategy C—agents have a 25% chance of accepting an offer of merger
Strategy D—agents will never merge

When agents have less than 10 energy units they know they are approaching death. Agents that have less than 10 energy units will search within a radius of five cells for others in their same lineage—that is, the same cooperation strategy. The agent that is close to starvation will ask one of their lineage for help. Those that always share (Strategy A) will always say yes; Strategy B will only say yes with a 50% probability, and Strategy C only will say yes with a 25% probability. Those in Strategy D never ask for help, because help will never be given.

Upon the acceptance of an offer of a merger, the merging agent donates all of its resources to the agent that accepted the offer of merger, and then households merge together. The combined households will then have more total energy, and perhaps a greater potential for fissioning the following summer.

This method of merging has been observed ethnographically in the region. For example, during ethnographic interviews conducted in northern Mongolia in 2012, a recently merged household was encountered. Only one week before interviews a child had set their family's *ger* on fire. The family took their belongings and joined their herds with another household. The households would remain merged until they were able to acquire or build another *ger*, and accumulate enough resources to move out on their own once again.

The simulation stops when either: (a) the simulation reaches 500 ticks (timesteps or seasons); or (b) there are no more agents on the landscape. Those households that survive to the end of the simulation, via luck and compassionate neighbors, represent the propagation of a kin descent group. As illustrated in the figures that follow, the most dynamic results occur in the first few hundred ticks. However, the simulation was run to 500 ticks in order to show the stability of the strategies over the long term.

2. Results and Discussion

For this study we examined how the variable "patch variability" affects the population of agents following the four different strategies. Patch variability reflects the likelihood at any timestep that a portion of the productive winter landscape will be unproductive. The different portions of unproductive landscape modeled can be related to both winter severity and differences in landscape in two or more compared regions. Seven values for patch variability were examined, displayed in Table 1.

In addition to testing each of these values for patch variability we examined how each of the strategies fared when just one strategy was present per patch variability (for example, *only* strategy A was practiced), *versus* when all strategies were present simultaneously. In this way we can examine the direct effects of patch variability on one strategy, as well as the effects of different competing strategies and patch variability.

Table 1. Description of key parameter "patch variability" and what each of the values corresponds to. When patches are set to 0% all patches during winter can be productive, while each increment decreases the productivity by that percentage.

Variability	Description
0	During the winter all patches can be productive
5	During the winter, 5% of all patches can be unproductive
10	During the winter, 10% of all patches can be unproductive
15	During the winter, 15% of all patches can be unproductive
20	During the winter, 20% of all patches can be unproductive
25	During the winter, 25% of all patches can be unproductive
30	During the winter, 30% of all patches can be unproductive

While multiple parameters were written in to the simulation (such as how much energy can be gained from grass, how much lost when grass is dead, what percentage to reproduce) the main question in this research is: "How well do the different sharing practices cope with impact of variable weather (such as localized temperature and precipitation)?" The parameter patch variability takes the simulation window and every year makes patches unproductive according to the values in Table 1. This creates unpredictable patchiness of the environment. The list of other parameters in this simulation and their values is reported in Table 2.

Table 2. List of key parameters and values that were swept across in this simulation. To note the parameter "winter patch variability" was the key parameter varied, with most other variables set to 5 for consistency.

Parameter Name	Value
Ger reproduction likelihood	5%
Random Number seed	197, 312, 414, 599, 822
Number of initial agents	0, 5
Winter Patch Variability	0, 5, 10, 15, 20, 25, 30
Ger gain from food	5
Grass regrowth time	5
Energy loss from dead patches	5

In total, 1750 runs of the simulation were completed for this study. For each of the seven values for the key parameter of patch variability, 10 runs were done with each of the five random number seeds so that outliers could be accounted for. Two separate experiments were done: looking at how each of these strategies fares when it is the only strategy represented on the landscape, and examining how these strategies fare when each strategy is represented on the landscape at the same time.

2.1. Single Strategies

As displayed in Figure 1, when only one strategy is present, regardless of which strategy is represented, population reaches carrying capacity and the mean population curve follows a regular logistic growth curve [31]. The most striking difference in this graphic is the difference between Column A (100% sharing) and the rest of the columns (50%, 25% and 0% sharing). While the mean population curve for column A is similar to the mean population curves for each of columns B, C, and D, the variance around the mean is much more pronounced. This is true even in row 1, which represents 0% patch variability.

The means for each value of patch variability are reported in Supplementary Figures S1–S7 so that means could be compared. With the means graphed in the same graphs the similarities among strategies are even more apparent. While there is some difference, those differences are small. The differences become larger as patch variability becomes higher—by the time patch variability is 30% the detriment of the all-sharing strategy becomes apparent. If agents always share, overall populations are lower, while restricted sharing strategies have higher populations. But even the difference between all share and the other strategies is minimal. As we will see below, this is in contrast to when each strategy is represented at the same time on the landscape.

Hegmon [6] found in her simulation of Hopi food sharing strategies that 100% cooperation was rarely the optimal strategy, but rather restricted sharing seemed to benefit the overall population the most. The results presented here compare positively with Hegmon's findings. While the mean of each of the sharing strategies reported here is similar, the variance in the 100% sharing strategy suggests that sharing with no restrictions could be detrimental, even in favorable conditions. While the mean of the all-share strategy is similar to all the other strategies (Figures S1–S7), the variance (Figure 1) belies the fact that an all-share strategy could have highly unpredictable outcomes. The tighter variance around the mean in the other strategies suggests that those strategies would have more predictable outcomes.

Hegmon also suggests that hoarding (here represented at 0% sharing) is only a good option in the years of the worst productivity. When looking at graphs S1–S7 there appears to be no functional difference between any of the strategies, so this finding is not necessarily echoed in our results at this stage.

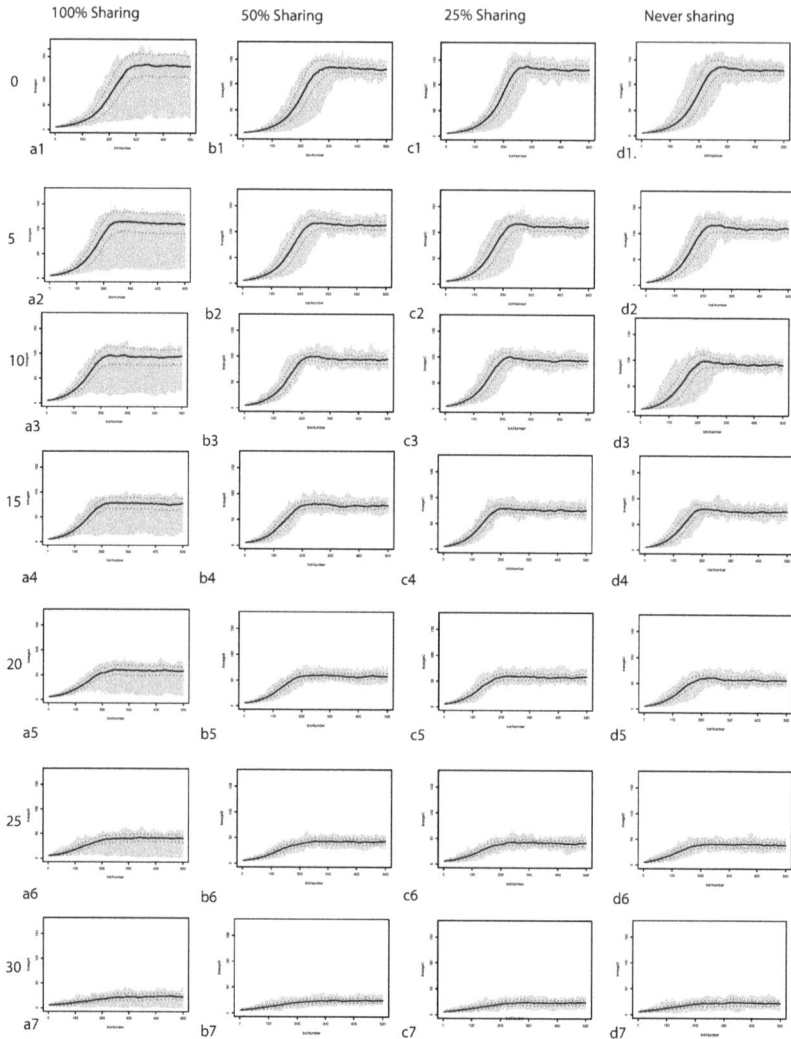

Figure 1. Figure showing how each individual strategy responds to environmental pressures when no other lineage is present. Each tile is as follows: Columns marked A correspond to the 100% sharing strategy. Columns marked B correspond to the 50% sharing strategy. Columns marked C correspond to the 25% sharing strategy. Columns marked D correspond to 0% sharing strategy. Row 1 is 0% winter patch variability. Row 2 is 5% winter patch variability. Row 3 is 10% winter patch variability. Row 4 is 15% winter patch variability. Row 5 is 20% winter patch variability. Row 6 is 25% winter patch variability. Row 7 is 30% winter patch variability. Thus, tile c3 is the 25% sharing strategy under 10% patch variability. Y-axis goes from 0 to 150 households, X axis goes from 0 to 500 ticks. Red-dotted line corresponds to the standard deviation from the mean, while the gray lines show each strategy. Black central line corresponds to the mean of each strategy.

208

2.2. Multiple Strategies

Here we examine how populations respond to environmental stressors when each of the different strategies coexist in the same landscape. At the beginning of the simulation five agents of each strategy are seeded on the landscape. Experiments followed the same trajectory as above: with seven variables for patch variability and five random number seeds.

First of note is the scale: when only one strategy is represented the sum of that strategy is higher than the sum of that individual strategy when there are multiple strategies present. In Figure 1 the scale is set to 150 agents, while in Figure 2 the scale is set to 60 agents. Because of this, in Figure 2 the variability might seem higher than it is when compared to Figure 1, but variance around the mean is only ever approximately 40 agents in both Figures 1 and 2 (Figure 1 strategy A excluded).

Comparing the means of each strategy against one another on one graphic provides more helpful information. In Figures 3–9 each of the mean strategies are graphed on top of one another without the variance surrounding the mean as in Figures 1 and 2. This allows us to directly compare the mean strategies without surrounding noise.

Figure 3 shows how each strategy fared against one another when the environment did not have any variability. To note, the 100% sharing strategy is never the best performing strategy. In these runs of the simulation, hoarding (0% sharing) is the highest performing strategy early in the simulation, while through time those *gers* that subscribe to a hoarding strategy decrease in number. The strategy of sharing 50% of the time, however, is very stable, and eventually becomes the most populous strategy.

In a situation of stable population we may expect to see a convergence upon the mean as agents coalesce upon stable landscapes. A population under stress, however, will see a wide range of variation around the mean as agents attempt to maximize their resource acquisition while dealing with a volatile landscape (as seen above when only one strategy is represented). While the landscape in these runs of the simulation does not have year-to-year variability, the use of the land will create barren patches for five timesteps. Thus, early on *gers* that do not share do well on the landscape because there is little environmental impetus for sharing. With a predictable environment from year-to-year, independence can be a viable strategy. However, as the simulation progresses and *gers* create barren patches on the landscape from over-use, sharing can help *gers* avoid the variable productivity in the landscape they themselves have created.

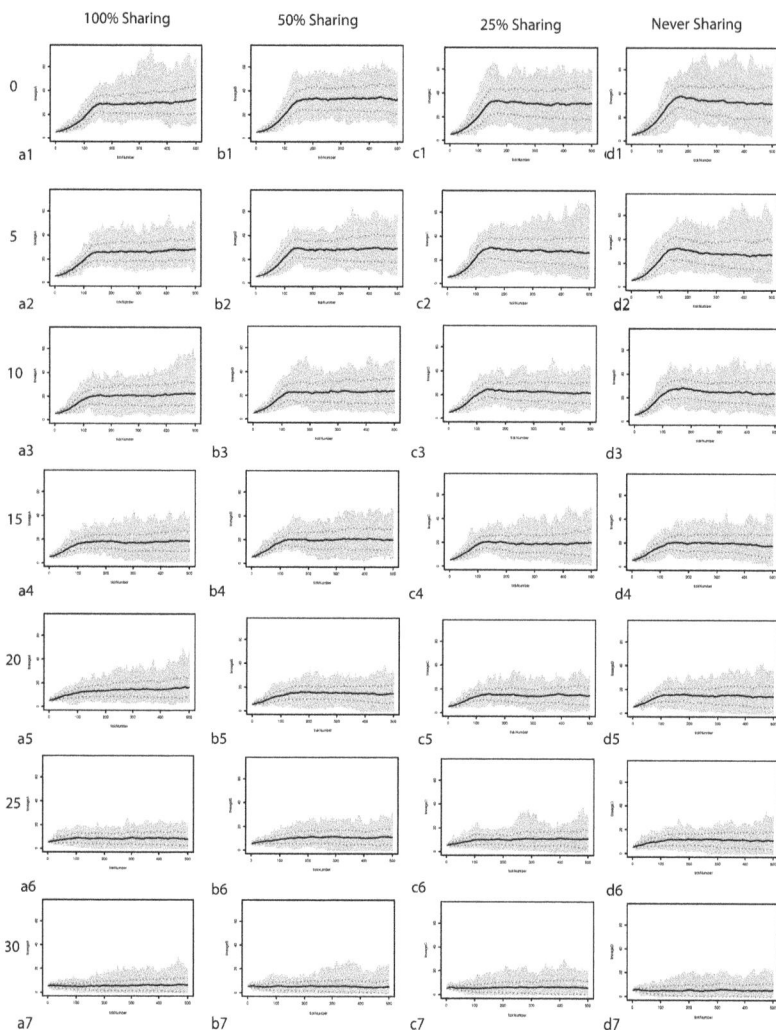

Figure 2. Figure showing how each individual strategy responds to environmental pressures when all other lineages are present. Each tile is as follows: Columns marked A correspond to the 100% sharing strategy. Columns marked B correspond to the 50% sharing strategy. Columns marked C correspond to the 25% sharing strategy. Columns marked D correspond to 0% sharing strategy. Row 1 is 0% winter patch variability. Row 2 is 5% winter patch variability. Row 3 is 10% winter patch variability. Row 4 is 15% winter patch variability. Row 5 is 20% winter patch variability. Row 6 is 25% winter patch variability. Row 7 is 30% winter patch variability. Thus, tile c3 is the 25% sharing strategy under 10% patch variability. Y-axis goes from 0 to 150 households, X-axis goes from 0 to 500 ticks. Red-dotted line corresponds to the standard deviation from the mean, while the gray lines show each strategy. Black central line corresponds to the mean of each strategy.

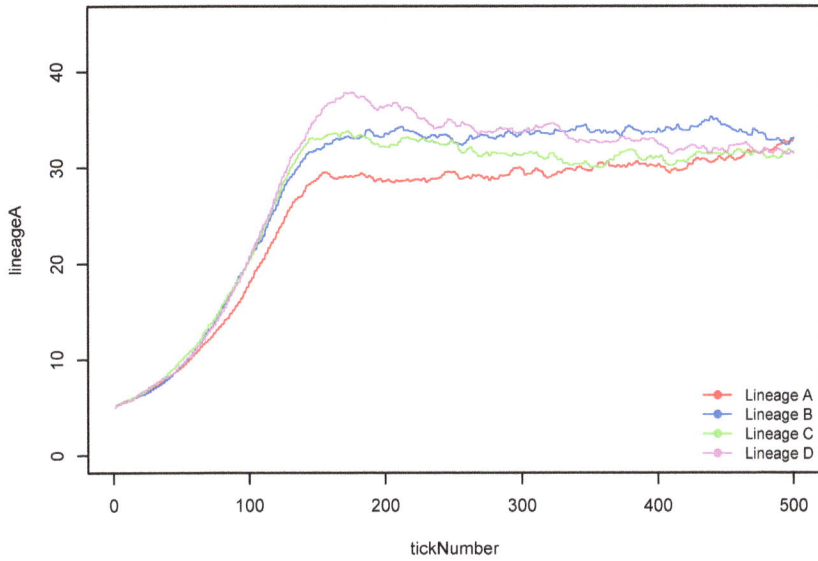

Figure 3. Means of each of the strategies for 0% patch variability. Means correspond to Row 1 of Figure 2. This figure reflects those runs when all strategies were present in the simulation.

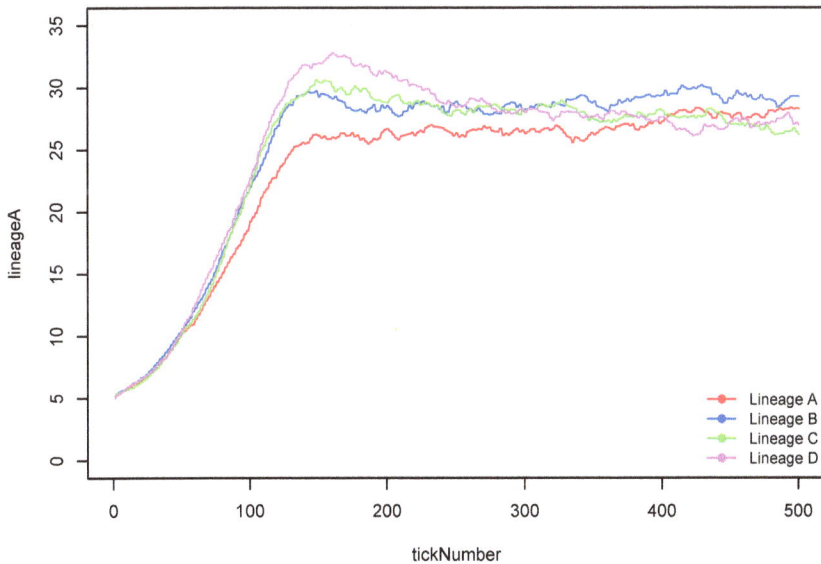

Figure 4. Means of each of the strategies for 5% patch variability. Means correspond to Row 2 of Figure 2. This figure reflects those runs when all strategies were present in the simulation.

Figure 4 follows a similar trajectory to Figure 3, with 100% sharing never being the best performing strategy of the four strategies, no sharing performing the best early on, and restricted sharing performing the best toward the end of the simulation. Figure 5, however, begins to diverge from Figures 3 and 4. In this figure the winter landscape had 10% variability. The sharing strategies are each fairly stable, reaching their own respective carrying capacities of 20 to 25 households on the landscape. In these runs of the simulation hoarding (0% sharing) is early on the highest performing strategy. However, this strategy has high variability, likely due to the unpredictability of the landscape, and the similar effect of overuse. However, as only 10% of the landscape is variable (due to the environment), independent *gers* can make a living on the landscape with the simple rules created for this simulation.

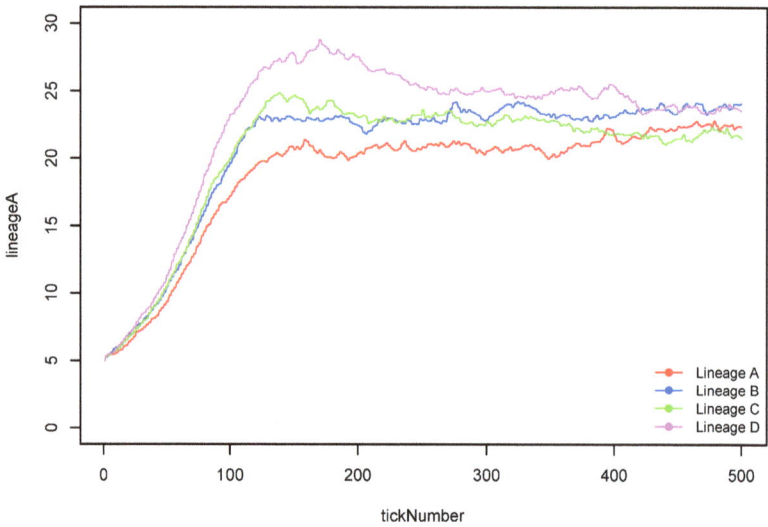

Figure 5. Means of each of the strategies for 10% patch variability. Means correspond to Row 3 of Figure 2. This figure reflects those runs when all strategies were present in the simulation.

Once the environmental unpredictability of the landscape reaches 15%, hoarding is no longer the strategy with the highest population, and will only become optimal again when the landscape's carrying capacity becomes very low (unpredictability of 25%). In Figure 6 we can see that the means of the restricted sharing strategies (50% and 25% sharing) perform the best. Early in the simulation the 25% sharing strategy has the highest mean, while later in the simulation the 50% sharing strategy has the highest mean. This holds true for Figure 7 as well. When the environmental landscape exhibits 20% unpredictability in winter patches, restricted sharing strategies perform well. Note, however, that in the final years of these simulations, the mean of the 100% sharing strategy performs well, while the other strategies remain relatively stable.

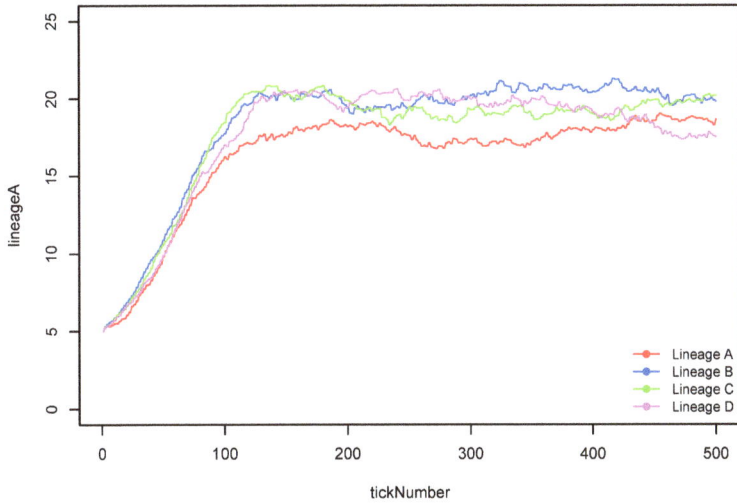

Figure 6. Means of each of the strategies for 15% patch variability. Means correspond to Row 4 of Figure 2. This figure reflects those runs when all strategies were present in the simulation.

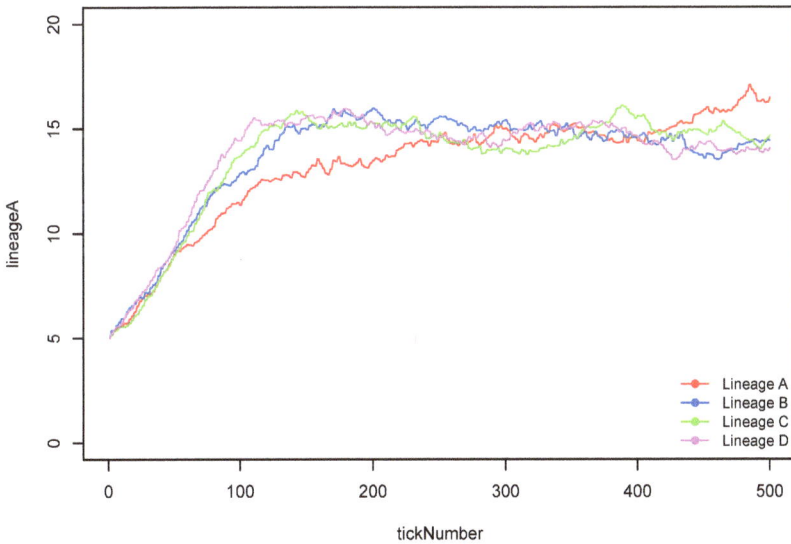

Figure 7. Means of each of the strategies for 20% patch variability. Means correspond to Row 5 of Figure 2. This figure reflects those runs when all strategies were present in the simulation.

In Figure 8 hoarding once again is the highest performing strategy. While above we suggest that hoarding is a good strategy when the landscape is productive enough that sharing is not necessary, Figure 8 echoes Hegmon's [6] finding that hoarding is a viable strategy when the landscape is so poor that sharing will be detrimental for the overall population. Please note, however, that the difference in this graph between the restricted sharing strategies and the hoarding strategy is one household. In fact, many of the differences are rather small. Over the long term, however, even small differences in survivability (small adaptive advantages) may impact decision making.

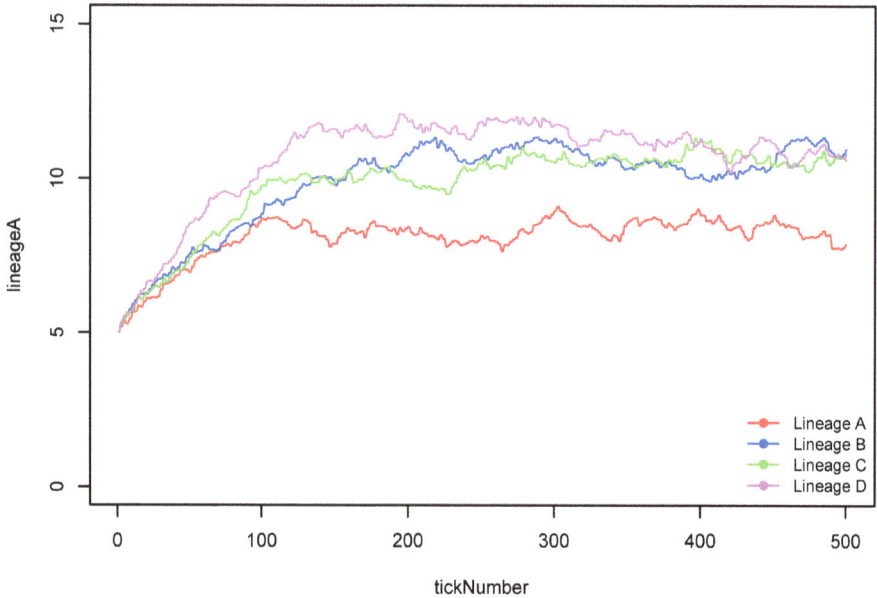

Figure 8. Means of each of the strategies for 25% patch variability. Means correspond to Row 6 of Figure 2. This figure reflects those runs when all strategies were present in the simulation.

In Figure 9, when the landscape exhibits 30% unpredictability in winter patches, the averages of all of the four strategies are within one household. However, the 25% sharing strategy seems to have the highest mean on average. These results, when compared with Figure 2(c7) show that this strategy also has the least variance (and thus might have the most predictable outcome).

Hegmon [6] found in her simulations that the all-share strategy was never the optimal strategy, and that hoarding is an optimal strategy for a population when the environment is highly unpredictable. These findings are comparable to our study results, although we show that there is little necessity for sharing in a highly predictable landscape. Only when the landscape becomes changed due to use, or the environmental predictability becomes great, do sharing strategies become necessary.

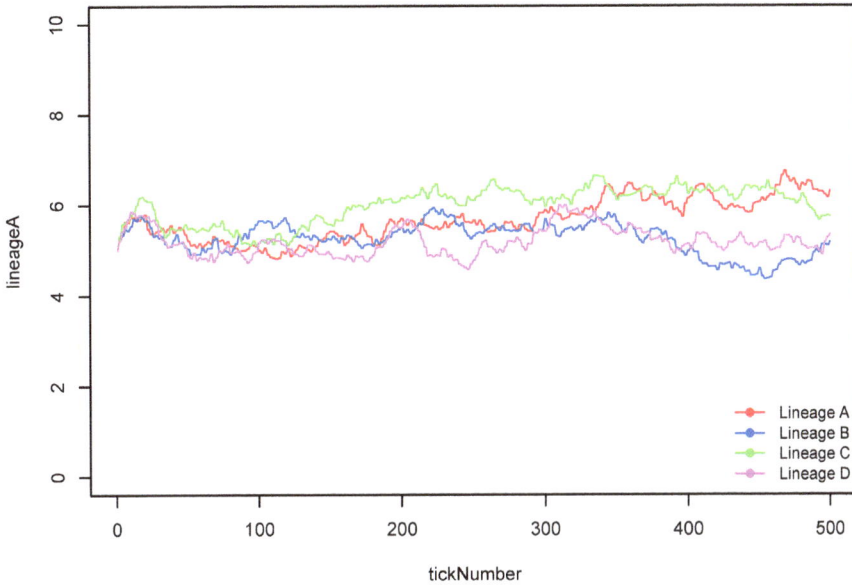

Figure 9. Means of each of the strategies for 30% patch variability. Means correspond to Row 7 of Figure 2. This figure reflects those runs when all strategies were present in the simulation.

Comparing Figures 1 and 2 we can see that some similar patterns are apparent—an "all share" strategy never outperforms the other strategies, but there appears to be little functional difference among the other strategies. Each strategy reaches the logistic population curve (the carrying capacity) in Figure 1, but in Figure 2 there is greater variability. When comparing the means in Figures 3–9 we see that restricted sharing seems to be the most beneficial strategy when environmental conditions are unpredictable.

For a final means of comparison, we examined the statistical difference among the strategies with a Kolomgorov-Smirnov analysis. Kolomgorov-Smirnov analyses allow for direct comparability of each of the simulated means to see if there are statistical differences between each of the strategies. We simplified these data into five time slices: 100 ticks (50 years), 200 ticks (100 years), 300 ticks (150 years), 400 ticks (200 years), and 500 ticks (250 years). Further, we compared the pair-wise difference between the following means: Strategy A to Strategy B, Strategy B to Strategy C, Strategy C to Strategy D, and Strategy A to Strategy D.

Frequency differences as well as p-values to the 0.05 level are reported in Tables 3 and 4. Table 3 corresponds to Figure 1 (single strategies modeled) while Table 4 corresponds to Figure 2 (all strategies present).

As can be seen in Table 3, when only one lineage is represented, 31 of the 140 K-S statistic values show clear statistical significance in their difference. In Table 4 we can see that 21 of the 140 values show clear statistical significance in their difference. Thus we can say that in 22% of the cases when only one lineage is represented there are real differences in the number of surviving households on the landscape, while when all lineages are represented 15% of the cases show real differences in the number of surviving households on the landscape. It is worth noting, the strongest difference is between strategy A (all share) and strategy D (no share) in both solo lineages and all lineages, with 17 and 11 cases showing statistical significance respectively. Little difference is seen in the restricted sharing strategies (50% and 25%) potentially showing that both of these are viable in most years and may be functionally the same.

In Table 3, the highest and most significant variation seems to be related to reaching the environmental carrying capacity, which generally is reached between 100 and 200 ticks. Most other times variation is not significant except between extreme strategies in less variable landscapes. In Table 4, however, variation is related to the end of the simulation, potentially showing that as sharing strategies stabilize the differences among them become more pronounced.

From Figure 1 through Figure 8 and Tables 3 and 4 we may be able to interpret that during years of middling unpredictably, those households that do not freely share their resources with everyone (but do share with a select few) are likely to have their caloric needs met, are likely to reproduce, and are likely to survive into the next year. The significance in variation among the strategies suggests that there are real differences in all sharing, restricted sharing, and hoarding and, potentially, that individuals using those strategies would be able to see how well their strategy compared to other strategies. These findings are also echoed in Crabtree [1]. Only during exceptional years would households want to horde their resources, potentially insuring their own survival at the detriment of others.

Table 3. Results from Kolomgorov-Smirnov analysis on single lineage values. Data is simplified into five time slices: 100, 200, 300, 400 and 500 ticks. K-S values that show significance above a *p*-value of 0.05 are highlighted blue and show "Sig" in the significance column. This means that there is a large difference in the mean values for the lines during that tick for those variability values. Of note are the 17 K-S values that show as significant between Strategy A (all share) and Strategy D (no share).

Tick	Variability	A to B Diff	p (0.05)	Sig?	B to C Diff	p (0.05)	Sig?	C to D Diff	p (0.05)	Sig?	A to D Diff	p (0.05)	Sig?
100	0	−0.0039	0.4643	Not	0.9322	0.4541	Sig	0.9450	0.4450	Sig	1.8733	0.4554	Sig
200	0	0.5399	0.2277	Sig	0.6984	0.2139	Sig	0.0349	0.2068	Not	1.2732	0.2210	Sig
300	0	−0.0134	0.1683	Not	−0.0834	0.1670	Not	0.0234	0.1673	Not	−0.0735	0.1685	Not
400	0	−0.0727	0.1686	Not	−0.1445	0.1687	Not	0.0008	0.1698	Not	−0.2164	0.1697	Sig
500	0	−0.0786	0.1698	Not	−0.0581	0.1690	Not	−0.0617	0.1699	Not	−0.1984	0.1707	Sig
100	5	−0.3384	0.4049	Not	0.7353	0.4006	Sig	2.4859	0.3830	Sig	2.8827	0.3875	Sig
200	5	0.1088	0.1994	Not	0.1921	0.1951	Not	0.2645	0.1898	Sig	0.5654	0.1942	Sig
300	5	−0.0642	0.1797	Not	−0.0695	0.1801	Not	−0.0898	0.1813	Not	−0.2236	0.1809	Sig
400	5	−0.0031	0.1819	Not	−0.0457	0.1814	Not	−0.1501	0.1830	Not	−0.1989	0.1835	Sig
500	5	0.0090	0.1821	Not	−0.0575	0.1815	Not	−0.1158	0.1829	Not	−0.1644	0.1834	Not
100	10	−0.3478	0.3802	Not	0.3518	0.3787	Not	1.4646	0.3684	Sig	1.4686	0.3699	Sig
200	10	0.1627	0.2015	Not	0.0639	0.1981	Not	−0.0319	0.1982	Not	0.1947	0.2017	Not
300	10	−0.0182	0.1975	Not	−0.0199	0.1972	Not	−0.0402	0.1984	Not	−0.0783	0.1987	Not
400	10	−0.0318	0.1990	Not	−0.0070	0.1987	Not	−0.0108	0.1994	Not	−0.0496	0.1997	Not
500	10	−0.0758	0.1975	Not	−0.0634	0.1984	Not	−0.0421	0.2001	Not	−0.1813	0.1992	Not
100	15	0.9795	0.3766	Sig	0.6125	0.3629	Sig	0.4403	0.3560	Sig	2.0323	0.3699	Sig
200	15	0.1259	0.2215	Not	0.1127	0.2175	Not	0.0984	0.2154	Not	0.3369	0.2195	Sig
300	15	−0.0415	0.2173	Not	−0.0891	0.2179	Not	−0.0830	0.2208	Not	−0.2136	0.2201	Not
400	15	−0.0941	0.2185	Not	−0.0195	0.2189	Not	−0.0988	0.2211	Not	−0.2124	0.2207	Not
500	15	−0.0948	0.2189	Not	−0.0888	0.2202	Not	0.0077	0.2219	Not	−0.1758	0.2206	Not
100	20	1.1205	0.4076	Sig	0.3596	0.3919	Sig	0.0860	0.3871	Not	1.5661	0.4030	Sig
200	20	0.2353	0.2575	Not	0.0125	0.2525	Not	0.1005	0.2504	Not	0.3484	0.2554	Sig
300	20	−0.0510	0.2490	Not	−0.1482	0.2504	Not	−0.0548	0.2529	Not	−0.2540	0.2515	Sig
400	20	−0.2497	0.2539	Not	0.0424	0.2554	Not	0.0390	0.2538	Not	−0.1684	0.2523	Not
500	20	−0.0855	0.2531	Not	0.0381	0.2523	Not	−0.1037	0.2528	Not	−0.1511	0.2536	Not
100	25	0.1657	0.4510	Sig	0.0865	0.4458	Not	0.9166	0.4370	Sig	1.1688	0.4422	Sig
200	25	−0.0232	0.3144	Not	0.1499	0.3108	Not	0.0877	0.3089	Not	0.2144	0.3125	Not

Table 3. *Cont.*

Tick	Variability	A to B Diff	p (0.05)	Sig?	B to C Diff	p (0.05)	Sig?	C to D Diff	p (0.05)	Sig?	A to D Diff	p (0.05)	Sig?
300	25	0.0601	0.2992	Not	0.0379	0.2960	Not	−0.2026	0.3007	Not	−0.1046	0.3039	Not
400	25	−0.1161	0.2979	Not	−0.0560	0.2994	Not	0.0682	0.3011	Not	−0.1039	0.2996	Not
500	25	0.0443	0.2981	Not	−0.1278	0.2980	Not	−0.1582	0.3048	Not	−0.2418	0.3048	Not
100	30	0.4897	0.5757	Not	0.1157	0.5568	Not	0.6519	0.5490	Sig	1.2573	0.5681	Sig
200	30	0.3456	0.4389	Not	0.0173	0.4243	Not	0.1977	0.4234	Not	0.5606	0.4380	Sig
300	30	−0.1855	0.4030	Not	0.1829	0.3969	Not	−0.2921	0.4030	Not	−0.2946	0.4089	Not
400	30	−0.2528	0.3964	Not	−0.0964	0.3984	Not	0.1441	0.4008	Not	−0.2051	0.3988	Not
500	30	0.0848	0.3964	Not	−0.1236	0.3911	Not	−0.2040	0.4021	Not	−0.2429	0.4073	Not

Table 4. Results from Kolomgorov-Smirnov analysis on multiple present lineage values. Data is simplified into five time slices: 100, 200, 300, 400 and 500 ticks. K-S values that show significance above a *p*-value of 0.05 are highlighted blue and show "Sig" in the significance column. This means that there is a large difference in the mean values for the lines during that tick for those variability values. Of note are the 11 K-S values that show as significant between Strategy A (all share) and Strategy D (no share).

Tick	Variability	A to B Diff	p (0.05)	Sig?	B to C Diff	p (0.05)	Sig?	C to D Diff	p (0.05)	Sig?	A to D Diff	p (0.05)	Sig?
100	0	0.2004	0.4380	Not	0.3571	0.4245	Not	−0.2796	0.4234	Not	0.2779	0.4369	Not
200	0	0.2692	0.3461	Not	0.0617	0.3356	Not	0.3047	0.3290	Not	0.6357	0.3397	Sig
300	0	0.0921	0.3421	Not	−0.0769	0.3376	Not	0.0921	0.3374	Not	0.1073	0.3418	Not
400	0	0.0600	0.3396	Not	−0.1675	0.3375	Not	−0.0582	0.3414	Not	−0.1657	0.3435	Not
500	0	−0.4464	0.3346	Sig	0.0393	0.3384	Not	−0.2786	0.3426	Not	−0.6857	0.3388	Sig
100	5	0.3933	0.4242	Not	0.2278	0.4091	Not	0.0064	0.4060	Not	0.6275	0.4212	Sig
200	5	−0.0263	0.3664	Not	0.1897	0.3594	Not	0.2578	0.3515	Not	0.4212	0.3587	Sig
300	5	−0.0153	0.3687	Not	0.1085	0.3630	Not	−0.1126	0.3637	Not	−0.0193	0.3694	Not
400	5	0.0124	0.3617	Not	−0.0632	0.3588	Not	−0.2068	0.3645	Not	−0.2577	0.3673	Not
500	5	−0.1759	0.3586	Not	−0.3916	0.3658	Sig	0.0195	0.3732	Not	−0.5480	0.3662	Sig
100	10	0.2237	0.4490	Not	0.2567	0.4320	Not	0.0896	0.4148	Not	0.5700	0.4324	Sig
200	10	0.0212	0.4163	Not	0.2597	0.4034	Not	0.2083	0.3841	Not	0.4891	0.3976	Sig
300	10	0.0690	0.4135	Not	0.1053	0.4029	Not	−0.1896	0.3943	Not	−0.0153	0.4051	Not
400	10	−0.1458	0.4068	Not	−0.1439	0.4062	Not	0.0799	0.3974	Not	−0.2098	0.3979	Not
500	10	−0.0832	0.3995	Not	−0.4201	0.4036	Sig	−0.2177	0.4058	Sig	−0.7211	0.4017	Sig
100	15	−0.0535	0.4671	Not	0.2600	0.4522	Not	−0.2272	0.4570	Not	−0.0206	0.4718	Not

Table 4. *Cont.*

Tick	Variability	A to B Diff	p (0.05)	Sig?	B to C Diff	p (0.05)	Sig?	C to D Diff	p (0.05)	Sig?	A to D Diff	p (0.05)	Sig?
200	15	−0.1695	0.4438	Not	0.1588	0.4341	Not	0.0890	0.4338	Not	0.0783	0.4435	Not
300	15	0.2532	0.4448	Not	−0.1974	0.4332	Not	0.3611	0.4346	Not	0.4169	0.4462	Not
400	15	0.1459	0.4397	Not	−0.3176	0.4334	Not	0.2244	0.4405	Not	0.0528	0.4468	Not
500	15	−0.1826	0.4389	Not	0.1344	0.4302	Not	−0.5178	0.4441	Sig	−0.5659	0.4525	Sig
100	20	0.5377	0.5536	Not	0.3391	0.5278	Not	0.3074	0.5129	Not	1.1842	0.5394	Sig
200	20	0.6620	0.5029	Sig	−0.2177	0.4868	Not	0.0377	0.4922	Not	0.4821	0.5082	Not
300	20	−0.0424	0.4934	Not	−0.4549	0.5017	Not	0.1634	0.5097	Not	−0.3339	0.5016	Not
400	20	−0.1135	0.5049	Not	0.3205	0.4952	Not	−0.2821	0.4951	Not	−0.0752	0.5048	Not
500	20	−0.7894	0.4904	Sig	0.0886	0.5042	Not	−0.1860	0.5075	Not	−0.8868	0.4937	Sig
100	25	−0.9495	0.6491	Sig	0.5460	0.6308	Not	−0.0947	0.6066	Not	−0.4983	0.6257	Not
200	25	0.2414	0.6260	Not	−0.3957	0.5986	Not	0.5060	0.5851	Not	0.3517	0.6131	Not
300	25	0.1750	0.6084	Not	−0.3469	0.5827	Not	0.1769	0.5774	Not	0.0051	0.6033	Not
400	25	−0.2866	0.6258	Not	0.4883	0.5893	Not	−0.3176	0.5732	Not	−0.1159	0.6107	Not
500	25	0.6869	0.6352	Sig	−0.1668	0.5859	Not	−0.2757	0.5880	Not	0.2444	0.6372	Not
100	30	1.0337	0.8347	Sig	−0.8822	0.8270	Sig	0.5212	0.8451	Not	0.6727	0.8526	Not
200	30	0.1821	0.8129	Not	0.1558	0.7978	Not	−0.0795	0.7978	Not	0.2585	0.8129	Not
300	30	0.1819	0.8001	Not	−0.1592	0.7975	Not	0.0087	0.8072	Not	0.0313	0.8098	Not
400	30	−0.6953	0.8201	Not	0.7294	0.8172	Not	−0.4789	0.8101	Not	−0.4448	0.8131	Not
500	30	−0.6060	0.8066	Not	0.1113	0.8242	Not	0.1235	0.8177	Not	−0.3712	0.8000	Not

219

2.3. Discussion

Winterhalder and Leslie [32] have shown that long-term stochastic processes may affect how individuals react to environmental conditions and how they approach risk. In their model, demographic response to an unpredictable environment will, by nature, be nonlinear. For example, people cannot predict exactly how many children to have so that four children will grow into adulthood. The results of our above analysis echo those of Winterhalder and Leslie and show that individuals may indeed seek risk when environments are highly unstable in order to have the chance of surviving, and may be risk-averse when environments are stable. The high levels of variance observed in the model presented in this paper are at least partially reflective of the unpredictable, highly unstable environments in which this simulation occurs. While Hegmon [6] found that restricted sharing will be the most beneficial strategy for overall populations (restricted sharing should decrease variance), Winterhalder and Leslie's findings may highlight why highly variance will be beneficial in unpredictable environments. People may need to try multiple strategies to survive.

Powers and Lehman [2] found that sharing increases the carrying capacity of a system. Such a result is potentially visible in our results as well. When environmental pressures become great, and households group together, the environmental pressures can become mitigated by the social sharing strategy. However, despite sharing strategies lessening environmental pressures, households are never outside of those environmental pressures, and the use of the landscape creates environmental pressures as well due to patch degradation.

Pastoralists have long been blamed for environmental degradation from overgrazing [33]. The "tragedy of the commons" theory states that unmonitored common-pool resources, as is the case in Mongolia with individual ownership of herds, but not land, leads to irresponsible usage of resources. However, critics of this theory point to various formal and informal social adaptations that oversee and regulate resource use [34]. The same cooperation and sharing networks modeled here may parallel the social networks ensuring sustainable resource utilization through traditional ecological knowledge.

The problem of common-pool resources is evident in the model. When agents land on patches they extract the resources from those patches, and must wait multiple timesteps until those patches regenerate. It is possible that all winter patches in one area could become used during one timestep, causing future households to have no opportunities for productive patches. If agents land on dead patches they are charged energy. Once agents have fewer than 10 energy stored, those agents with a sharing strategy must rely on other agents in their network for survival. In this way we can see how agents react to a simulated tragedy of the commons. Once resources are over-exploited in an area, households must call upon their networks for help. As we see in this simulation, agents are doubly burdened by both simulated *dzuds* and by simulated resource over-use. Those agents that are able to rely on their greater social network fare better overall than those agents with no social network when both climatic and overuse pressures affect the environment.

One final issue addressed by this model is the poor resolution of the archaeological record. While research is ongoing in household studies in Mongolia (e.g., [12]) most studies in Mongolia have focused on monumental archaeology. This is coupled with poor resolution of household archaeology (centimeters of deposition equating to centuries of occupation). Consequently, our understanding of the past can be blurred. Simulations, therefore, help us to address these gaps in our knowledge.

Notably missing from our study is a goodness-of-fit exercise between the model and the real settlement patterns [1,35]. This is due in part to there not being many complete archaeological datasets in the region to do goodness-of-fit tests against yet. Consequently, we must make do and use models as a way to inform our understanding of the limited archaeological information available at this time.

This model, while not meant as a reproduction of reality, presents a plausible scenario based on developed theory and hopes to address key questions of how semi-nomadic Mongolians address local weather events, such as drought and heavy winters. While this model is highly simplified, it presents a plausible suite of directions that people in this highly unpredictable environment could face. Therefore the outcome of our study can be used to make some conclusions of a much more complicated system.

3. Conclusions

The mobility of Mongolia's pastoralists presents a unique case rather different than the settled, Ancestral Pueblos investigated by Hegmon [6,7]. Household units, which are moving frequently anyway, can fission and fusion without large disruptions to the social, economic or political order. Rather than reaching a breaking point, temporary solutions can mitigate risk and catastrophe, followed by a return to the normal order.

So which of the above cooperation strategies works best for Mongolia? This is a tricky question to answer with a single straightforward answer. All of Mongolia is hit by *dzuds*, but they do not impact different regions of the country equally; one area will be more susceptible to them than others for various natural and socio-cultural reasons. For instance, the weather in southern Mongolia's Gobi Desert is quite different than that of northern Mongolia's Taiga-Mountain-Steppe ecotones. Therefore, which strategy is most beneficial may vary geographically as well as temporally. Additionally, the availability of other risk-mitigating adaptations is different by region. There may be many more types of wild resources available in the northern ecotones than in the more homogenous steppe or desert zones in central and southern Mongolia. In regions where it is more difficult to fall back on wild resources, this may place much more importance on social or kin networks to mitigate risk. This might be seen archaeologically in Mongolia by looking at facets of the ritual landscape as a reflection of the strength of social and kin networks [12].

Ger Grouper is a very simplified model. However, this "wrong" model (*sensu* [29]) is useful in that it helps us to understand how individuals might react to catastrophic events. We began with a highly simplified model to examine how variables interact with one another, so that in future we can truly examine the effects of variables in a realistic setting. Future development of this model will include bringing real world variables into the model. The rates of environmental catastrophes (e.g., *dzuds* and droughts) can be reconstructed using historical weather data which can then be added to create a more realistic "patchy" element to the model. In addition, realistic GIS landscapes can be created based on real locations within Mongolia and the surrounding regions. As more detailed archaeological and paleoenvironmental data become available, the parameters of the model will improve. The results from multiple regions can then be compared, illuminating any differences in socially adaptive risk management responses due to environmental variation. The Ger Grouper model was designed to work at a landscape-scale compatible with the annual seasonal rounds of mobile pastoralists in Mongolia. Agent-based-modeling, when implemented at this scale, will allow for explicit connections between computer-aided models and archaeological project design.

Acknowledgments: We would like to thank Luke Premo, Camilla Kelsoe and Bryan Hanks as well as the participants and organizers of the Complex Systems in Prehistoric Research symposium at the European Conference on Complex Systems for feedback on earlier versions of this model and paper. Clark recognizes funding from NSF Doctoral Dissertation Improvement Grant #BCS-1236939. Crabtree recognizes funding from NSF Graduate Research Fellowship #DGE-080667 and the Chateaubriand Fellowship Program. Crabtree built the model Ger Grouper so any faults in the model design belong exclusively to her faulty logic.

Author Contributions: Both authors contributed equally to this work. Julia K. Clark directs fieldwork in Mongolia and provided the background and guidance for the logic of the model. Stefani A. Crabtree created the model and ran the statistics.

Conflicts of Interest: The authors declare no conflict of interest.

References

1. Crabtree, S.A. Inferring ancestral Pueblo social networks from simulation in the central Mesa Verde. *J. Archaeol. Method Theory* **2015**, *22*, 144–181.
2. Powers, S.T.; Lehmann, L. The co-evolution of social institutions, demography, and large-scale human cooperation. *Ecol. Lett.* **2013**, *16*, 1356–1364.
3. Gregory, C.A. *Gifts and Commodities*; Academic Press Inc.: London, UK, 1982.

4. Kohler, T.A.; van West, C.R. The calculus of self interest in the development of cooperation: sociopolitical development and risk among the northern Anasazi. In *Evolving Complexity and Environment: Risk in the Prehistoric Southwest*; Tainter, J.A., Tainter, B.B., Eds.; Santa Fe Institute Studies in the Sciences of Complexity, Santa Fe Institute and Oxford University Press: New York, NY, USA, 1996; pp. 171–198.

5. Sahlins, M.D. *Stone Age Economics*; Aldine de Gruyter: New York, NY, USA, 1972.

6. Hegmon, M. Risk reduction and variation in agricultural economies: A computer simulation of Hopi agriculture. In *Research in Economic Anthropology*; Issac, B.L., Ed.; JAI Press: Greenwich, CT, USA, 1989; pp. 89–121.

7. Hegmon, M. The risks of sharing and sharing as risk reduction: Interhousehold food sharing in egalitarian societies. In *Between Bands and States*; Gregg, S.A., Ed.; Center for Archaeological Investigations, Occasional Paper No. 9; Southern Illinois University: Carbondale, IL, USA, 1991; pp. 309–329.

8. Fernandez-Gimenez, M.E. The role of nomadic pastoralists' ecological knowledge in rangeland management. *Ecol. Adapt.* **2000**, *10*, 1218–1326.

9. Müller, B.; Lindstädter, A.; Frank, K.; Bollig, M.; Wissel, C. Learning from local knowledge: Modeling the pastoral-nomadic range management of the Himba, Namibia. *Ecol. Appl.* **2007**, *17*, 1857–1875.

10. Cribb, R. *Nomads in Archaeology*; Cambridge University Press: Cambridge, UK, 1991.

11. Dyson-Hudson, R.; Dyson-Hudson, N. Nomadic pastoralism. *Annu. Rev. Anthropol.* **1980**, *9*, 15–61.

12. Clark, J.K. Modeling Late Prehistoric and Early Historic Pastoral Adaptation in Northern Mongolia's Darkhad Depression. Ph.D. Dissertation, University of Pittsburgh, Pittsburgh, PA, USA, 2014. unpublished.

13. Costopoulos, A.; Lake, M.; Gupta, N. Introduction. In *Simulating Change–Archaeology into the Twenty-First Century*; Costopoulos, A., Lake, M., Eds.; University of Utah Press: Salt Lake City, UT, USA, 2010; pp. 1–12.

14. Lake, M.W. Trends in archaeological simulation. *J. Archaeol. Method Theory* **2014**, *21*, 258–287.

15. Cioffi-Revilla, C.; Rogers, J.D.; Latek, M. The MASON households world model of pastoral nomad societies. In *Simulating Interaction Agents and Social Phenomena: The Second World Congress*; Takadama, K., Cioffi-Revilla, C., Deffuant, G., Eds.; Springer: Tokyo, Japan, 2010; pp. 193–204.

16. Cioffi-Revilla, C.; Rogers, J.D.; Wilcox, S.P.; Alterman, J. Computing the steppes: Data analysis for agent-based modeling of polities in Inner Asia. In Proceedings of the 104th Annual Meeting of the American Political Science Association, Boston, MA, USA, 28–31 August 2008.

17. Houle, J.-L. "Socially integrative facilities" and the emergence of societal complexity on the Mongolian steppe. In *Monuments, Metals and Mobility: Trajectories of Complexity in the Late Prehistory of the Eurasian Steppe*; Hanks, B.K., Linduff, K.M., Eds.; Cambridge University Press: Cambridge, UK, 2009; pp. 358–377.

18. Wright, J. The adoption of Pastoralism in Northeast Asia: Monumental transformation in the Egiin Gol Valley, Mongolia. Ph.D. Dissertation, Harvard University, Cambridge, MA, USA, 2006. unpublished.

19. Crabtree, S.A.; Clark, J.K.; Harris, K.; Bullion, E.; Wason, C.; Neyroud, M.; Cooper, C.M.; Bayarsaikhan, J. Jade mining in the Darkhad: Changing pastoral economies in northern Mongolia. *Archaeol. Ethnol. Anthropol. Eurasia* **2015**, *43*. in press.

20. Endicott, E. *A History of Land Use in Mongolia: The Thirteenth Century to the Present*; Palgrave Macmillan: New York, NY, USA, 2012.

21. Fernandez-Gimenez, M.E. Land use and land tenure in Mongolia: A brief history and current issues. In *US Department of Agriculture Forest Service Proceedings RMRS*; USDA Forest Service: Fort Collins, CO, USA, 2006; pp. 30–36.

22. Houle, J.-L. Emergent Complexity on the Mongolian Steppe: Mobility, Territoriality, and the Development of Nomadic Polities. Ph.D. Dissertation, University of Pittsburgh, Pittsburgh, PA, USA, 2010. unpublished.

23. Vainshtein, S. *Nomads of South Siberia: The Pastoral Economies of Tuva*; Cambridge University Press: Cambridge, UK, 1980.

24. Batima, P.; Bold, B.; Sainkhuu, T.; Bavuu, M. Adapting to drought, zud and climate change in Mongolia's rangelands. In *Climate Change and Adaptation*; Learly, N., Adejuwon, J., Barros, V., Burton, I., Kulkarni, J., Lasco, R., Eds.; Earthscan: London, UK, 2008; pp. 197–210.

25. Fernandez-Gimenez, M.E. *Restoring Community Connections to the Land: Building Resilience through Community-Based Rangeland Management in China and Mongolia*; CABI: Wallingford, UK, 2011.

26. Janes, C.R. Failed development and vulnerability to climate change in central Asia: Implications for food security and health. *Asia Pac. J. Public Health* **2010**, *22*, 236S–245S.

27. Fitzhugh, B.; Philips, S.C.; Gjesfjeld, E. Modeling variability in hunter-gatherer information networks: An archaeological case study from the Kuril Islands. In *The Role of Information in Hunter-Gatherer Band Adaptations*; Whallon, R., Lovis, W., Hitchcock, R., Eds.; Cotsen Institute of Archaeology, University of California: Los Angeles, CA, USA, 2011; pp. 85–115.

28. Clarke, D.L. Models and paradigms in contemporary archaeology. In *Models in Archaeology*; Methuen & Co., Ltd.: London, UK, 1972; pp. 1–60.

29. Box, G.E.P.; Draper, N.R. *Empirical Model-Building and Response Surfaces*; John Wiley & Sons: Hoboken, NJ, USA, 1987.

30. Kohler, T.A.; Varien, M.D. *Emergence and Collapse of Early Villages*; University of California Press: Berkeley, CA, USA, 2012.

31. Chamberlain, A.T. *Demography in Archaeology*; Cambridge University Press: Cambridge, UK, 2006.

32. Winterhalder, B.; Leslie, P. Risk-sensitive fertility: The variance compensation hypothesis. *Evol. Hum. Behav.* **2002**, *23*, 59–82.

33. Hardin, G. The tragedy of the commons. *Science* **1968**, *162*, 1243–1248.

34. Ostrom, E.; Burger, J.; Field, C.B.; Norgaard, R.B.; Plicansky, D. Revisiting the commons: Local lessons, global challenges. *Science* **1999**, *284*, 278–282.

35. Kohler, T.A.; Bocinsky, R.K.; Crabtree, S.; Ford, B. Exercising the model: Assessing changes in settlement location and efficiency, emergence and collapse of early villages: Models of central Mesa Verde archaeology. In *Emergence and Collapse of Early Villages*; Kohler, T.A., Varien, M.D., Eds.; University of California Press: Berkeley Oakland, CA, USA, 2012; pp. 153–164.

Simulating Littoral Trade: Modeling the Trade of Wine in the Bronze to Iron Age Transition in Southern France

Stefani A. Crabtree

Abstract: The Languedoc-Roussillon region of southern France is well known today for producing full-bodied red wines. Yet wine grapes are not native to France. Additionally, wine was not developed indigenously first. In the 7th century B.C. Etruscan merchants bringing wine landed on the shores of the Languedoc and established trade relationships with the native Gauls, later creating local viticulture, and laying the foundation for a strong cultural identity of French wine production and setting in motion a multi-billion dollar industry. This paper examines the first five centuries of wine consumption (from ~600 B.C. to ~100 B.C.), analyzing how preference of one type of luxury good over another created distinctive artifact patterns in the archaeological record. I create a simple agent-based model to examine how the trade of comestibles for wine led to a growing economy and a distinctive patterning of artifacts in the archaeological record of southern France. This model helps shed light on the processes that led to centuries of peaceable relationships with colonial merchants, and interacts with scholarly debate on why Etruscan amphorae are replaced by Greek amphorae so swiftly and completely.

Reprinted from *Land*. Cite as: Crabtree, S.A. Simulating Littoral Trade: Modeling the Trade of Wine in the Bronze to Iron Age Transition in Southern France. *Land* **2016**, *5*, 5.

Niketas then asked for some wine and poured a cup for Baudolino. "See if you like this. It's a resinous wine that many Latins find disgusting; they say it tastes of mold." Assured by Baudolino that this Greek nectar was his favorite drink, Niketas settled down to hear his story. — *from* Baudolino [1]

1. Introduction

Understanding the choices that people made in the past is difficult, if not impossible, without written sources directly telling us why people chose specific courses of action. Yet it is these choices that led to the archaeological record; today we can see the aggregate of these decisions. The following model presents a simple case of examining prehistoric economies. Through using an agent-based model on a heterogeneous population it is suggested that the economy of this area was driven by the choices of Gauls as consumers, and not by the availability of goods; this work articulates with longstanding debates in the prehistory of France.

This research specifically asks the question: what caused the complete switch in wine amphorae from Etruscan to Greek styles in the Languedoc when clearly both groups were present on the landscape? This model aims to examine the abrupt transition from Etruscan amphorae to Greek amphorae as discovered by Py [2] and reported in Figure 1 by modeling strictly local processes. A pattern oriented modeling approach [3] was used to examine the overall process and validate the model with the archaeological record. Validation in this model is via a complete shift in artifact types from Etruscan to Greek amphorae—output from the simulation is directly compared against output from the archaeological record. This research is one of the first forays into formal modeling of the archaeological record in France, thus this article represents both the utility of agent-based modeling for examining the prehistory in France, and also acts as a first step for more complex models on French prehistory.

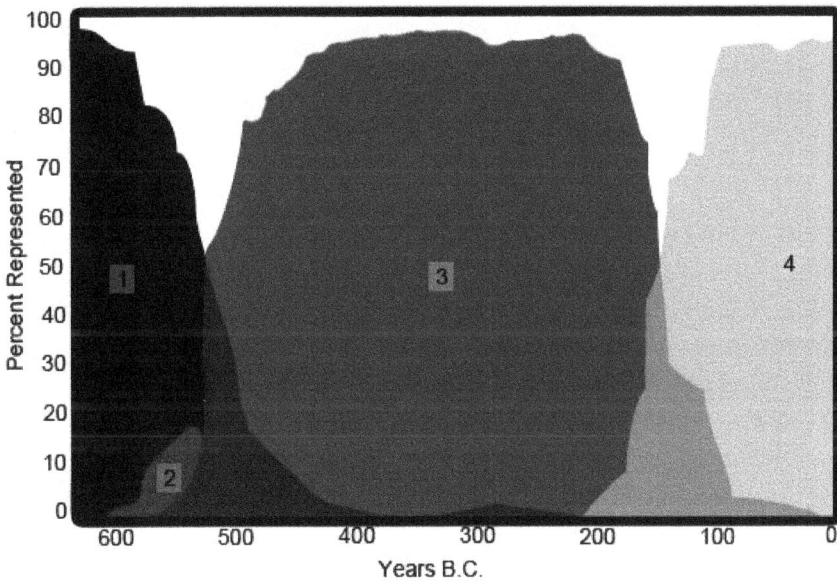

Figure 1. Redrawn from Py [2], curves of artifact percentages through time. (1) represents Etruscan amphorae, which make up almost 100% of the assemblage at that time. (2) represents archaic Greek amphorae, which have a small percentage of the assemblage. (3) represents Greek amphorae. Note that while Greek amphorae are the dominant form of wine vessels after 500 B.C. that amphorae of Etruscan type (1) are still present into the 3rd millennium B.C. (4) represents Roman amphorae, which are not examined in this paper.

The article is organized as follows. First, a brief background situates the research and the research question. Next, methods and then the model is presented; please note that much of the model detail is described in the supplementary ODD protocol to allow for a streamlined description. Results follow, focusing on those outputs that directly facilitate comparison with observed archaeological phenomena. Following the results of the model, a lengthy discussion of the cultural history of southern France is presented, showing exactly how this model articulates with research in this area. Much of the data on this region is published in French; thus this article provides a summary of the culture history in English, advancing understanding for this area for the Anglophone audience. Finally, the archaeological data are discussed in conjunction with the results presented here, and suggestions to future directions are presented.

1.1. Background

The Languedoc-Roussillon region of southern France (Figure 2) is well known today for producing full-bodied red wines. Yet wine grapes were introduced in antiquity. In the 7th century B.C. wine-bearing Etruscan merchants landed on the shores of the Languedoc and established trade relationships with the native Gaulish elite. In fact, most wine consumed in southern France was not even grown by Gauls, instead being imported to Gaulish settlements [4,5]. In complement to this, some argue that certain colonial settlements were so large they outstripped their local carrying capacities, and thus had to import grain and other comestibles [5]. Complex economic partnerships linking Gauls to Etruscan and later to Greek merchants were essential, yet these trade relationships had far-reaching effects for the household economies of both indigenous and colonist populations. "Greek [colonist] towns in general and Greek houses in particular, constitute evidence of a new type of materialism, individualism and consumer display, where patron-client relations were negotiated in semi-public homes, in which creators of wealth were linked to local and international business opportunities" [6]. In this paper, I examine how the trade relationships between colonist merchants and the indigenous Gaulish elite facilitated the development of complex economies and created distinctive artifact patternings in the archaeological record.

This model examines three processes: (1) the arrival of wine-bearing merchants in Gaul; (2) the establishment of trade relationships between these merchants and native Gauls; and, (3) the replacement of Etruscan wine amphorae by Greek wine amphorae. To understand the establishment of trade between Gauls and colonial merchants we need to understand why Gauls would engage in trading grain for wine. Then, to understand the replacement of one amphora type by another, we need to examine the choices made by the Gaulish elite at home. After all, Gauls were the agents of demand in the supply chain. As in the above quote, the preference for one type of wine over another would influence how people would choose to consume wine. Agent-based modeling is a perfect method for examining how local decisions could affect overarching patterns of artifact distributions.

The agent-based model presented here looks at the distribution of artifact types over time across a simplified landscape. By reducing the model to a few key parameters, I am able to directly examine how a preference for one type of wine over the other might affect archaeological assemblages. This model articulates with current debates over the nature of trade within this region. As I discuss below, ethnic identities in the past are difficult to identify, but patterns of artifacts across space and through time can be identified. This simple agent-based model acts as a first step to understanding economies in prehistory and sets up studies that can further examine land use in the past.

Existing models for the interaction between Gaulish inhabitants and colonial traders along the littoral (the region abutting the Mediterranean) of southern France are descriptive. According to Py [7] the paradigms underlying research by proto-historians working in these contexts can be summarized as follows: indigenous Gauls living along the littoral zone were forced to abandon some of their traditional practices, such as semi-nomadic pastoralism, to generate the agricultural surplus required to develop their economies and engage in trade with outsiders [7]. Yet these descriptive models have not been formally tested; thus, the research here formally examines how early colonialism can create distinctive economic partnerships and artifact patterns.

The terms "colonist," "colonizer," and "colonialism" come with academic baggage. To avoid confusion, and differentiate the colonization in southern Gaul from Colonialism in the 1600s–1900s, I will use the term "settled nonlocal merchants," "settled merchants," or simply "merchants" henceforth to refer to the Etruscan and Greek merchants. Settled, because in general the colonizers who arrived in southern Gaul settled in colonies, or in already established Gaulish settlements, as is argued for the Etruscans at Lattara [5]. Nonlocal, because the first wave of Etruscans and Greeks were born in other areas. (Through time this becomes debatable, as later generations of merchants may have been born in Gaul [5].) And finally I use the term merchants, because the Etruscans and Greeks who came to southern Gaul are characterized by engaging in trade with the locals.

How the development of agricultural surplus could lead to trade relationships with merchants is directly examined in the model presented here. By creating multiple parameters related to flows of exchange and the ability to extract resources (discussed in more detail below) and sweeping across values for these parameters I can directly examine existing conceptual models for southern Gaul. To state it simply, this model directly examines how trade affects the survivability of agents on the landscape and allows for the examination of the percent of different artifacts on the landscape. I examine this model in two steps: (1) a simple model allowing for the exchange of wine for grain; and (2) a model that allows for two types of merchant populations, Etruscans and Greeks, to trade with the Gauls for grain.

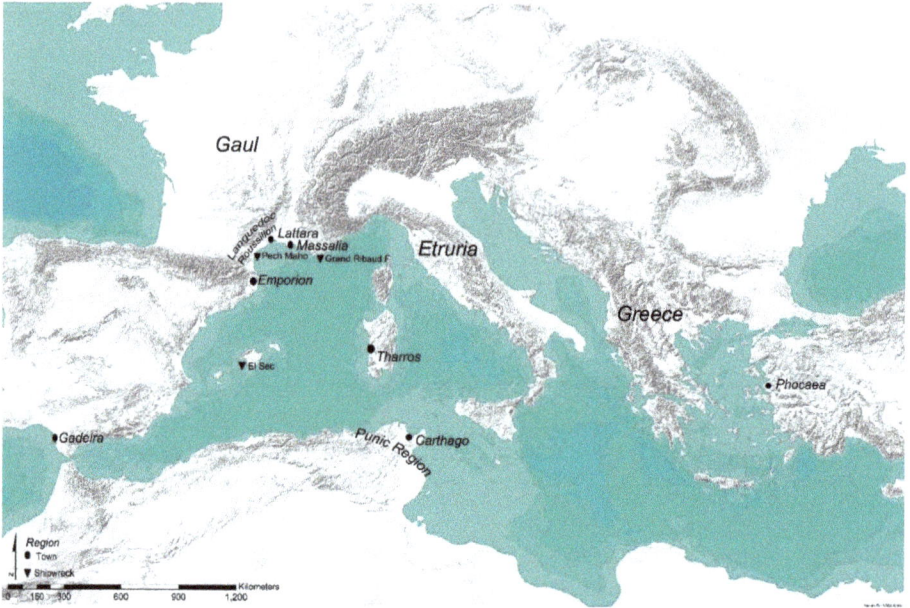

Figure 2. Area of interest for this study. This study specifically examines the development of viticulture and trade in the Languedoc Roussillon region, but map includes surrounding areas of interest to this study. Here I show those cities that are specifically mentioned in this manuscript, as well as the three shipwrecks mentioned that show integration of ethnic identities.

1.2. Methods

The agent-based model developed in this paper was created in NetLogo [8], though could have easily been created in any other modeling platform; figures were created in R [9]. The modeling framework consists of a simple resource extraction model coupled with a trade model (see below). Each timestep of the model represents one year and the model is run for 500 timesteps. Two types of output are generated: populations of agents (Gauls and merchants) and populations of artifacts (Etruscan and Greek wine amphorae).

This model is meant to reproduce patterns for validation. While no reliable population estimates exist for this area, patterns of agent survival are helpful in calibrating whether or not exchange of grain for wine would have enabled merchant survival in prehistory. Patterns of artifacts, however, are more reliable in this study area. Output of the quantity of Etruscan and Greek amphorae are compared against real archaeological patterns of artifacts (Figure 1) to determine if local processes could have led to the archaeological record.

1.3. The Model

Here I ask two questions: (1) could visiting merchants have survived in the littoral without farming grain? and, (2) can a transition in the number and type of amphorae be generated through modifying a simple set of parameters? I examine these questions through the simple agent-based model detailed below. Following I describe the base of the model to provide a background for the questions answered in this paper, then I detail each of the models.

1.3.1. The Landscape

The landscape is 80-cells by 80-cells wide, creating a total of 6400 cells for the simulation window. In this model the landscape is created in three portions: the sea to the south (2400 cells), the littoral region abutting the sea (320 cells; light green in Figure 3), and the rest of the land (3680 cells; medium green in Figure 3). Grain (energy) can only be grown on green patches.

At simulation instantiation a random 33% subset of the farming landscape is unproductive. Regrowth "clocks" are set on each cell randomly between 0 and 60 years and the patches regenerate during this time. While the model presented here does not use realistic paleoproductivity estimates [10], the random generation of unproductive cells creates a patchy environment that farming Gauls likely faced when they began cultivating wheat. As stated above, the conceptual model used by proto-historians [7] suggests that Gauls abandoned semi-nomadic pastoralism to create surplus for trade. It is likely that not all Gauls would have abandoned this way of life immediately, suggesting that some parts of the landscape would still be in use for pastoralism and foraging. Moreover, lanscape productivity may have been effected by generations of landscape use before settled farming took hold. Thus it is reasonable to expect that not all of the land was available for farming right away. Further, agent actions degrade the landscape (see below) which makes Gaulish agents need to learn to be able to farm, reproduce, and trade.

The decision to abstract the landscape to a rectangular space was made to enable an examination of the simple process of exchange without having to model multiple historical details. Archaeology and historical study has been ongoing in this region for decades. An agent-based model would not be able to encapsulate all of the specifics of the historical record of this region. Moreover, as this is the first agent-based model to be made in this region, it was determined that it would be best to create a highly simplified model with the goal of adding complexities later.

1.3.2. The Agents

There are two main types of agents in this model: Gaulish agents and Merchant agents. In this model, agents correspond to the economic production unit of a household [10]. The composition of households may have been slightly different for Gauls than for the Etruscan and Greek visiting merchants, and may have differed depending on social status. For example, on arrival in southern France many of the visiting merchants were likely single men who later may have married locally to create a family or returned home and brought their families from their home countries to the west [5]. For simplicity, in this model it is assumed that agents are independent economic production units. As such, each agent produces goods specific for its type: Gauls produce grain, and Etruscans and Greeks produce wine.

In this model households can be of varying size, and this is tied to production (see below). It is assumed here that the basic household may begin with only one agent—for example, when merchants land, a household consists of one merchant. As households increase their grain storage, they can support more individuals. Then, as households fission and split their grain storage, they can support fewer individuals within their own household from their storage. So household size fluctuates as storage fluctuates, and as daughter households bud off of the parent household. This is explored below in the discussion of grain consumption rates.

To examine how the trade of grain for wine helped the survival of visiting merchants, we need to understand consumption rates of grain in the Gaulish world. Gras [11] identified average consumption rates of roughly six hectoliters (*hl*) of grain per year for adults. I use this as a base value for consumption by the agents, with four *hl* of grain as the base for juveniles. In the simulation, if a Gaulish agent has below 10 *hl* of grain, the household can only support one individual. This scales up as agents store more grain (Table 1). Average annual yields of fields have been suggested to be up to eight *hl* per hectare [5], so I use this upper bound to calibrate consumption rates and field productivity in the simulation. To calculate the size of family farms I use estimates by White [12] who reports that small farms in the Roman republic, which used similar farming techniques, were between 18 and 108 *iugera* or 4.5 to 27 hectares during the 5th century B.C. (contemporaneous to this study). The amount of grain harvested also scales with the size of the family; a small family can harvest from 5 hectares, while a large family can harvest up to 15 hectares. This is explained below in Table 1.

Wine cultivation, however, does not scale with a larger family. In this simulation individuals can harvest 10 amphorae of wine per cell and do not create more viticulture cells with increased family size. Rather, an agent owns one cell of wine production. While amphorae in antiquity varied in size, in this simulation I assume that the amphorae are the standard Attic size of roughly 50 liters of wine per transport amphora [13]. When I discuss trade rates below, the optimum trade is 40 hl of grain for 5 hl (10 amphorae) of wine.

Table 1. How storage level affects the number of individuals in a household and their consumption rates. This enables agents to increase their family size, and thus the productivity of their land, as well as increasing the ability to trade. However, once an agent trades, its storage level will be cut in half (as half is donated to the daughter household) decreasing the household size in the process. Merchants have a higher storage level because they cannot grow their own food, and thus need to plan more to be able to raise daughter households.

Storage Level Merchants	Storage Level Gauls	Size of Plots Gauls	Size of Harvest Gauls	Consumption Rates Gauls and Merchants	Corresponding Number of Individuals per Household
< 45 hl	⩽ 10 hl	5 ha	40 hl	6 hl	1
⩾ 46, <50 hl	>10, ⩽ 30 hl	5 × 1.5 (7.5 ha)	60 hl	12 hl	2
⩾ 50, <60 hl	⩾ 31, ⩽ 40 hl	5 × 2 (10 ha)	80 hl	16 hl	3
⩾ 61, <70 hl	⩾ 41, ⩽ 50 hl	5 × 2.5 (12.5 ha)	100 hl	20 hl	4
⩾ 71 hl	⩾ 51 hl	5 × 3 (15 ha)	120 hl	24 hl	5

Consumption rates are tied to various parameters, including the basic consumption of grain (6 hectoliters per year, per adult) plus the quantity of grain required for planting and harvesting (see below). While farming yield, amount consumed, and exchange rate are all parameterized, in this run of the simulation these parameters were fixed for simplicity. Fixed parameters are reported in Table 2. Of note, planting calories and harvest calories are both set to 4 hectoliters. Gauls would have needed to store seed to plant their fields each year, and planting would be energetically costly. Thus the parameter "planting calories" encapsulates both the stored grain, and the cost to plant a large field. Harvests, on the other hand, are known to come in at once and need to be harvested rapidly before the grain falls off the stalk. Thus Gauls likely relied on neighbors (and potentially slaves, see Discussion) to help with harvest, and may have fed them to help with this cost. Further, some grain that grew may be lost in harvest, due to improper techniques, harvesting too late, or storing improperly. Thus "harvest" encapsulates the costs associated with harvest and storage. Swept parameters are reported in Table 3.

At the beginning of the simulation—here set to year 0, but corresponding to roughly year 700 B.C.—Gaulish agents are distributed randomly on the land portion of the landscape. Each Gaulish agent is created with a storage of grain set to 20 hectoliters. The initial number of Gaulish households is set to 150. Colonist agents are seeded on the landscape during their birth years (Table 2) with 60 hl of grain in storage, and the initial number of colonists is set to 100.

In this model agents have yearly basic metabolic needs which are met by consuming grain (Table 1). If agents get to zero energy, they die. There is an additional parameter, "life expectancy," that ensures agents—the natal household—do not live too long. If an agent reaches above the number of timesteps set by "life-expectancy" they have a 50% chance of dying every timestep. (Note that agents can die before that due to lack of resources.) In this sweep life expectancy was set to 80 timesteps since birth; while this is likely a high estimate for antiquity, this allowed many agents to die of "natural" causes (e.g., having too few grain) before being killed off by the simulation. Reproduction in this model is via fission (see Supplementary materials). Daughter households form near their parent household, and storage is divided evenly between daughter and parent households.

Consuming wine decreases harvest costs (and is consumed at a rate of one unit per year). Elsewhere, beer parties are used as a form of payment to help in collective labor [14]. Alcohol mobilizes workers at work-parties, and was likely used in Gaul for harvest, since crops would mature and need to be harvested quickly. Historians have suggested that beer parties indeed aided in Gaulish grain harvest [15]. For this simulation I apply the concept of beer parties and assume that consuming wine would decrease costs to the harvester. Therefore, having wine is beneficial for farming agents, as it makes harvesting less costly for them.

Table 2. Fixed parameters used in this simulation. Many of these were tested in earlier sweeps, which are not reported here.

Parameter Name	Value	Explanation
Grain Storage (Gauls at birth)	20 *hl*	Amount of grain per Gaul when seeded on landscape
Grain Storage (Merchants at arrival)	60 *hl*	Amount of grain per merchant when seeded on landscape
Wine Storage (Merchants at arrival)	20 amphora	Amount of wine per merchant when seeded on landscape
Number Gauls Seeded	150	Number of Gauls at start of simulation
Number merchants seeded (both types)	100	Number of merchants upon arrival
Life expectancy	80	Year after which agent has 50% probability of mortality per timestep
Etruscan arrival	Year 34	Year Etruscans arrive
Greek arrival	Year 100	Year Greeks arrive
Grain harvest amount	20	Amount of grain (in hectoliters) harvested per farmed cell
Wine harvest amount	10	Amount of wine (in amphora) harvested per cultivated cell
Planting calories	4 *hl*	How much it costs to plant each year
Harvest calories	4 *hl*	How much it costs to harvest each year
Wine decay rate	1 amphora/yr	How much wine rots per year
Wine drinking rate	1 amphora/yr	How much wine an agent can consume per year, per type
Reproduction	3%	Probability of reproduction per timestep
Probability of selling wine (merchants)	5%	Probability a merchant will be able to sell wine each time step
Probability of buying wine (Gauls)	1%	Probability a colonist will be able to buy wine each time step

Table 3. Parameters swept across in two models. Grain Trade Rate was swept across in first model. Preference was swept across in second model.

Parameter Name	Values Swept Across	Explanation
Grain Trade Rate (examined in part 2.1)	20:10; 30:10; 40:10; 50:10, 60:10	Amount of hectoliters of grain traded per 10 amphorae of wine
Preference (examined in part 2.2)	0, 10, 20, 30, 40, 50, 60, 70, 80, 90, 100	Weighted value for when two types of wine are available. Explained further in Table 3

Agents trade wine for grain, and trade is costly. Both the wine and grain traded would need to be transported between exchanging agents, so agents are charged calories for the trade of these goods across the simulation in a manner similar to Crabtree [16]. Further, wine was likely an elite drink, and so the trade of grain for wine could only be accomplished by the elite. Thus, agents must account for costs when trading. In this model agents calculate the distance between themselves and their trading partner. The agent that is buying is charged 0.25 *hl* per cell traveled. This, then, ties to the agent's move algorithm.

In this model cells degrade after 5 years of farming use; cells become productive again after up to 5 years lying fallow, set randomly. If a Gaulish agent's farm cell has become unproductive, the agent must move to another cell. When Gauls move, they will look at their most recent trading costs and assess how costly they were. If the trading costs were greater than 1/4 of the gain in storage, the agent will move to a productive cell closer to the merchant settlements. If the costs were less than 1/4 of the agent's grain storage, the agent will simply look for another productive cell in a radius of 10 cells to begin a new farm. The Gaulish agent is charged 1 *hl* to move to a new farm.

Trade in this model is simple, but occurs both from the Gaulish side and from the Merchant side (see Supplementary Material: ODD Protocol, Scheduling). Gaulish agents trade before merchant agents do (demand for goods comes first). When Gaulish agents have stored twice the trade rate in the simulation (in Section 2.2 set to 40 *hl*) they may choose to trade for wine. This threshold is so that if an agent reproduces (dividing energy equally) it will have 40 energy to divide between itself and offspring after trading energy for wine; this threshold minimizes agent-death. Merchant agents require grain to survive and reproduce; thus a merchant will always trade for grain when approached by a Gaulish agent asking for wine. The agent that instantiates trade pays the cost for trading as described above [16]. After Gaulish agents trade and complete their scheduling, merchant agents trade.

Second, merchant agents trade wine for grain. When colonist agents have greater than 10 wine-units, they ask a Gaulish agent to trade following the above logic. The merchant agent asks a Gaulish agent to trade; the Gaulish agent then has a

50% probability of accepting this trade. If the trade is accepted, the merchant agent pays the cost of trading (0.25 energy multiplied by the number of cells separating it from the Gaulish agent).

Following I now describe the differences in each model, building from the simplest base model that examines the trade of grain for wine with one type of merchant-agent, to the more complicated model that examines the trade of two types of wine for grain. I additionally discuss the results from running sweeps of each model-type.

2. Results

2.1. Base Model

In this section I use the base model to establish the trade rate of grain to wine to be used in the subsequent model. While future applications of this model may enable agents to barter for an appropriate trade rate [17], this model sought to reduce variables, so a global exchange rate was determined in this first step. This model examines the verbal model as explained by Py [2], that Etruscan merchants arrived in Gaul and influenced an intensification of agriculture in the area, with Gaulish people creating surplus to engage in trade for wine with the Etruscans. Here I specifically examine population of Etruscan agents, since their survival depends on their ability to trade with Gaulish agents. In this model I calibrate the amount of grain traded, which then feeds into the following models. For this model I specifically ask:

Could Gauls have generated enough surplus to feed visiting merchants, while still enabling their own survival?

Here only farming Gauls and Etruscan merchants exist, so only grain (energy) and one type of wine are traded.

Five exchange rate values were examined (Table 2): a rate of 2:1, 3:1, 4:1, 5:1, and 6:1. Value of 1:1 and 7:1 were examined; at 1:1, Etruscan merchants died out quickly (as they do in 2:1), while at 7:1 Gaulish agents died out quickly, which caused the simulation to stop. The basal amount of trade each year is 10 amphorae of wine, so the amount of grain scales accordingly (e.g., 40:10, which equals 40 *hl* of grain for 5 *hl* of wine). In summary, colonist agents cannot survive unless they trade wine for grain. Figure 3 reports the response of population to these trade values.

In Figure 3a. I examine the trade of 20 hectoliters of grain for 10 amphorae of wine. Note that merchant-agents die out almost immediately. In Figure 3b. the trade rate (3:1) is more favorable to merchant populations, and their population trajectory reflects this. Note there is large variance around the mean. In Figure 3c. the trade rate (4:1) is increasingly favorable for merchant populations, with their population trajectories more-or-less overlapping by year 300. In Figure 3d. the trade rate (5:1) is again favorable to merchant populations, and the two population trajectories have significant overlap, as with Figure 3c. However, the variance around the mean is larger in Figure 3d. than Figure 3c. In Figure 3e. Gaulish agent populations begin to die out due to the unfavorable trade rate (6:1). This may reflect the trade-rates that some merchants attempted to achieve reported by Diodorus Siculus [18]. This poor trade rate negatively effects merchant populations as well; with fewer Gauls to trade with, the quantity of available grain diminishes, decreasing merchant population.

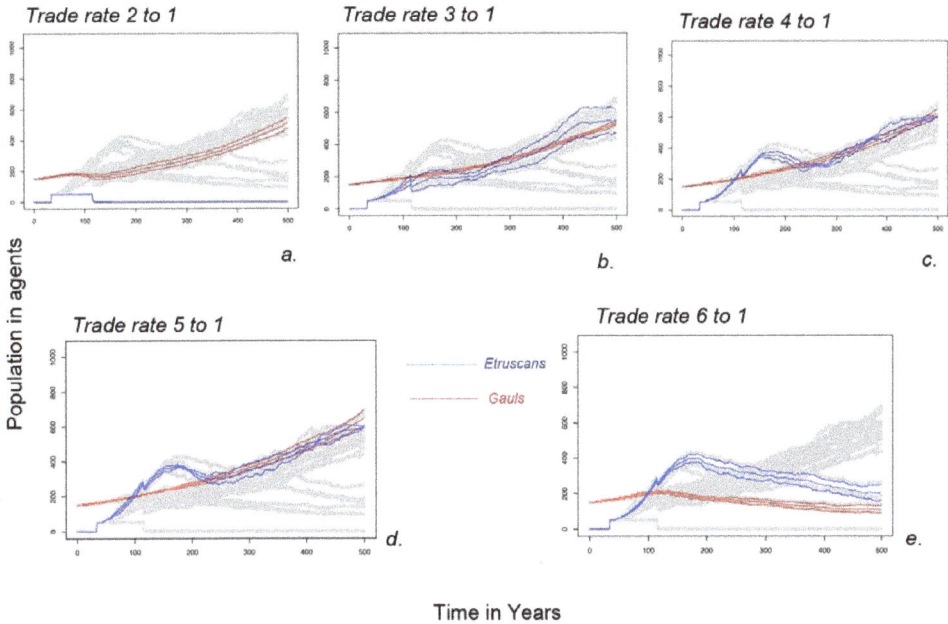

Figure 3. Population response in the simulation, tied to the consumption of grain (grey lines). Solid lines denote the mean population for the scenario being examined, dotted lines denote one standard deviation of the mean above and below the mean. Scenarios a through e represented different trade rates examined in this simulation.

While many of the trade rates examined here would have enabled the survival of merchant populations in southern France, a trade rate of 40 *hl* of grain for 5 *hl* of wine (4:1) creates a favorable exchange rate for both merchant and Gaulish populations, while reducing variance around the mean (reducing path dependence). Thus a trade rate of 4:1 was set for the subsequent models examined in this paper.

Though this model is highly simplified, by using historically reported yield rates (8 *hl* of grain per hectare, with family farms from 4.5 to 27 hectares, consumption of 6 *hl* of grain annually per adult, and 4 *hl* per child, 50 *l* of wine per amphora) it shows that Gauls would have been able to grow enough grain to support themselves and a burgeoning economy. This has important ramifications, and will be discussed below (Discussion). Using these historical rates reported above for average field productivity and average farm size, it is completely feasible that a household would be able to produce enough grain for immediate consumption, storage, and trade. Then, through the trade of wine for grain, merchant populations were able to reproduce and grow their numbers, establishing colonies along the littoral, and engaging in long-term trade with Gaulish farmers. This model verifies Py's first hypothesis [7]. Next, I build on this simple model to examine how the inclusion of two different types of merchant populations effects the distribution of artifact types across the landscape, and the survivability of each type of agent.

2.2. Multiple-Colonist Model

At the beginning of this article I quote Eco [1], who illustrates the preference of one type of wine for another. While Eco writes of 12th century Italy, the preference for red wines from Etruria, or for wines that are "bitter" and "tasting of mold" would have governed purchasing tactics by prehistoric consumers. In this second simple model I show how these preferences create distinctive artifact patternings that can then be compared to real archaeological data. The model presented in this section builds on the simple trade model presented in Section 2.1. In this model Gaulish agents choose to trade for either Etruscan or Greek wine.

Gaulish agents will favor buying wine according to the parameter, "preference." Preference governs the choice between Etruscan and Greek wine, weighting the probability of choosing a Gaulish or Etruscan wine depending on the perceived value by Gauls. Of course, before Greek agents arrive, Gaulish agents will only purchase Etruscan wine, and thus preference has no effect. Preference can take many forms. Preference could be for the taste of the wine, the rarity of it (causing it to have higher prestige status), or in mimicking the elite [19]. When preference is set to 50, Gaulish agents have a 50% chance of choosing Etruscan or Greek wine (they don't prefer either, they just want wine). The closer the value is to 0, the more weighted it is in favor of Etruscans, while the closer it is to 100 the more weighted it is in favor of Greeks. These are explained below in Table 4.

Eleven preference values were swept across (reported in Tables 3 and 4) to examine how a simple change of preference could influence both the survival of agents on the landscape and the artifact assemblage across the landscape. Each of these models was run for a total 30 runs per preference value, creating a sweep of 330 runs. Results are reported in Figures 4 and 5. In these figures the average population per each preference value is reported along the left column while the average number of artifacts of each type through time is reported along the right column; solid lines indicate the mean of all runs, while the dotted lines indicate the high and low standard deviations around the mean. It should be noted, however, that even though Gaulish agents may prefer one type of wine over the other when they initiate purchase, each merchant agent initiates trade with a Gaulish agent after the Gaulish agent has finished its scheduling (see Supplementary Information). The Gaulish agent then has a 50% chance of choosing to trade with the merchant or not. Thus, while preference should affect the results, it should not completely control the assemblage types, and even when Gaulish agents prefer one type of wine over another, due to the logic in this simulation, merchant agents should be able to survive, albeit in low numbers, since merchants can initiate trade as well.

Table 4. Preference Values swept across in this study.

Preference	Explanation
0	Weight is entirely in favor of Etruscan Wine
10	Weight is strongly in favor of Etruscan Wine
20	Weight is in favor of Etruscan Wine
30	Weight is slightly in favor of Etruscan Wine
40	Weight is very slightly in favor of Etruscan Wine
50	There is no weighted preference between Greek or Etruscan wine.
60	Weight is very slightly in favor of Greek wine.
70	Weight is slightly in favor of Greek wine.
80	Weight is in favor of Greek wine.
90	Weight is strongly in favor of Greek wine.
100	Weight is entirely in favor of Greek wine.

In Figure 4, preference is set initially so that Gaulish agents prefer Etruscan wine (preference 0, Figure 4a,b). In this model, Greek agents have difficulty establishing trade relationships with native Gauls (Figure 4b) and die out essentially upon arrival (Figure 4a). The same occurs when preference is set to 10 (Figure 4c,d); when preference is set to 20, Greek agents survive slightly longer, but still die out (Figure 4e,f). This type of situation may be expected when a strong economic partnership develops between two entities, making it difficult, if not impossible, for a new competitor to enter the market. The new goods may be seen as "strange" (e.g., they may "taste of mold" [1]) and thus not desirable. Moreover, the new product may not offer anything better than the older products, and the lack of a relationship between the new sellers and the buyers may influence the sale of those products [20].

As we move down preference values in Figure 4, Greek agents are able to survive easier as the preference value approaches 40%. Yet even in Figure 4i the population of Etruscans holds strong even after Greek agents arrive. In Figure 4j it is evident that the slight preference for Etruscan wine over Greek wine influences the distribution of artifacts so that Etruscan amphorae are more prevalent.

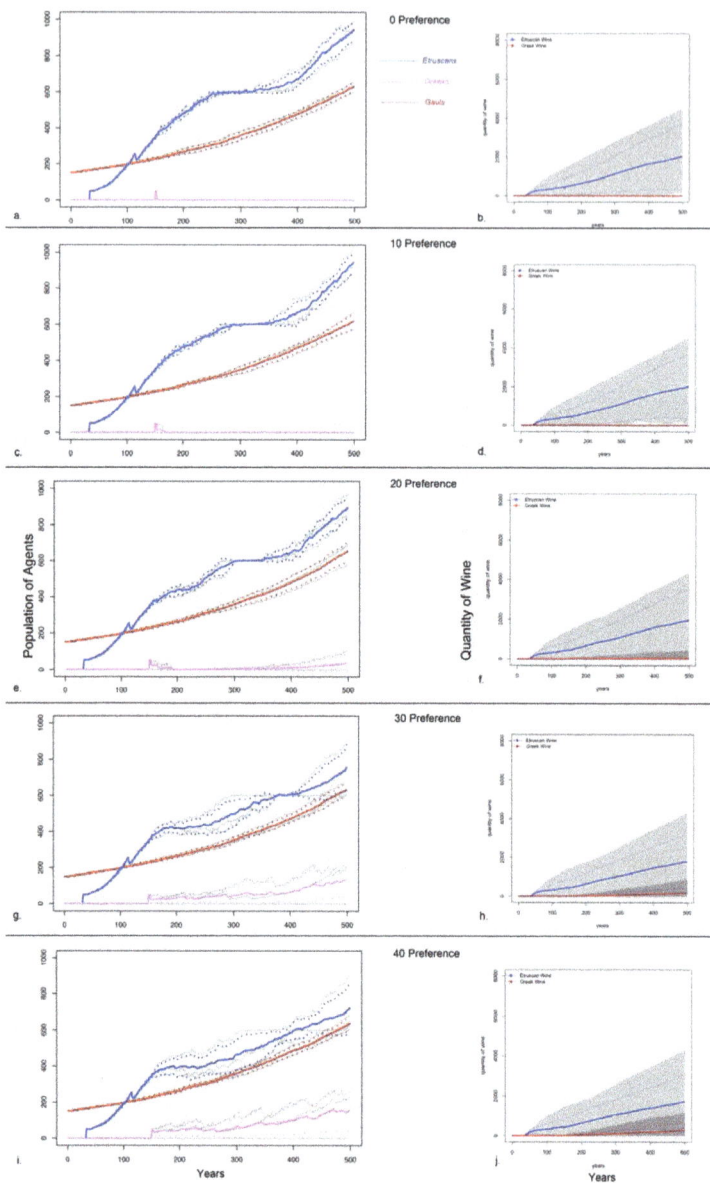

Figure 4. (a–j) Response of population and artifact type based on preference value, beginning with a preference of 0 (in favor of Etruscans) and ending at a preference value of 40 (almost equal preference, still in favor of Etruscans). Preference values are reported in the middle of each tile, corresponding to the values on the left and the right. Left side of tiled figure corresponds to population, while the right side corresponds to the artifact assemblage. Solid colored lines denote the mean, while dotted colored lines denote one standard deviation above and below the mean. Grey lines indicate overall variation of output in simulations.

In the next set of tiled figures, response of population when preference is set at 50% is examined (Figure 5a,b). When Gauls weight Etruscan and Greek wine evenly, both Etruscan and Greek wine are present. However, since Etruscans arrive sooner in this simulation (during year 34) they have a longer time to establish trade relationships with Gauls and reproduce along the littoral. Thus, when Greek agents arrive, Etruscan agents outnumber them. The low proportion of Greek wine in the assemblage shows that, while Greek merchants can (and do) trade wine for grain, the quantity reflects the challenge for Greek merchants to gain a foothold in the region.

When preference values begin to favor Greek merchants (Figure 5c,d) the average number of Etruscan agents and Greek agents stays similar, yet because Etruscans were on the landscape longer they maintain the majority of amphorae (Figure 5d). Only when preference reaches a value of 80, and Greek agents dramatically outpace Etruscan agents (Figure 5g) do the mean number of Greek amphorae begin to be more numerous than Etruscan amphorae (Figure 5h). When preference is set to a value of 90, the mean number of Etruscan amphorae levels out (Figure 5i) showing that the growth of grapes and trade of wine is at a strict replacement rate for the amphorae that are being discarded. Finally, when preference values are set to 100, we see both Etruscan population dying out (Figure 5k) and attrition of Etruscan vessels decrease their presence in the simulated assemblage (Figure 5l).

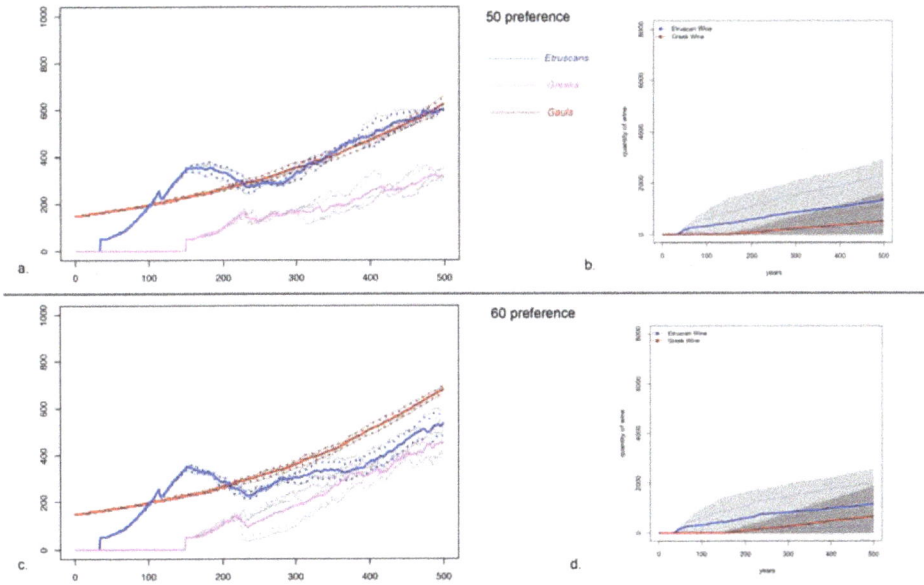

Figure 5. *Cont.*

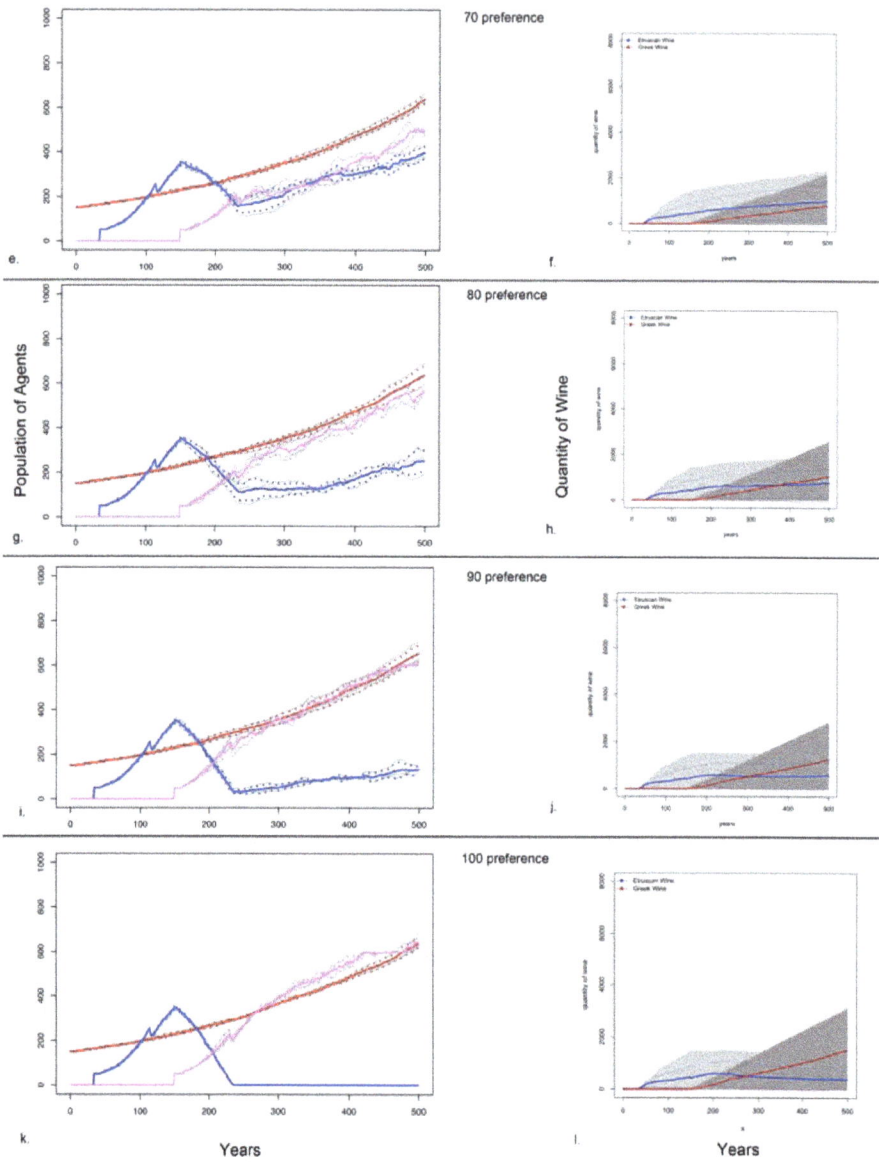

Figure 5. (a–l) Response of population and artifact type based on preference value, beginning at preference of 50 (no preference for Greek or Etruscan wine) and ending at 100 (preference for Greek wine). Preference values are reported in the middle of each tile, corresponding to the values on the left and the right. Left side of tiled figure corresponds to population, while the right side corresponds to the artifact assemblage. Solid colored lines denote the mean, while dotted colored lines denote one standard deviation above and below the mean. Grey lines indicate overall variation of output in simulations.

In Figures 4 and 5 each of the 11 preference values is displayed, demonstrating how preference affects both the distribution of artifacts on the landscape and the survival of each of the agents types on the landscape. Here a phase transition at preference value 30 is evident (Figure 4g,h), after which Greek agents are able to survive and trade. Below this value it is difficult for Greek agents to survive since Gaulish agents have a strong preference for Etruscan wine. Further, since Etruscans arrive first on the landscape, they are able to monopolize the market and establish their own territories.

The next phase transition occurs at a preference value of 70 (Figure 5e,f) where the mean quantity of Greek wine begins to approach that of Etruscan wine by year 400 (Figure 5f); in this scenario the mean population of both Etruscans and Greeks is quite similar from year 250 onward. Then, when preference values are set to 80 and 90 Greeks do well, yet Etruscans do not die out. While their populations diminish, they still exist. This is in stark contrast to when preference was set to 10 or 20; in those scenarios, since Etruscans had already established a monopoly on the economy, Greeks were unable to trade enough (or quickly enough) to reproduce. When preference values are set to 80 and 90, in contrast, Etruscans have already lived on the landscape long enough to create storage and establish trade relationships with the Gauls. They can weather a few years of bad trade relationships due to their longevity in the region. It is only when a preference value of 100 is used that Etruscan merchants completely die out. Indeed, their die off is precipitous and complete by year 250 (Figure 5k).

When Figures 4 and 5 are compared to Figure 1, we can see that it is Figure 5h–l that most closely resemble Figure 1; when Gauls "prefer" Greek wine, Greek wine amphorae begin to outnumber Etruscan amphorae even though Etruscans can still trade wine for grain. While it should be noted that the consumption/decay rate built into the simulation decreases the amphorae at a rate of 1 amphora per agent per year, these rates likely reflect use (and discard) in prehistory. Many amphorae were reused, yet many more were recycled when they became cracked or chipped (or perhaps even unfashionable).

Prehistoric France was littered with amphora sherds, and not necessarily because of the ravages of time. (Indeed, modern France is still littered with these amphora sherds). Instead, archaeological evidence suggests that humans who lived during the time examined here (~700 B.C. to ~100 B.C.) recycled amphorae as paving gravel, as chips for building walls, as fill for creating land (such as building an artificial hill), or as roof tiles for buildings [21]. If objects were cherished, they would be preserved; consequently, we should see those objects lasting for generations in the archaeological record. Yet if objects are not cherished, and instead are utilitarian, utilitarian objects that outlasted their utility (or became unfashionable) would be discarded. In the simulation amphorae are discarded when the wine is drunk, but also there is no inheritance when an agent dies. Consequently, when an agent dies all of its amphorae are discarded (metaphorically recycled into roof tiles or paving sherds). Thus the new generation will drive the demand for certain types of amphorae due to the practicality of recycling. The artifact curves examined in Figure 1, and recreated in Figure 5h–l, reflect an evolving preference by Gaulish consumers for Greek wine.

However, even though Greek wine replaces Etruscan wine in the simulation, the discard rate of Etruscan sherds in Figure 5h–l is much slower than in Figure 1. This suggests that the use and discard rate in prehistory is much faster than what was used in the simulation. In reality, Etruscan amphorae were used and discarded quickly, suggesting that these amphorae were not treasured objects, but utilitarian vessels that had more use when recycled than being reused in their original form.

This pattern—of a replacement of one type of amphora by another—has been mystifying archaeologists in this region for decades. The model presented here provides a way forward to examine how Gauls drove the economy, and created the archaeological assemblage seen today by their preference for one type of wine over another. Thus, pattern oriented modeling, where I sought to create a virtual artifact assemblage through simple rules of exchange, helps to illuminate the complex processes of prehistoric decision-making and prehistoric economies. Further, as will be discussed below, while archaeologists can identify the amphora, the objects the amphorae were traded *for* are missing. This model proves that merchant agents, if they did not engage in farming practices (which was likely, see below) could trade local farmers for grain, and through this trade they could survive along the littoral. Thus through creating patterns of artifacts, and examining how the trade of grain for wine effects the survival of merchant agents, I conclude that Gauls drove the economy, but their desire for luxury wine and their willingness to farm enough grain for trade enabled the survival of merchant agents (both Eturscan and Greek) along the littoral.

Of interest, however, is the fact that Etruscan merchants do not die out in Figure 5g,i. Rather, they persist until the end of the simulation. This is explored below with archaeological data on the persistence of different ethic identities of merchants well into the Iron Age in Southern France.

3. Discussion

When an entity creates a monopoly on a type of good, a dramatic and complete switch to another type of that good is suspect. From archaeological data what is clear is that Etruscan amphorae become replaced by Greek amphorae in a rapid amount of time. In the following data (Section 3.2, *Trade with Outsiders*) I suggest that this is not because Etruscan merchants were not present along the littoral. Rather, Etruscans *were* present, and their wares were represented *alongside* those brought by Greek merchants. If these were both present, how would one dominant type of amphora become completely replaced by another?

If one artifact type is technologically superior it may gain a higher quantity of the market share; it may be tempting to suggest that Greek wine, or at least Greek amphorae, were superior. However, it is not necessarily always the case that a technologically superior good will become dominant. Technologically superior goods can be expensive, and if other more readily accessible goods are still at hand, replacement makes little sense. Rather I suggest that it is the *desire* for a different type of material that creates the switch. Otherwise, both should be present, since some individuals will continue to use (consume) the older material.

In the above model I show phase transitions at 30% and 80%. Here, it is not necessary for 100% of Gauls to prefer one type of wine over another. Rather, when an individual prefers Greek wine 1 out of every 5 times, Grecian merchants are able to stake a stronghold along the littoral. Thus, this model demonstrates that it is not necessary for 100% of the populace to prefer buying one type of wine over another, but rather that critical transitions happen at percentages much less than 100%.

Further, recall that these results are for the demand side. Etruscan and Greek merchants can still approach Gaulish agents and ask to sell. While Gauls will only accept buying 50% of the time when they are approached, even then this pattern of complete artifact replacement is present. This shows how important the demand side is for the supply-demand chain of Gaulish consumption (discussed further below) and demonstrates the agency Gauls had in shaping their economies.

3.1. Archaeological Evidence: Mixing of Colonial Entities

Modern conceptions of nationalist trade ventures likely do not hold for trading in antiquity; Greek and Etruscan merchants likely coexisted and traded each others' wares. This is evident in recovered vessels from shipwrecks. De Hoz [22] notes that the El Sec shipwreck (Figure 2), dating to the 4th century B.C., contained a vast array of types of amphorae, 30 percent of which came from Samos (Figure 2), with Punic and Greek graffiti present on the recovered vessels [22,23]. This mixing is present in other contexts, such as the Grand Ribaud F shipwreck (Figure 2) where Etruscan and Greek goods are both represented [24], and on a lead tablet inscribed with both Greek and Etruscan text, recovered at Pech Maho in western Languedoc (Figure 2 [5,25]). The replacement of amphorae from Etruscan-type to Greek-type does not necessarily mean that ethnically identified Etruscans were no longer present in Gaul, or that Etruscans were no longer producing goods to trade. The replacement rather indicates that there was a cultural shift from wanting Etruscan wine vessels to wanting Greek wine vessels, and likely the contents within them, too. Etruscans and Greeks were present simultaneously, yet vessel-type changed rapidly. Understanding Gauls as drivers of the economy may help illuminate the transition to Roman amphorae that occurred much later (Figure 1).

3.2. Trade with Outsiders

The Gaulish littoral was not isolated, but had contact with traders well before the development of the complex exchange networks noted archaeologically. For example, Punic traders interacted with Gauls in the Languedoc since at least the 8th century B.C. [7,26]. However, these interactions were short and established no high-intensity trade relationships. Objects of Punic origin, including amphorae, vases, and glass objects are present in Gaul beginning in the 7th century B.C. However, no evidence for Punic settlement is present. While Punic boats likely made frequent trips across the Mediterranean to Languedoc [7], these interactions left ephemeral traces. Further, Villard [26] notes that Gauls in a small settlement in what would become Massalia likely had contact with merchants from Phocaea, an Ionian Greek city on the western coast of Anatolia, a half century before the founding of Massalia as a Greek city (contemporary Marseille, see Figure 2; Villard [26] places the foundation of Massalia between 600 and 596 B.C.)

Ceramics for the transportation and drinking of wine arrive in southern Gaul by the late seventh century B.C. These ceramics are composed primarily of Etruscan wine amphorae, although Etruscan *bucchero nero* pottery, as well as a small quantity of Greek ceramics (likely imported by Etruscans) are also present (Villard notes roughly 30 of these in Marseille [26]).

Once Massalia was founded, locally produced fine-wares called "Pseudo-Ionian" and "Grey Monochrome" began to be produced (although Villard remarks that the massaliote and imported ceramics are fundamentally the same, just made in different areas [26]). Some of these wares were traded to indigenous peoples. Villard [26] finds a wider range of fine-ware ceramic vessels in Massalia than in indigenous settlements; it appears that ceramics at indigenous sites include only wine-related vessels to complement indigenous bowl forms, while in Massalia ceramics take on more numerous forms. Additionally, wine begins being produced locally in the littoral, such as at Massalia by Greeks [27]. After c. 525 B.C., local imports of Etruscan amphorae fell off sharply as Massalian-produced amphorae replaced them [27]. However, Villard also notes that the imported amphorae from Greece are much more abundant in Massalia than locally made amphorae, postulating that "imported wine was more or less consumed where it arrived, even while locally grown and produced wine was largely exported [locally] into the indigenous market" [26] (my translation).

Thus, the pattern of trade between Gauls and visiting nonlocal merchants shows that Gauls received almost exclusively pottery related to drinking wine. These included amphorae and wine-drinking apparatus. Yet ceramic assemblages from other areas that traded with Etruscan and Greek merchants show a higher diversity of objects. If Gauls drove the demand for Greek wine, we may be able to expect to see in the archaeological record evidence for Gauls driving other areas of their economy. This is examined below.

3.3. Supply and Demand

Morel [28] states that contemporary trade between Etruscans and North Africans does not follow the same pattern as that between Etruscans and Gauls. Rather, southern Gaul's limited type of imports likely reflects consumer demand more than the range of artifacts available. Specifically, the artifact type "amphorette," a ceramic object used for storing high quality wines less than half the size of an amphora [29], makes up approximately half of the *bucchero nero* pottery in Carthage, almost 100% of pottery in Tharros (a city on the western coast of Sardinia, see Figure 2), but is "practically inexistent" in Gaul [28]. Table wine and wine amphorae are the objects the Gauls desired and do not reflect the variety of objects offered for trade by the Etruscans; rather the makeup of Gaulish assemblages reflect a cultural preference for drinking materials. As Dietler [15] states, Gauls "avidly adopted this foreign form of drink while at the same time rejecting other cultural borrowings."

The trade of amphorae seems to be one-way—evidence for Gaulish products in merchant settlements is thin—so secondary measures for identifying the goods traded are often used. For example, historians suggest that Massalia was so large it would have outstripped its local carrying capacity, and only through trade were inhabitants of Massalia able to eat [5]. Coupling this with primary sources, such as Strabo [30] who describes Massalian land as too poor to produce grain, and the suggestion that Etruscan and Greek traders would not have engaged in subsistence farming due to it being seen as below their station [31], grain was likely produced by Gauls and traded to settled merchants. However, this statement had never been tested formally. The model presented above illustrates how the trade of grain could have enabled merchant survival. Further evidence of ships bearing large quantities of grain are recorded as arriving in Greek and Etruscan homelands, and this grain likely came from Gaul [5].

Metal and salt are two other commodities likely to have originated in Gaul and traded to settled merchants. Copper, gold, iron, tin and silver are all found within France, and sources for these are noted in antiquity [4]. These metals would have been essential for the creation of objects during the Iron Age, and salt would have been essential for food preservation. Overland transalpine exchange of metals and salts from Gaul to northern Italy began in the early Bronze Age [4], so it is likely that Etruscan traders knew that minerals could be obtained in Gaul, thus influencing their decision to trade in Gaul.

As suggested above in discussion of harvest (Section 2.2), enslaved people were likely present along the littoral. Briggs [4] suggests that Etruscans commonly used enslaved people as servants, and that women and children especially would have been brought to the colonizer homeland as household slaves [4]. While "one of the most elusive of all prehistoric objects of exchange is human labour" [4], the importance of slaves in Etruscan households may suggest that Gaulish women and children were some of the "objects" that enabled the trade system to function [4,5]. Indeed Diodorus Siculus [18] writes that some brazen merchants would attempt exchanging one amphora of wine for one slave. (Though this anecdote relates to the first century BC, and this exchange value was likely not the norm.)

So, while wine amphorae are plentiful in Gaulish settlements [2], the objects for which they were traded remain elusive. Indirect evidence suggests that grain, metals, salt, and slaves were traded to the settled merchants. The model presented above intervenes in these debates. While the objects that were traded may be invisible, the survival of merchants along the littoral suggests that they were able to trade their goods for foodstuffs. This model shows that a simple economic model can enable merchant survival, and can lead to distinctive artifact patterns. While this model is highly simplified, it enables a first step into using agent-based modeling in Southern France, and will be expanded upon the in future to examine expanded economies (such as the trade of metal or salt) and the aggregation of Gauls into oppida.

4. Future Directions

I began this article by proposing that a simple preference for one type of wine over another could cause the empirical artifact distribution recognized by Py [2] and reported in Figure 1. To do that, historically-based farming production rates were employed on a simplified landscape to enable the intensification of agriculture and the trade of surplus wine for grain. In this we can examine landscape use in antiquity and see how it could lead to the establishment of complex economies in the past.

Results in this model showed that when Gaulish agents did not prefer one type of wine over another (when preference was set to 50%) that both Etruscan and Greek wine were present in the simulation, but that Etruscan wine was more common due to being present in the area longer. When preference was set to 20 (Table 4) or below it was very difficult for Greek merchants to trade wine for grain and to exist on the landscape. Additionally, when preference was set to 70 or higher (Table 4), Greek wine supplanted Etruscan wine as the more common type in the simulation (after Greek merchant arrival). However, it was only when preference was set to a value of 100 that Etruscan died off. Even when this occurred, however, Etruscan wine amphorae were still present for the remainder of the simulation due to a slow use and decay rate.

These findings have important implications for the archaeological record. First, these results suggest that when Greek merchants arrived in southern Gaul that their product was found as desirable. If it was not, the archaeological record may reflect those results in Figure 4f–h. Instead, Figure 1 resembles most closely Figure 5i–k, where Greek wine arrived and became common along the littoral. In these figures Etruscan amphorae make up the early assemblage, but are quickly supplanted by a second type of amphora. In the simulation, not only were Greek wines seen as desirable, but upon their arrival they were preferred by Gauls and became the largest part of the assemblage. However, these results also suggest that artifacts can have a long uselife. Archaeological assemblages may not reflect the presence of a population, but may reflect instead the storage and use of those artifacts after a population moves on.

The work begun here is ongoing, as this simple model was a first step in establishing an agent-based model for the development of colonial interactions in southern France. As mentioned above, multiple other types of resources besides grain were traded for wine. While these scenarios are not examined in this publication, this model is being developed to enable the trade of two types of wine for two types of resources—grain and metal. Future research will examine how the incorporation of diverse resources effects the survival of agents on the landscape and the distribution of materials on the landscape. Research is also being pursued into using realistic GIS dataplanes in the simulation, instead of using a simple patchy and regenerating landscape, as was used in the model presented here. This will enable the development of aggregation models based on least-cost path analysis to help agents trade resources across the landscape and establish settlements at optimal locations to enable trade.

5. Conclusions

What drove the preference of Greek over Etruscan wine? Was it the desire for a less expensive product? Was it because Gauls liked the taste of Greek wine better? Did Greek merchants treat Gaulish farmers better than their Etruscan counterparts had? These are not questions that can be answered with an agent-based model, but would rather need to be examined through the archaeological record and through primary texts. However, the model presented here enables us to begin to ask these open questions, since we now know through systematic analysis that preference can drive artifact assemblages. Gauls preferred Greek amphorae, and likely the contents within them, over Etruscan amphorae, and it was through this demand that the artifact assemblage changed so rapidly and completely. If Etruscan amphorae signaled wealth or prestige, archaeologists should see them much later in the archaeological record. Instead they are discarded and recycled to make way for new Greek ceramics.

Debates about the causes of the complete replacement of Etruscan amphorae by Greek amphorae, as reported in Figure 1, are longstanding for this area. This research directly intervenes in these debates. The importance of this work is that the replacement event might be understood from internal, rather than external, processes.

While further studies would need to take into account economic decisions—such as Greek amphorae being less costly to produce—this work begins these debates and allows for a thorough and systematic study of the distribution of wine types across the littoral. Further, this simple model shows that using a modeling approach can help shed light on complex processes. This model provides a useful tool to support the hypothesis that it is the demand for wine that drove these artifact patterns, not necessarily the availability of products [26]. Gaulish people were the creators of the economy of southern France, and their preferences drove what we see in the archaeological record.

This model is meant as a first step toward understanding the complexities of early colonist interactions in southern France, as well as a first step toward understanding how France became a viniculture powerhouse. The modern wine industry in France has roots that date back to the founding of the wine trade between Etruscans, Gauls and Greeks, and it is through the development of this complex economy that the wine industry exists today [27]. Even though this model may be simple, it helps advance our understanding of local populations as drivers of the economy of a globalizing antique world.

Acknowledgments: I would like to thank Laure Nuninger, François Favory, Tim Kohler, and Rachel Opitz for guidance in developing this model. Further, I would like to thank Andrew Duff, Kathryn Harris, James Millington and John Wainwright, as well as two anonymous reviewers who helped to edit this work. This work also is indebted to each member of the TransMonDyn research group, who have commented on, guided, and listened to countless versions of this project. This work is also indebted to Michel Py and Michael Dietler, whose foundational work in understanding colonial interactions along this region influenced this model. Crabtree recognizes funding from the NSF Graduate Research Fellowship #DGE-080667 and the Chateaubriand Fellowship Program for helping fund this work. Crabtree created this model, wrote this paper, and did the analyses herself; any errors or mistakes are strictly her own.

Conflicts of Interest: The author declares no conflict of interest.

References

1. Eco, U. *Baudolino*; Harcourt, Inc.: New York, NY, USA, 2000.
2. Py, M. *Culture, Économie Et Société Protohistoriques Dans La Région Nimoise*; Publications de l'École française de Rome: Rome, Italy, 1990.
3. Grimm, V.; Revilla, E.; Berger, U.; Jeltsch, F.; Mooij, W.M.; Railsback, S.F.; Thulke, H.-H.; Weiner, J.; Wiegand, T.; DeAngelis, D.L. Pattern-Oriented modeling of agent-based complex systems: Lessons from ecology. *Science* **2005**, *301*, 987–991.
4. Briggs, D.N. Metals, salt, and slaves: economic links between Gaul and Italy from the eighth to the late sixth centuries BC. *Oxf. J. Archaeol.* **2003**, *22*, 243–259.
5. Dietler, M. *Archaeologies of Colonialism: Consumption, Entanglement, and Violence in Ancient Mediterranean France*; University of California Press: Los Angeles, CA, USA, 2010.

6. Bintliff, J. The Hellenistic to Roman Mediterranean: A proto-capitalist revolution? In *Economic Archaeology: From Structure to Performance in European Archaeology*; Kerig, T., Zimmermann, A., Eds.; Rudolf Habelt GmbH: Bonn, Germany, 2013; pp. 285–292.

7. Py, M. *Les Gaulois du Midi, de la fin de l'Age du Bronze à la Conquête Romaine*; Nouvelle édition revue et augmentée, collection Les Hespérides, Errance; 2012.

8. Wilensky, U. *NetLogo*; Center for Connected Learning and Computer-Based Modeling, Northwestern University: Evanston, IL, USA, 1999; Available online: http://ccl.northwestern.edu/netlogo/ (accessed on 12 December 2015).

9. R Core Team. *R: A Language and Environment for Statistical Computing*; R Foundation for Statistical Computing: Vienna, Austria, 2013. Available online: http://www.R-project.org/ (accessed on 12 December 2015).

10. *Emergence and Collapse of Early Villages*; Kohler, T.A., Varien, M.D., Eds.; University of California Press: Berkeley and Los Angeles, CA, USA, 2012.

11. Gras, M. *Trafics Tyrrhéniens Archaïques*; Bibliothèque des Ecoles Français d'Athènes et de Rome 258; Ecole Française de Rome: Rome, Italy, 1985.

12. White, K.D. *Roman Farming*; Cornell University Press: Ithaca, NY, USA, 1970.

13. Cahill, N. *Household and City Organization at Olynthus*; Yale University Press: New Haven, CT, USA, 2002.

14. McAllister, P. *Xhosa Beer Drinking Rituals: Power, Practice and Performance in the South African Rural Periphery*; Carolina Academic Press: Durham, NC, USA, 2006.

15. Dietler, M. Driven by drink: The role of drinking in the political economy and the case of Early Iron Age France. *J. Anthropol. Archaeol.* **1990**, *9*, 352–406.

16. Crabtree, S.A. Inferring social networks from aggregation in, and simulation of, the central Mesa Verde. *J. Archaeol. Method Theory* **2015**, *22*, 144–181.

17. Cockburn, D.; Crabtree, S.A.; Kobti, Z.; Kohler, T.A.; Bocinsky, R.K. Simulating social and economic specialization in small-scale agricultural societies. *J. Artif. Sci. Soc. Simul.* **2013**, *16*, 4.

18. Siculus, D. *The Library of History*; Volume III, Books 4.59–8; Translated by C.H. Oldfather, Loeb Classical Library; Harvard University Press: Cambridge, MA, USA, 1939.

19. Hashim, M.N.; Che, R.R. Consumer ethnocentrism: The relationship with domestic products evaluation and buying preferences. *Int. J. Manag. Stud.* **2004**, *11*, 29–44.

20. Mazzeo, M.J. Product choice and oligopoly market structure. *RAND J. Econ.* **2002**, *33*, 221–242.

21. Twede, D. Commercial amphoras: The earliest consumer packages? *J. Macromark.* **2002**, *22*, 98–108.

22. De Hoz, J. El Sec: les graffites mercantiles en Occident et l'éprave de l'El Sec. *Revue des Etudes Anciennes* **1987**, *89*, 117–130.

23. Koehler, C.G. Corinthian A and B Transport Amphoras. PhD Dissertation, Princeton University, Princeton, NJ, USA, 1978.

24. Rouillard, P. *Les Grecs et la peninsula ibérique du VIIIe au IVe siècle avant Jésus-Christ*; Broccard: Paris, France, 1992.

25. Chadwick, J. The Pech Maho lead. *Zeitschrift für Papyrologie und Epigraphik* **1990**, *82*, 161–166.

26. Villard, F. La Céramique Grecque de Marseille (VIe-IVe siècle); Essai D'histoire économique; Thèse pour le doctorat ès lettres, présentée à la faculté des lettres de l'université de Paris. Editions e. de Boccard: Paris, France, 1960.

27. McGovern, P.E.; Luley, B.P.; Rovira, N.; Mirzoian, A.; Callahan, M.P.; Smith, K.E.; Hall, G.R.; Davidson, T.; Henkin, J.M. Beginning of viniculture in France. *Proc. Natl. Acad. Sci. USA* **2013**, *110*, 10147–10152.

28. Morel, J.P. *Le Commerce Etrusque en France, en Espagne, et en Afrique*; L'Eturia Mineraria: Atti del XII Convegno di studi etruschi e etalica; Firenze-Populonia-Piombino, 16–20 giugno 1979; LS Olschki: Florence, Italy, 1981; pp. 463–508.

29. Vallat, P.; Cabanis, M. Le site de "Champ Chalatras" aux Martres-d'Artière (Puy-de-Dôme) et les premiers témoins archéologiques de la viticulture gallo-romaine dans le bassin de Clermont-Ferrand (Auvergne). Revue Archéologique du centre de la France. 2009.

30. Strabo, J. *Geography*; Volume II, Books 305; Translated by H.L. Jones, Loeb Classical Library; Harvard University Press: Cambridge, MA, USA, 1923.

31. Wood, E.M. Agriculture and slavery in classical Athens. *Am. J. Anc. Hist.* **1983**, *8*, 1–47.

Landscape Epidemiology Modeling Using an Agent-Based Model and a Geographic Information System

S. M. Niaz Arifin, Rumana Reaz Arifin, Dilkushi de Alwis Pitts, M. Sohel Rahman, Sara Nowreen, Gregory R. Madey and Frank H. Collins

Abstract: A landscape epidemiology modeling framework is presented which integrates the simulation outputs from an established spatial agent-based model (ABM) of malaria with a geographic information system (GIS). For a study area in Kenya, five landscape scenarios are constructed with varying coverage levels of two mosquito-control interventions. For each scenario, maps are presented to show the average distributions of three output indices obtained from the results of 750 simulation runs. Hot spot analysis is performed to detect statistically significant hot spots and cold spots. Additional spatial analysis is conducted using ordinary kriging with circular semivariograms for all scenarios. The integration of epidemiological simulation-based results with spatial analyses techniques within a single modeling framework can be a valuable tool for conducting a variety of disease control activities such as exploring new biological insights, monitoring epidemiological landscape changes, and guiding resource allocation for further investigation.

Reprinted from *Land*. Cite as: Arifin, S.M.N.; Arifin, R.R.; de Alwis Pitts, D.; Rahman, M.S.; Nowreen, S; Madey, G.R.; Collins, F.H. Landscape Epidemiology Modeling Using an Agent-Based Model and a Geographic Information System. *Land* **2016**, *4*, 378–412.

1. Introduction

Spatial epidemiology, *medical geography*, and *geographical epidemiology* are all effectively synonymous terms for the study of the geographical distribution of disease spread or population at risk [1–3]. A closely related research field, *landscape epidemiology*, studies the patterns, processes, and risk factors of diseases across time and space. It describes how the spatio-temporal dynamics of host, vector, and pathogen populations interact within a permissive environment to enable transmission [4–6]. The emergence and spread of infectious diseases in a changing environment require the development of new methodologies and tools. As such, disease dynamics models on geographic scales ranging from village to continental levels are increasingly needed for quantitative prediction of epidemic outcomes and design of practicable strategies for control [7,8].

Understanding a landscape epidemiology system requires more than an understanding of the different types of individuals (host, vector, and pathogen) that comprise the system. It also requires understanding how the individuals interact with each other, and how the results can be more than the sum of the parts. In this regard, agent-based models (ABMs), also known as individual-based models (IBMs), have become very popular in recent years. ABMs are computational models for simulating the actions and interactions of autonomous agents with a view to assessing their effects on the system as a whole. An ABM often exhibits emergent properties arising from the interactions of the agents that cannot be deduced simply by aggregating the properties of the agents. Thus, an ABM can be a very practical method of analysis of the dynamic consequences of agents for a landscape epidemiology model.

In recent years, despite the proliferation of spatial models which acknowledge the importance of spatially explicit processes in determining disease risk, the use of spatial information beyond recording spatial location and mapping disease risk is rare [9]. Although numerous recent tools have been developed using geographic information systems (GIS), global positioning systems (GPS), remote sensing and spatial statistics, there is still a lack of and hence a serious need to develop efficient and useful tools for research, surveillance, and control programs of vector-borne diseases (VBDs).

In this paper, we present a landscape epidemiology modeling framework by integrating an established spatial ABM of malaria with a GIS (preliminary results of integrating an earlier version of the ABM with a GIS were described in a conference paper in [10]). Malaria is one of the largest causes of global human mortality and morbidity. According to the World Health Organization (WHO), half of the world's population (about 3.4 billion people) are currently at risk of malaria, with about 207 million cases and an estimated 627,000 deaths in 2012 [11]. The ABM describes the population dynamics of the malaria-transmitting mosquito species *Anopheles gambiae*. To account for three output indices and five scenarios (that represent two coverage levels of the two interventions being modeled), a total of 750 simulations are run for two years, and the average results are reported in this paper. Using spatial statistics tools, hot spot analysis is performed for all scenarios and two output indices in order to determine the statistical significance of the simulation results. Additionally, we have applied ordinary kriging with circular semivariograms on all three output indices considering all the scenarios. To allow the viewers for an improved spatial analysis perspective, the kriged maps are presented along with other results for a better insight for the unmeasured (*i.e.*, not simulated) locations on the maps.

Besides being useful for simulation modelers in different branches of science and engineering, this work can provide important insights from the epidemiological perspective, and thus would be valuable for epidemiologists, disease control managers, and public health officials for research as well as in practical fields. In particular, we believe that the insights gained through this study can assist these stakeholders in refining further research questions and surveillance needs, and in guiding control efforts and field studies. Additionally, although the landscape epidemiology modeling framework described in this paper utilizes an ABM of malaria-transmitting mosquitoes, it is applicable to a wider range of other infectious VBDs (e.g., dengue, yellow fever, *etc.*), and hence may find its use in a much wider scenario.

Although the work presented in this paper builds upon a previous work [10] of a subset of the authors, is new and different (from the previous work) in a number of dimensions. In particular, the current work presents the following new features:

- Use of improved models: Although the current paper builds upon the exploratory ideas presented in [10], much improved versions of both the core model and the spatial agent-based model (ABM) have been used for this paper. Over the last few years, we have developed several versions of the core model and the corresponding ABMs. The earlier versions, including the one used for [10], mostly dealt with exploratory features [12–14]. Many of those results were not tested using the verification & validation (V&V) and replication features/techniques of the models.

  On the other hand, the version described and used in this paper reflects the most recent updates in an attempt to enrich the models with features that reflect the population dynamics of *An. gambiae* in a more comprehensive way, as described in [15,16]. Since the most recent ABM is tested using the rigorous V&V and replication techniques, the results presented in this paper entail much higher confidence from both the epidemiological and the simulation perspectives. A summary of major improvements incorporated in the current ABM used for this paper is presented in Table 1.

- Modeling malaria-control interventions: From an epidemiological point of view, one of the most important roles of modeling is to quantify the effects of major malaria-control interventions such as insecticide-treated nets (ITNs) or long-lasting impregnated nets (LLINs), indoor residual spraying (IRS), larval source management (LSM), *etc.* Recent malaria control efforts have seen an unprecedented increase in their coverages. Impact of these interventions, often applied and assessed in isolation and in combination, is the focus of investigation of numerous recent and ongoing studies. In this study, the combined impacts of LSM and ITNs have been evaluated. Notably, the scope of the work in [10] did not cover the study of mosquito control interventions and hence, naturally, no results thereof were reported therein. To this end, the scope of the current study is much broader and more meaningful from the epidemiological perspective.

- Reporting aggregate measures by replicating all simulations: Replicability of the *in silico* experiments and simulations performed by various malaria models bear special importance. Replication is treated as the scientific gold standard to judge scientific claims and allows modelers to address scientific hypotheses [17,18]. In agent-based modeling and simulation (ABMS), replication is also known as model-to-model comparison, alignment, or cross-model validation. It falls under the broader subject of V&V. As highlighted by recent simulation research, most simulation models (including the one presented in the current paper) that involve substantial stochasticity should conduct sufficient number of replicated runs, and some form of aggregate measures of these replicated runs should be reported as results (as opposed to reporting results from a single run). Sufficient number of replications is required to ensure that, given the same input, the aggregate response can be treated as a deterministic number, and not as random variation of the results. This allows modelers to obtain a *more complete* statistical description of the model variables.

Table 1. Updated features for the models used for this paper. Each row represents a specific model feature. The second column refers to the exploratory features from the previous versions [12–14]. The third column refers to the most recent features from [15,16], which are used for this paper. Resource-seeking includes both host-seeking and oviposition. For fecundity, N indicates a normal distribution with *mean* and *standard deviation*. LSM and ITNs refer to the two interventions, larval source management and insecticide-treated nets, respectively.

Feature	Previous Versions	Current Versions
Combined interventions	No	Yes
Coverage scheme for ITNs	Not applicable	Complete coverage
Egg development time	Constant	Temperature-dependent
Fecundity (eggs per oviposition)	Constant	$N(170, 30)$
Interventions modeled	None	LSM, ITNs
Modeling human population	No	Yes (static)
Replication of simulations	No	Yes
Resource-seeking	Anytime	Only at night
Stage transitions	Anytime	Only in permitted time-windows
Time step resolution	Daily	Hourly

Since the spatial ABM involves considerable stochasticity in the forms of probability-based distributions and equations, performing sufficient number of replicated runs is extremely important for validation of the results. In the ABM, mosquito agents' decisions are often simulated using random draws from certain distributions. These sources of randomness are used to represent the diversity of model characteristics, and the behaviour uncertainty of the agents' actions, states, *etc.* For example, when a host-seeking mosquito agent searches for a blood meal in a ITN-covered house, a 20% ITN coverage would mean that it may find a blood meal with a probability of 0.2, which can be simulated using random draws from a uniform distribution. The randomness has significant impact on the results of the simulation, and different simulation runs can therefore produce significantly different results (due to a different sequence of pseudo-random numbers drawn from the distributions). As a consequence, in this study, 50 replicated runs for all simulations are performed, and their averages are reported.

- Kriging analysis: In addition to hot spot analysis, spatial analysis has been conducted using ordinary kriging with circular semivariograms for all scenarios for all the output indices using ArcGIS 9.3 [19]. For the entire study area, kriging analysis produces predicted values for unmeasured (*i.e.*, not simulated) spatial locations, which are derived from the surrounding weighted measured values. Interpolation (prediction) for spatial data for all the three output indices is performed by kriging.

These new dimensions allowed us to present new results in this paper, which entail much higher confidence from both the epidemiological and the simulation perspectives.

2. Experimental Section

2.1. The Core Model

In this section, we present a brief overview of the conceptual biological core model (hereafter referred to as *the core model*) from which the spatial agent-based model (ABM) was developed. The core model describes the population dynamics of *An. gambiae*, which is regarded as one of the most efficient mosquito species that transmits malaria. Due to its pivotal role in malaria transmission, modeling its population dynamics can assist in finding factors in the mosquito life cycle that can be targeted to decrease malaria transmission to a lower level. The *An. gambiae* complex, a closely related group of eight named mosquito species found primarily in Africa, includes three nominal species, *An. gambiae*, *An. coluzzii*, and *An. arabiensis* that are among the most efficient malaria vectors known (in this paper, the terms 'vector' and 'mosquito' are used interchangeably). The model described in this paper has been designed specifically around the mosquito *An. gambiae*. While the respective ecologies and involvement in malaria transmission among other members of the *An. gambiae* complex differ in important ways, this model could effectively apply to all three and even to many of the several dozen other major malaria vectors in the world.

257

The complete *An. gambiae* mosquito life cycle consists of aquatic and adult phases, as shown in Figure 1. The *aquatic* phase (also known as the *immature* phase) consists of three aquatic stages: Egg (E), Larva (L), and Pupa (P). The *adult* phase consists of five adult stages: Immature Adult (IA), Mate Seeking (MS), Blood Meal Seeking (BMS), Blood Meal Digesting (BMD), and Gravid (G) (the term *gravid* denotes the *egg-laying* stage). The development and mortality rates in all eight stages of the life cycle are described in terms of the aquatic and adult mosquito populations.

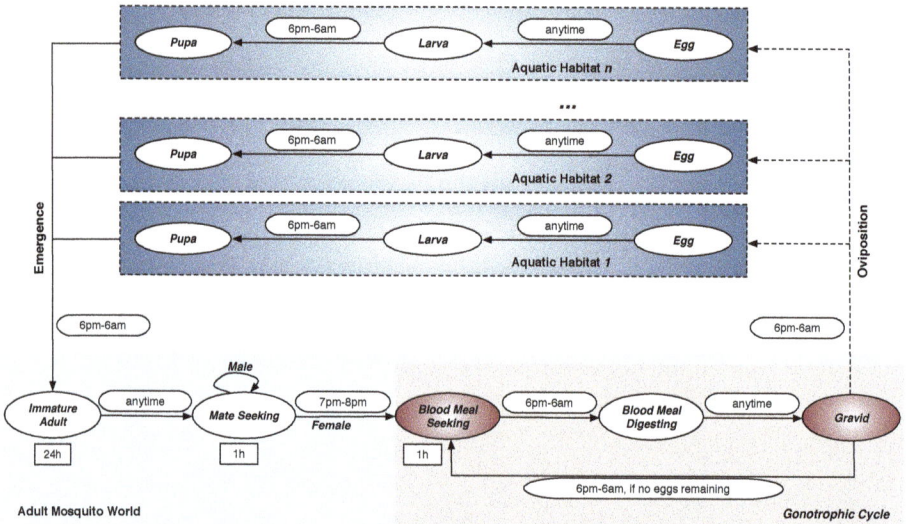

Figure 1. Life Cycle of Mosquito Agents. The *An. gambiae* mosquito life cycle consists of *aquatic* and *adult* phases. The *aquatic* phase consists of three aquatic stages: Egg (E), Larva (L), and Pupa (P). The *adult* phase consists of five adult stages: Immature Adult (IA), Mate Seeking (MS), Blood Meal Seeking (BMS), Blood Meal Digesting (BMD), and Gravid (G). Each oval represents a stage in the model. Stages in which agents move through the landscape are marked in red. The rectangles represent durations for the fixed-duration stages. The symbol *h* denotes hour. Permissible time transition windows (from one stage to another) are shown next to the corresponding stage transition arrows as rounded rectangles. Note that adult males, once reaching the *Mate Seeking* stage, remain forever in that stage until they die; adult females cycle through obtaining blood meals (in *Blood Meal Seeking* stage), developing eggs (in *Blood Meal Digesting* stage), and ovipositing the eggs (in *Gravid* stage) until they die. By Arifin *et al.* [16], used under a Creative Commons Attribution 4.0 International License.

The core model addresses several important features of the *An. gambiae* life cycle, including the development and mortality rates in different stages, the aquatic habitats, oviposition, *etc.* Another important feature, vector senescence, is adopted to account for the age-dependent aspects of the mosquito biology, and implemented using density- and age-dependent larval and adult mortality rates. Further details about the core model can be found in [16].

The *Anopheles* mosquitoes need to access human blood meals (in houses) and aquatic habitats (various water bodies) to complete their life cycle. Thus, the houses and aquatic habitats can be termed as important *ecological resources* for the mosquitoes. These resources have a direct impact on the spatial heterogeneity of the landscapes being modeled, and their availability has long been recognized as a crucial determinant for malaria transmission [20].

2.2. Aquatic Habitats and Oviposition

The core model assumes simplistic, homogeneous aquatic habitats for all mosquitoes. All habitats are uniform in size and capacity (this assumption is relaxed for this study by including five different types of habitats with varying habitat capacities, as described in Table 2), and the water temperature of a habitat is assumed to be the same as the air temperature. To account for the combined seasonality factor, each aquatic habitat is set with a carrying capacity that can be *below* or *above* a baseline capacity, representing low or high precipitation/rainfall, respectively. The carrying capacity essentially represents the *density-dependent oviposition* mechanism by regulating an age-adjusted biomass that the habitat can sustain.

Table 2. Feature types and counts for the ABM. A total of 975 aquatic habitats and 941 houses are used. The last column represents the assigned capacity per feature.

Type	Count	Assigned Capacity
Pool	4	2000
Puddle	13	1000
Pit latrine	395	500
Borehole	4	300
Wetland	559	10
House	941	5

Oviposition is the process by which gravid female mosquitoes lay new eggs. The oviposition behavior of *An. gambiae* mosquitoes can be affected by a variety of factors, as demonstrated by several studies [21–28]. In the core model, all larvae are categorized into different age groups, or cohorts, according to the common age of the cohort. The model keeps track of the age-adjusted biomass in each aquatic habitat, which is defined as the sum of the eggs, the pupae, and the one-day old equivalent larval population in the habitat (for details, see [16]).

2.3. The Spatial Agent-Based Model (ABM)

The spatial ABM, described in detail in [14,15], simulates the life cycle of the mosquito vector *An. gambiae* by tracking attributes relevant to the vector population dynamics for each individual mosquito. It is developed in the Java [29] object-oriented programming (OOP) language using the Eclipse Software Development Kit (SDK, Version: 3.5.2, freely available from [30]). In this section, we present a brief overview.

The three major components of the spatial ABM are the mosquito agents, their environment, and rules. *An. gambiae* mosquitoes are modeled as autonomous *agents* with explicit spatial locations (however, once *within* a cell, a mosquito agent's spatial location does not vary until it moves to another cell). An agent's life in the ABM evolves in artificial, well-defined environments modeled as landscape environments. A landscape environment can be thought of as a medium on which the agents operate and with which they interact. Agents have internal attributes (states) to store relevant attributes and data represented by discrete variables. The major attributes of a mosquito agent include its age, life cycle stage, environment, spatial location, movement counter, id (identifier), sex (gender), available eggs counter, egg batch identifier, *etc.* Some attributes (e.g., id, sex) may remain fixed throughout the agent's lifespan in the ABM, while others (e.g., age, life cycle stage, spatial location) may change through interaction with the environment and/or other agents.

In the ABM, *An. gambiae* mosquitoes are the only dynamic agents (humans are included as static agents, *i.e.*, human agents do not move in space). A new mosquito agent begins its life cycle in the aquatic phase as an egg, and then proceed through larva and pupa stages. When the aquatic phase completes, the agent emerges as an adult mosquito into the adult phase, and advances through the five adult stages (see Figure 1). To account for the limited flight ability and perceptual ranges of *Anopheles* mosquitoes, the cell resolution in the selected landscape is chosen as 50 m × 50 m, yielding a total area of ≈25 km² (for this study). Note that at every time step of the BMS and G stages, the agent needs to search the cell-based landscape by moving from one cell to another until the desired resource is found (the *search* event is guided by several flight heuristics, as described in Section 2.5). A male adult mosquito, after reaching the MS stage, stays in this stage for the rest of its life. The stage transitions (from one stage to another), development rates, and mortality rates are governed by rules as described by the core model. The number of eggs that a gravid mosquito agent can lay is governed by the density-dependent oviposition rules (see [16] for details). New agents, in the form of eggs, possess the same spatial locations as that of the aquatic habitat in which they are oviposited.

The GIS-processed data layers are synthesized in the spatial ABM with a landscape-based approach, where each *landscape* comprises discrete and finite-sized cells (grids). A landscape is used to represent the coordinate space necessary for the spatial locations of the environments and the adult mosquito agents. Resources, in the forms of aquatic habitats and houses, are contained within a landscape. Each cell, with its spatial attributes, may represent a specific habitat environment (human or aquatic), or be part of the (adult) mosquito environment. Landscapes are topologically modeled as 2D torus spaces with a *non-absorbing (periodic) boundary* (with a *non-absorbing (periodic) boundary*, when mosquitoes hit an edge of a landscape, they re-enter it from the edge directly opposite of the exiting edge, and thus are not killed due to hitting the edge).

2.4. Event Action List (EAL) Diagram

In order to capture the major daily events of a simulation for the ABM in a standard, canonical manner, a new type of descriptive diagram, called the *Event Action List (EAL)* diagram, is proposed and presented. It depicts the simulation *events* (occurring on a daily basis), the corresponding *actions* triggered by those events, and the *list(s)* of agents (data structures) affected by them. In an EAL diagram, each event represents a biological phenomenon, and the corresponding action represents the programmatic task(s) performed by the simulation. Optionally, some list(s) of agents may be modified as a direct result of the performed action. Thus, an EAL diagram summarizes the daily events of the simulation model by listing all major events, actions, and lists. For example, when the simulation is started, it needs to create initial adult mosquito agents. The biological phenomenon *"create initial adults"*, termed as an event, is realized by the (simulation) action *"add agents"*; this event-action pair affects the list of adult agents in the simulation. An EAL diagram for the ABM is shown in Figure 2.

2.5. Flight Heuristics for Mosquito Agents

In the spatial ABM, movement of adult female mosquito agents in a landscape is restricted: they move only when in BMS and G stages (marked in red in Figure 1) in order to seek for resources. Since each landscape comprises discrete and finite-sized (50 m × 50 m) cells, the landscape-based modeling approach appeared to be especially suitable to capture the details of the resource-seeking process. In summary, the resource-seeking process is modeled with random non-directional flights with limited flight ability and perceptual ranges until the agents can perceive resources at close proximity, at which point, the flight becomes directional.

A mosquito agent's neighborhood is modeled as an eight-directional *Moore* neighborhood. The maximum distance that an agent may travel in a day is controlled by a *movement counter*, which is reset to 5 at the beginning of each day for a moving agent (thus, the counter controls the maximum daily range of movement, which translates to $250\sqrt[2]{2}$ m). The flight heuristics, depicted in the form of flow-charts in Figure 3, are described below.

261

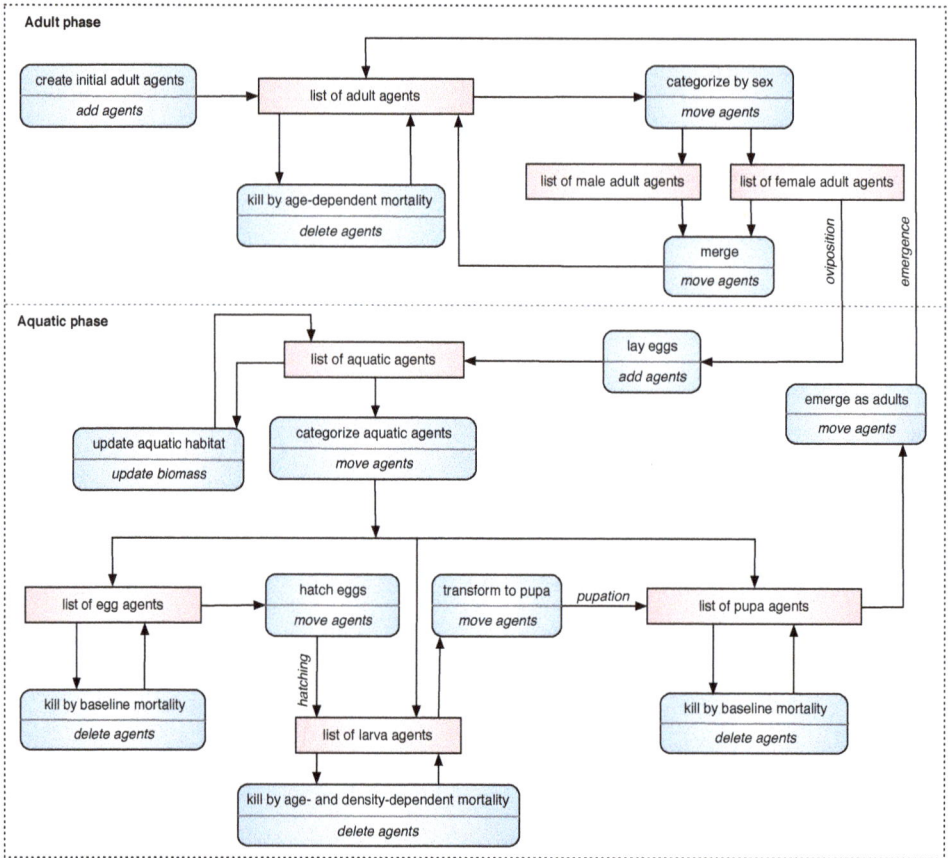

Figure 2. An Event Action List (EAL) diagram for the ABM. Each *squashed rectangle* represents an event-action pair, in which the *event* is denoted at the upper-half, and the *action* is denoted at the lower-half. Each *rectangle* represents the *list(s)* (data structures) of agents affected by the event-action pair.

The host-seeking event starts when a female adult mosquito agent enters the BMS stage and searches for a human blood meal in a house. If the current cell contains a house, it immediately gets a blood meal, and enters the BMD stage to digest the meal, rest, generate new eggs, and eventually enter the G stage to search for an aquatic habitat (if the current cell contains multiple houses, one is chosen at random). If the current cell does not contain any house, a new search event starts as follows. First, the agent's *movement counter* is checked. If the agent is permitted to move, its *Moore* neighborhood *M* is checked. If *M* contains multiple cells that have houses, a random cell *C* (from these cells) is selected, and the agent moves to cell *C*. If cell *C* contains multiple houses, a random house is selected, the agent gets a blood meal, and continues as before. However, if the current cell and its *Moore* neighborhood do not contain any house, the agent starts a random flight and moves randomly into one of the adjacent eight cells (following a previous study [31], the probability of a random move into a diagonally-adjacent cell is set as half that of moving into a horizontally- or vertically-adjacent cell).

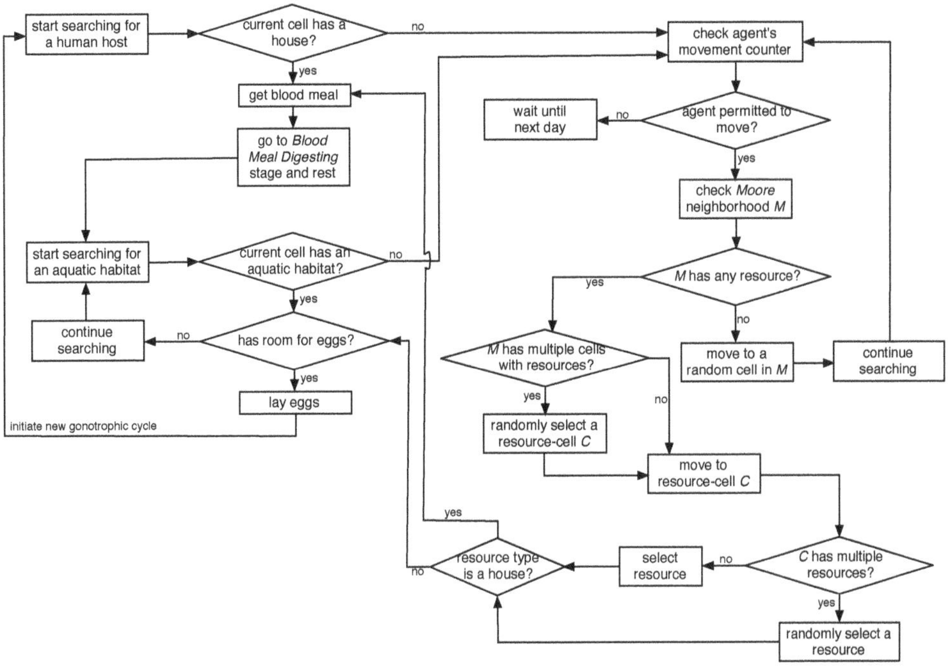

Figure 3. Flight heuristics for mosquito agents.

In an oviposition event, an agent searches for an aquatic habitat. If the current cell contains an aquatic habitat, it's current capacity is checked to see if it has any remaining capacity for new eggs, in which case, the agent lay the eggs (again, if the current cell contains multiple aquatic habitats, one is chosen at random). Once all of the eggs are laid, it goes to the BMS stage, thus initiating a new gonotrophic cycle. If the current cell does not contain any aquatic habitat, the search continues in the same fashion as described above.

As evident from the above, in case of a directional flight, if multiple resources (houses or aquatic habitats) are found within a single cell, a random resource is selected. Note that this strategy can be easily extended/modified for future work to select a resource based on some preference criterion, e.g., to select the house which has the fewest number of mosquitoes visited or to select the aquatic habitat which has the largest remaining capacity.

As evident from the above, in case of a directional flight, if multiple resources (houses or aquatic habitats) are found within a single cell, a random resource is selected. Note that this strategy can be easily extended/modified for future work to select a resource based on some preference criterion, e.g., to select the house which has the fewest number of mosquitoes visited or to select the aquatic habitat which has the largest remaining capacity.

2.6. The Study Area

The study area is located within a subsection of the Siaya and Bondo Districts (Rarieda Division, Nyanza Province) in western Kenya. It comprises a village which is selected from a set of 15 villages with an area of approximately 70 km^2. The greater area is locally known as *Asembo*, which covers an area of 200 km^2 with a population of approximately 60,000 persons [32]. It lies on Lake Victoria and experiences intense, perennial (year-around) malaria transmission [33]. The primary reason for selecting Asembo is the availability of relevant data from the Asembo Bay Cohort Project [34] and the Asembo ITN project [32]. In a series of 23 articles, these studies reported important public health findings from a successful trial of ITNs in western Kenya [35]. The study area is shown in Figure 4: Figure 4A shows the boundary and administrative units for Kenya, Figure 4B shows the selected data layers within the village cluster, and Figure 4C shows the selected village cluster in Asembo, Kenya.

The ABM, without explicit parallelization or multiple runs, can handle a landscape with finite maximum dimensions. Hence, a subset of villages with 95 × 96 cell dimensions is selected for all simulation runs in this study, as outlined by the polygons in Figure 4B,C.

2.7. GIS Processing of Data Layers

ArcGIS Desktop 10 [36] is used to produce, process, and analyze the relevant data layers. Different types of water features and villages (including houses) are identified, extracted and projected to the *Arc 1950 UTM Zone 36S* projection system for all over Kenya. The selected water features include rivers, lakes, wetlands, wells/springs, falls/rapids, lagoons, *etc.* Each water feature type is assigned a unique ID.

The selected features are scaled down to a village cluster around Asembo. Water features for different types of aquatic sites are included. Since the spatial ABM deals with spatial features at the habitat levels, the study area is further scaled down to village and household levels, and then to subsets of villages levels. Some of the water features are ranked by precedence by sub-grouping the water source data layers based on their attributes. Similar types of water features in the same data layer are combined.

The selected data layers are then converted to the raster format, with a cell resolution of 50 m × 50 m. All point shapefiles for aquatic habitats and houses are converted using the *Point to Raster* tool. Since pit latrines are usually found inside the household boundaries, the shapefile for pit latrines is created from the shapefile for houses. It is possible to have more than one feature type within a single cell. In these cases, to calculate the number of features (of each type) in each cell, the summation of value fields of the corresponding data features is used. Finally, the raster files are converted to the ASCII format, and are ready to be used as input to the spatial ABM.

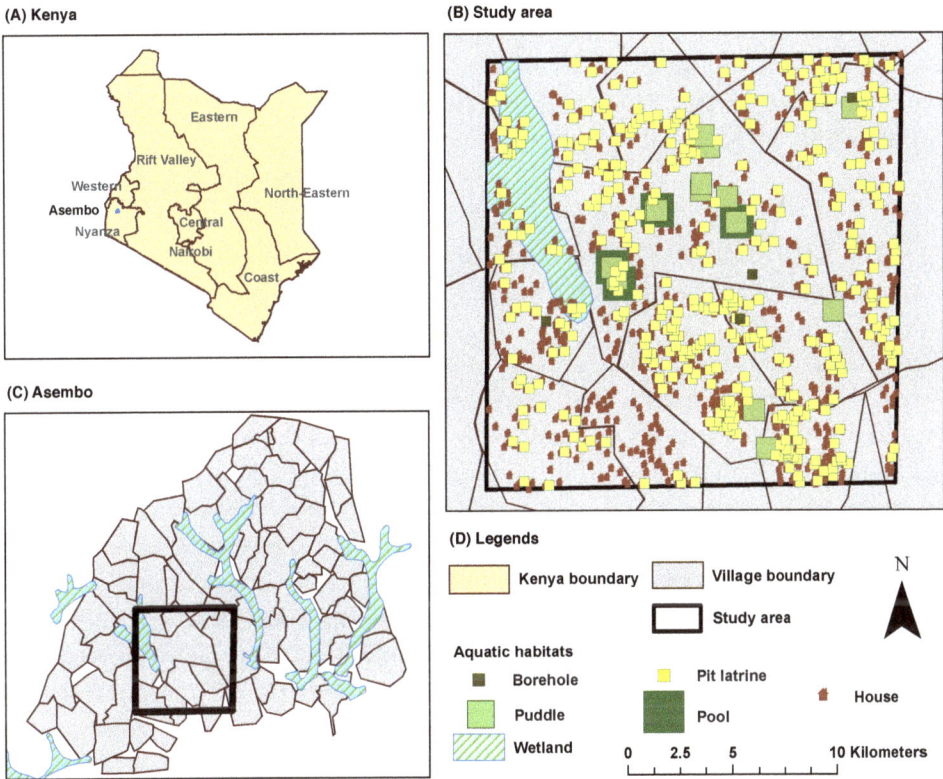

Figure 4. The Study Area. (**A**) Kenya Boundary and Administrative Units (Provinces); (**B**) Study Area with Selected Data Layers; the outlined polygon represents a subset of villages selected for the simulation runs in this study; (**C**) Village Cluster in Asembo; (**D**) Legends.

2.8. Feature Counts

A total of 975 aquatic habitats, categorized into five different types, are identified in the selected area as follows: (1) pools (large); (2) puddles (small); (3) pit latrines; (4) boreholes; and (5) wetland. Boreholes, also known as borrow pits, have significant potential as breeding sites in the area. They represent man-made holes or pits in the ground when local people use clay or soil for building houses, making pots, *etc.*, thereby leaving depressions in the ground that easily get filled with rain water. Pit latrines are very common to the households in the area. The wetland represents a stretch of marsh lying to the northwest corner of the area which is dominated by herbaceous plant species.

As mentioned before, each aquatic habitat is set with a predefined *carrying capacity* (CC), which regulates the aquatic mosquito population that the habitat can sustain, and reflects the habitat heterogeneity (e.g., in terms of productivity) to some degree (see [15] for details). A total of 941 houses, each having a mean of five occupants, are also identified. These feature counts and their assigned values are summarized in Table 2. Note that for wetland, which covers multiple cells in the northwest corner of the study area, the same CC value is assigned to each cell.

2.9. Vector Control Interventions

The last decade (2000–2010) of worldwide malaria control efforts has seen an unprecedented increase in the coverage of vector (mosquito) control interventions for malaria, with ITNs/LLINs, IRS, and LSM as the front-line vector control tools [37]. These interventions are often applied in isolation and in combination, and their impacts have been investigated by numerous early and recent studies [38–45]. In addition to these time-tested, established tools, new and novel intervention tactics and strategies such as new drugs, vaccines, insecticides, improved surveillance methods, *etc.*, are also being investigated [46]. Some of the promising approaches include genetically engineered mosquitoes through sterile insect technique (SIT) or release of insects containing a dominant lethal [47,48], fungal biopesticides that increase the rate of adult mosquito mortality [49], the development of genetically modified mosquitoes (GMMs) or transgenic mosquitoes manipulated for resistance to malaria parasites [50], transmission blocking vaccines (TBVs) which are intended to induce immunity against the malaria parasites [51], *etc.*

As mentioned before, the combined impacts of two vector control interventions (LSM and ITNs) are evaluated for this study. Both interventions have been extensively used as intervention tactics to reduce and control malaria in sub-Saharan Africa, as reported by numerous early and recent studies [37,39,41,43]. LSM (also known as source reduction) is one of the oldest tools in the fight against malaria. It refers to the management of aquatic habitats in order to restrict the completion of immature stages of mosquito development. ITNs, particularly LLINs, are considered among the most effective vector control strategies currently in use [39,52]. ITNs offer direct personal protection to users as well as indirect community protection to non-users (through insecticidal and/or repellent effects). For this study, LSM refers to the permanent elimination of targeted aquatic habitats. For ITNs, the *household-level complete coverage* scheme is used, which ensures that if a house is covered, all persons in the house are protected by bed nets; the two other relevant variables, killing effectiveness and repellence, are both fixed at 50%. *Killing effectiveness* refers to an increased mortality (increased probability of death of a mosquito), toxicity, or killing efficiency due to the insecticidal killing effects of the ITNs; the insecticide kills the mosquitoes that come into contact with the ITNs. *Repellence* refers to the insecticidal excito-repellent properties of the ITNs which repel the blood meal seeking mosquitoes; it adds a chemical barrier to the physical one, further reducing human-mosquito contact and increasing the protective efficacy of the ITNs (see [15,16] for details).

Four different scenarios are constructed by using two coverage (C) levels of *low* (20%) and *high* (80%). For a specific coverage, aquatic habitats and houses which will be covered by the corresponding intervention are selected by using random sampling. The actual numbers of objects covered approximate the desired coverage levels. A baseline scenario (with no intervention) is also added for comparison. The scenarios are summarized in Table 3.

Table 3. Scenarios obtained by applying the two vector control interventions LSM and ITNs. A total of 975 aquatic habitats and 941 houses are used to calculate the desired coverage (C) levels of *low* (20%) and *high* (80%). The first column denotes the scenario (interventions applied). The actual coverage (C) levels are given in the last two columns for aquatic habitats and houses covered in the landscape, respectively.

Scenario	Coverage (C) %	
	% Aquatic Habitats Covered	% Houses Covered
Baseline	0	0
$LSM_{Low} - ITNs_{Low}$	$208/975 = 0.21$	$204/941 = 0.22$
$LSM_{Low} - ITNs_{High}$	$215/975 = 0.22$	$751/941 = 0.8$
$LSM_{High} - ITNs_{Low}$	$774/975 = 0.79$	$195/941 = 0.21$
$LSM_{High} - ITNs_{High}$	$781/975 = 0.8$	$736/941 = 0.78$

2.10. Simulations

For each of the five scenarios (*Baseline*, $LSM_{Low} - ITNs_{Low}$, $LSM_{Low} - ITNs_{High}$, $LSM_{High} - ITNs_{Low}$, and $LSM_{High} - ITNs_{High}$), 50 replicated simulation runs are performed and the average results are reported (in order to rule out any stochasticity effects). Each simulation runs for 730 days (2 years) (in this paper, all time units related to the simulation runs refer to *simulated time* as opposed to *physical time* or *wall clock time*; thus, a 2 years run indicates a virtual simulation run within the computer which represents an imitation of operations in the real-world for the same time duration), and reaches a steady state (equilibrium) at around day 50. Interventions are applied on day 100 and continued up to the end of the simulation.

Initially, all simulations start with 1000 female adult mosquito agents (no male agents). Each female agent is assigned to a randomly-selected aquatic habitat. The maximum daily range of movement for mosquito agents is set to 5 cells per day, which translates to $250\sqrt[2]{2}$ m. Biological aging (senescence) of the mosquitoes is assumed. The ABM implements age-specific mortality rates for the adult mosquitoes and the larvae (*i.e.*, the probability of death for mosquito agents increases with their age).

2.11. Output Indices

Mosquito abundance is the primary output index of the ABM. However, the spatial model also allows us to explore some spatial indices by overlaying these on the entire landscape. These indices capture the spatial heterogeneity of various objects (aquatic habitats and houses) in the landscape. Some of these indices are generated as *cumulative aggregates* at the end of each simulation run, and represent measures on a *per object* basis. The output indices are listed below:

1. Mosquito Abundance: represents a spatial *snapshot* of the female adult mosquito population distribution at the end of simulations (see Figures 5 and 6)
2. Oviposition Count per Aquatic Habitat: for each aquatic habitat x, it represents the *cumulative* number of female adult mosquitoes which have oviposited (laid eggs) in x; depicted spatially at the end of simulations by overlaying on top of the aquatic habitats (see Figures 7 and 8)
3. Blood Meal Count per House: for each house y, it represents the *cumulative* number of blood meals successfully obtained by female adult mosquitoes in y; depicted spatially at the end of simulations by overlaying on top of the houses (see Figures 9 and 10)

Note that for all output indices, the average measures of 50 replicated simulation runs are reported (in order to rule out any stochasticity effects). The spatial indices are sampled across all daily time steps throughout the entire simulations. The output maps are produced by overlaying the *averaged* indices on top of the relevant data layers.

All output indices are mapped using the graduated symbology. The graduated symbol renderer is one of the common renderer types used to represent quantitative information. Using a graduated symbols renderer, the quantitative values for the output indices are separately grouped into ordered classes, so that higher values cover larger areas on the map. Within a class, all features are drawn with the same symbol. Each class is assigned a graduated symbol from the smallest to the largest.

2.12. Hot Spot Analysis

Using spatial statistics tools, hot spot analysis (spatial cluster analysis) is performed for all scenarios for the last two indices (*oviposition count per aquatic habitat* and *blood meal count per house*) in order to determine whether a specific value is statistically significant or not [53]. In hot spot analysis, if a higher value is surrounded by similar magnitude of other higher values, it is considered a hot spot (with 95% or 99% confidence intervals). The cold spots are determined using the same principle. The values (or cluster of values) between the statistically significant hot spots and cold spots are considered as random samples of a distribution. The hot spot analysis tool calculates the Getis-Ord Gi* statistic (z-scores and p-values) for each feature in a dataset [36]. Z-scores are measures of standard deviations, and define the confidence intervals (in this case, 95%–99%). A p-value represents the probability that the observed spatial pattern was created by some random process.

The *null hypothesis* for pattern analysis essentially states that the expected pattern is just one of the many possible versions of complete spatial randomness. If the z-score is within the 95%–99% confidence interval or beyond, the exhibited pattern is probably too unusual to be of random chance, and the p-value will be subsequently small to reflect this. In this case, it is possible to reject the null hypothesis and proceed to determine the cause of the statistically significant spatial pattern. On the other hand, if the z-score lies below the 95% confidence interval, the p-value will be larger, the null hypothesis cannot be rejected, and the pattern exhibited is more likely to indicate a random pattern. Thus, a high z-score and small p-value for a feature indicates a significant hot spot. Conversely, a low negative z-score and small p-value indicates a significant cold spot.

2.13. Kriging Analysis

Kriging, also known as *Gaussian process regression*, is a popular method of interpolation (prediction) for spatial data. It is an interpolation technique in which the surrounding measured values are weighted to derive a predicted value for an unmeasured location. Weights for the measured values depend on the distance between the measured points, the prediction locations, and the overall spatial arrangement among the measured points [54]. Various kriging techniques provide a framework for predicting values of a variable of interest at unobserved locations given a set of spatially distributed data, incorporating spatial autocorrelation and computing uncertainty measures around model predictions [55,56].

In recent years, kriging has been extensively used in public heath and epidemiology modeling for variable mapping to interpolate estimates of occurrence of a variable or risk of disease [57–59]. For example, de Carvalho Alves and Pozza characterized the spatial variability of common bean anthracnose using kriging and nonlinear regression models [60]. Alexeeff *et al.* evaluated the accuracy of epidemiological health effect estimates in linear and logistic regression when using spatial air pollution predictions from kriging and land use regression models [61]. For malaria modeling, the Malaria Atlas Project (MAP) [62] developed several Bayesian geostatistical kriging models for spatial prediction of *Plasmodium falciparum* prevalence, estimated human populations at risk, vector distribution, *etc.*, generating malaria maps of many endemic countries in sub-Saharan Africa [63–65].

The basic idea of kriging is to predict the value of a function at a given point by computing a weighted average of the known values of the function in the neighborhood of the point. To this end, kriging is closely related to the method of regression analysis. The data represent a set of observations of some variable(s) of interest, with some spatial correlation. Usually, the result of kriging is the expected value, referred to as the *kriging mean* and the *kriging variance* computed for every point within a region of interest. If kriging is done with a known mean, it is then called *simple kriging*. On the other hand, in *ordinary kriging*, estimating the mean and applying (simple) kriging are performed simultaneously.

Kriging uses semivariogram functions to describe the structure of spatial variability. A semivariogram is one of the significant functions to indicate spatial correlation in observations measured at sample locations, and plays a central role in the analysis of geostatistical data using kriging. The effect of different semivariograms on kriging has also been a focus of interest in different branches of the literature (e.g., [66]). In this paper, spatial analysis is conducted using ordinary kriging with circular semivariograms for all scenarios for all the output indices using ArcGIS 9.3 [19]. We note that similar analyses have also been conducted for other insects in the literature (e.g., for fig fly [67]).

3. Results

In this section, we describe the results by categorizing them according to the output indices. For the output indices and scenarios (see Table 3), simulation results are presented along with hot spot analysis and kriging results. For clarity, houses and pit latrines are not shown in the output maps. Each scenario (in the output maps) represents the average results of 50 replicated simulations.

3.1. Mosquito Abundance

The mosquito abundance maps are shown in Figure 5. These maps depict the *mosquito abundances* index, which represent a spatial snapshot of the female adult mosquito population distribution at the end of simulations. Figure 5A shows the abundance map for the baseline scenario (in which no intervention was applied). Figure 5B depicts the symbols used in the maps: it shows the village boundary, different types of aquatic habitats, and the graduated symbols for abundances. Note that for the aquatic habitats, the symbol sizes vary according to the assigned carrying capacities of the habitats (see Table 2). The symbol sizes for abundances also vary depending on the magnitudes. Figure 5C–F show the abundance maps for the four different scenarios with control interventions LSM and ITNs having two coverage levels: $LSM_{Low} - ITNs_{Low}$, $LSM_{Low} - ITNs_{High}$, $LSM_{High} - ITNs_{Low}$, and $LSM_{High} - ITNs_{High}$, respectively. The corresponding kriged maps for mosquito abundance are illustrated in Figure 6.

As shown in Figure 5, with increasing coverage levels of both interventions, the mosquito abundances are significantly reduced, as evident from the progressively lower number of "Above 40" symbols (which denote the highest abundances) in the series of figures. The changes are more clear and evident from the kriged maps (Figure 6).

It is interesting to note that ITNs are more effective in reducing abundances than LSM (compare Figure 5D,E as well as the kriged maps in Figure 6D,E): covering 80% of the houses has more impact than removing a total of 80% different types of the aquatic habitats. This is partially due to the fact that the household-level complete coverage scheme (used for ITNs, see Experimental Section) prohibits a blood meal-seeking female mosquito to obtain a blood meal from any person in any house which is covered by ITNs. As coverage of ITNs increases, more houses fall within the range of coverage, and the probability of finding an unprotected human in another house (during the blood meal-seeking stage) decreases. Thus, with increasing coverage of ITNs, abundances are reduced more effectively.

The *low* (20%) coverage levels for both interventions do not produce significant reduction in abundances, as evident from the baseline and $LSM_{Low} - ITNs_{Low}$ maps (compare Figure 5A,C). In general, higher abundances are observed near the pools (which have the highest carrying capacities) and in the north east and the south east portions of the map.

When either of the interventions has a *high* (80%) coverage level, abundances are significantly reduced, as evident from the $LSM_{Low} - ITNs_{High}$ and $LSM_{High} - ITNs_{Low}$ maps. For these two scenarios, the highest abundances observed are significantly lower than the baseline (compare Figure 5D,E with Figure 5A). However, for the $LSM_{Low} - ITNs_{High}$ scenario higher abundances do not always coincide with the spatial locations of aquatic habitats with higher carrying capacities, while for the other scenario this expected trend is observed for some cases.

Not surprisingly, when both interventions have *high* (80%) coverage levels, abundances are reduced to the lowest level, as evident from the $LSM_{High} - ITNs_{High}$ map shown in Figure 5F. For this scenario very few higher abundances are observed; these occur at greater distances from the spatial locations of aquatic habitats with higher carrying capacities, since most of them are eliminated by LSM.

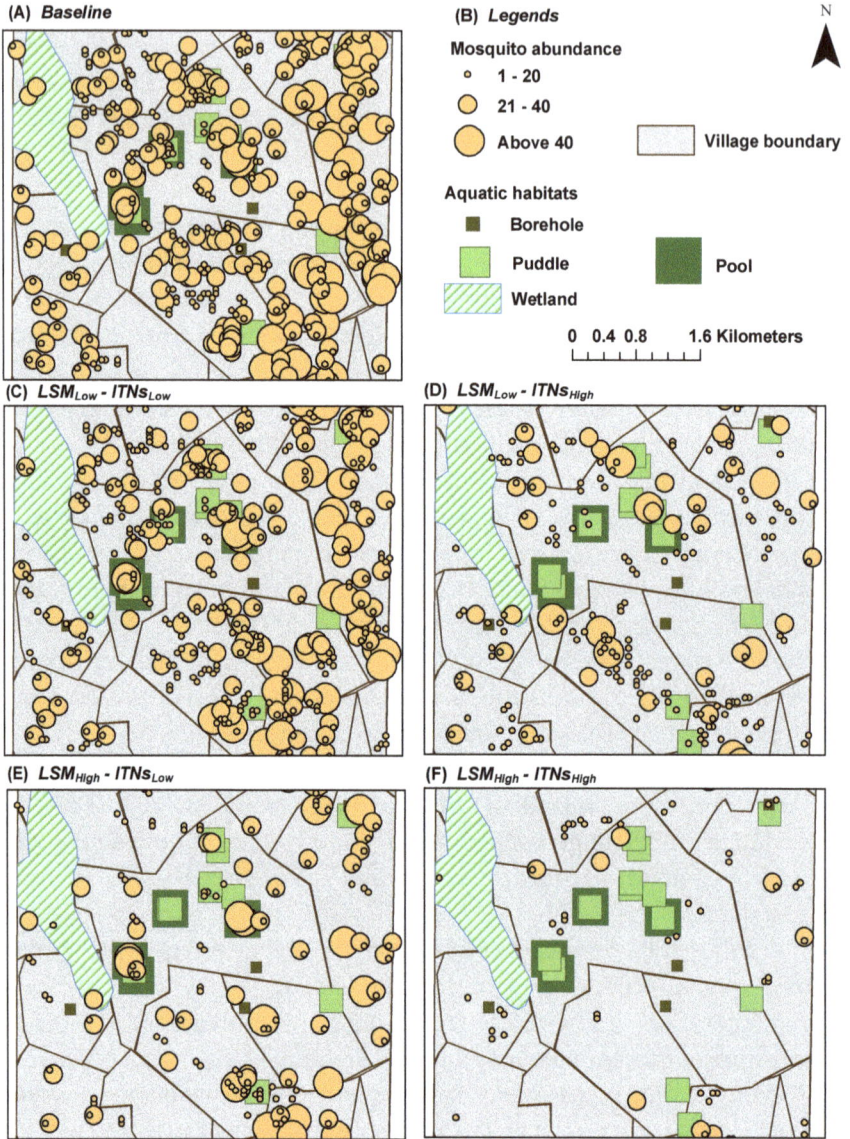

Figure 5. Maps for all scenarios for the *mosquito abundances* index. Each scenario represents the average results of 50 replicated simulations. (**A**) Abundance map for baseline; (**B**) Legends: symbol sizes are proportional to the carrying capacities of the aquatic habitats (see Table 2); graduated symbol sizes are proportional to the magnitudes of abundances. For clarity, houses and pit latrines are not shown; (**C**) Abundance map for $LSM_{Low} - ITNs_{Low}$; (**D**) Abundance map for $LSM_{Low} - ITNs_{High}$; (**E**) Abundance map for $LSM_{High} - ITNs_{Low}$; (**F**) Abundance map for $LSM_{High} - ITNs_{High}$.

(A) *Baseline*

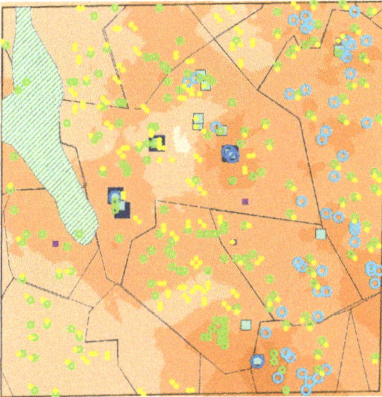

(B) *Legends*

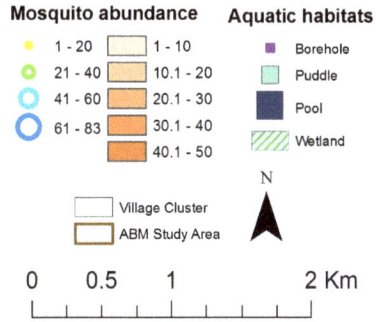

(C) *LSM$_{Low}$ - ITNs$_{Low}$*

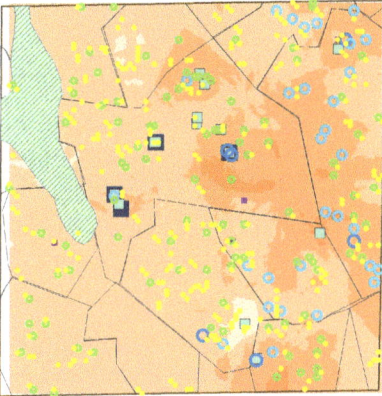

(D) *LSM$_{Low}$ - ITNs$_{High}$*

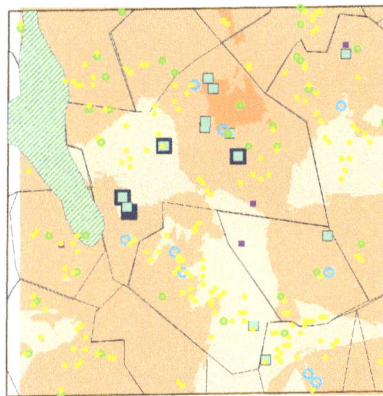

(E) *LSM$_{High}$ - ITNs$_{Low}$*

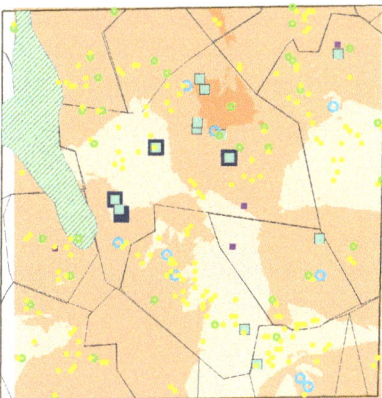

(F) *LSM$_{High}$ - ITNs$_{High}$*

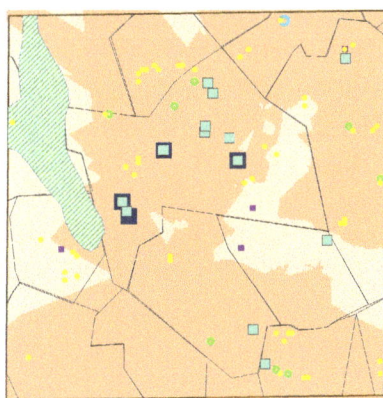

Figure 6. Kriged maps for all scenarios for the *mosquito abundances* index. (**A**) Kriged abundance map for baseline; (**B**) Legends; (**C–F**) The four intervention scenarios.

3.2. Oviposition Count per Aquatic Habitat

Results for the *oviposition count per aquatic habitat* index are shown in Figure 7. These maps depict the cumulative number of female adult mosquitoes which have oviposited (laid eggs) in the aquatic habitats, as well as the predicted hot spots and cold spots identified by hot spot analysis. For the five scenarios, oviposition counts for the aquatic habitats are placed into three ordered classes of 1–$20,000$, $20,001$–$50,000$ and *above* $50,000$ using the same quantitative scale, and are shown using graduated symbols. Hot spots and cold spots are spatially clustered using two confidence interval (CI) levels of 95% and 99%. The legends denote the color-coding for the classes, the hot spots, the cold spots, and the CIs.

Figure 7A shows a higher frequency of higher values for the *oviposition count per aquatic habitat* index in the baseline map. Significant number of these appear to be statistically significant, and hence considered as hot spots. Notable clustering of lower values can also be seen over the wetland area (where each cell is assigned a tiny CC), which are categorized as cold spots.

Figure 7C shows a drop in frequency of higher values in the $LSM_{Low} - ITNs_{Low}$ map for the same index, about half of which are considered as hot spots. In addition, more cold spots can be seen over the wetland area. Both of these results can be explained as the effects of *low* coverage levels for both interventions.

When either of the interventions has a *high* coverage level, frequencies of higher values are further reduced, as evident from the $LSM_{Low} - ITNs_{High}$ and $LSM_{High} - ITNs_{Low}$ maps in Figure 7D,E, respectively. For the $LSM_{Low} - ITNs_{High}$ scenario, some moderate oviposition counts become statistically significant, fewer hot spots are detected, and most of the cold spots are eliminated from the wetland area. On the other hand, the $LSM_{High} - ITNs_{Low}$ scenario has higher frequencies of higher oviposition counts, hot spots, and cold spots. These observations confirm to our previous results (for abundances) that ITNs are more effective in reducing oviposition counts than LSM. As before, when both interventions have *high* coverage levels, frequencies of higher oviposition counts, hot spots, and cold spots are reduced to the lowest level, as evident from the $LSM_{High} - ITNs_{High}$ map shown in Figure 7F.

Similar deductions can be made from the kriged maps presented in Figure 8. For example, when both interventions are applied with higher coverages (Figure 8F), areas with the light blue and green colors representing the two highest levels of oviposition counts are simply non-existent from the map, and the third one (light brown) is greatly diminished. This illustrates the drastic reductions in the oviposition counts.

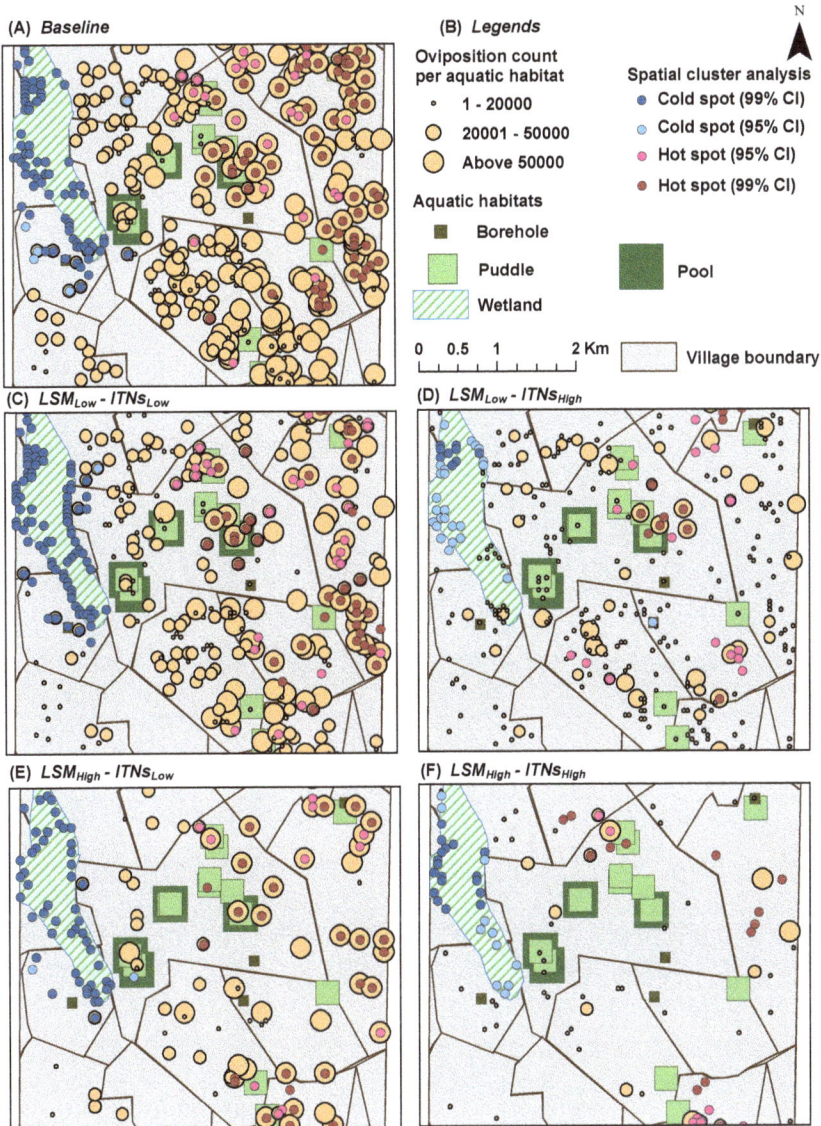

Figure 7. Maps for all scenarios for the *oviposition count per aquatic habitat* index. Each scenario represents the average results of 50 replicated simulations. Oviposition counts are categorized using the same quantitative scale, and are shown using graduated symbols which are proportional to the magnitudes. For clarity, houses and pit latrines are not shown. Hot spots and cold spots are spatially clustered using two confidence intervals (CIs) of 95% and 99%. (**A**) Baseline; (**B**) Legends; (**C–F**) The four intervention scenarios.

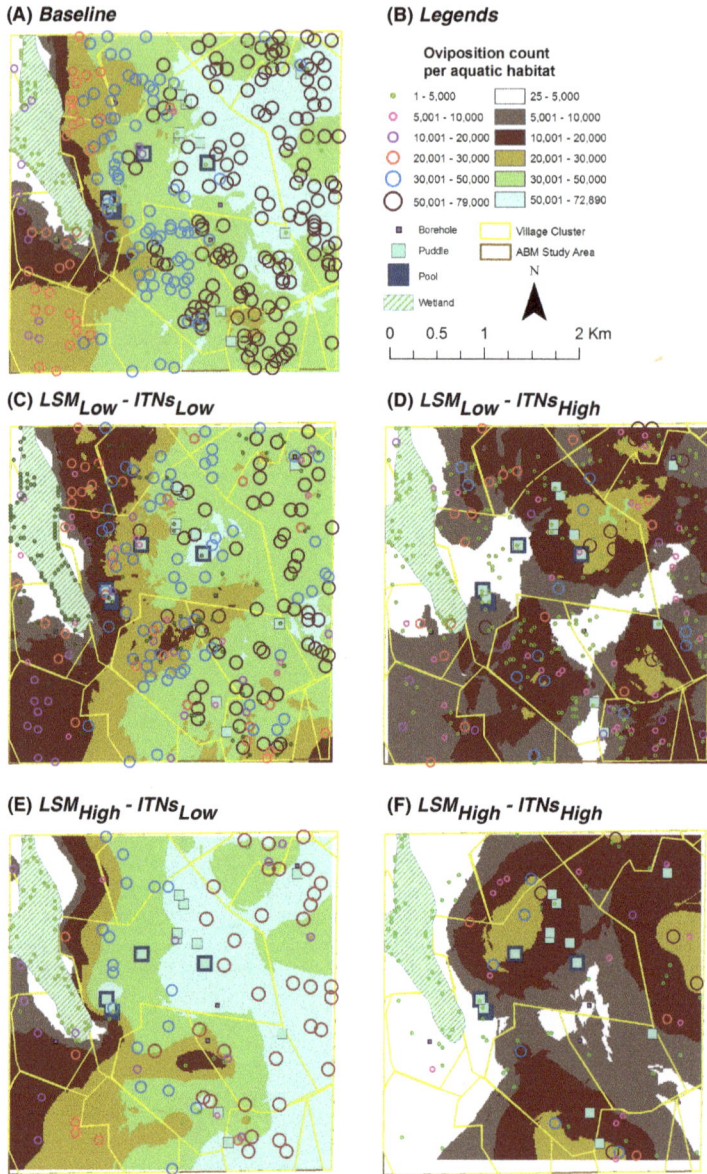

Figure 8. Kriged maps for all scenarios for the *oviposition count per aquatic habitat* index. (**A**) Baseline; (**B**) Legends; (**C–F**) The four intervention scenarios.

3.3. Blood Meal Count per House

Results for the *blood meal count per house* index are shown in Figure 9. These maps depict the cumulative number of blood meals obtained by female adult mosquitoes in the houses, as well as the predicted hot spots and cold spots identified by hot spot analysis. For the five scenarios, blood meal counts for the houses are placed into three ordered classes of 1–3000, 3001–9000, and *above* 9000 using the same quantitative scale, and are shown using graduated symbols. Hot spots and cold spots are spatially clustered using two confidence interval (CI) levels of 95% and 99%. The legends denote the color-coding for the classes, the hot spots, the cold spots, and the CIs.

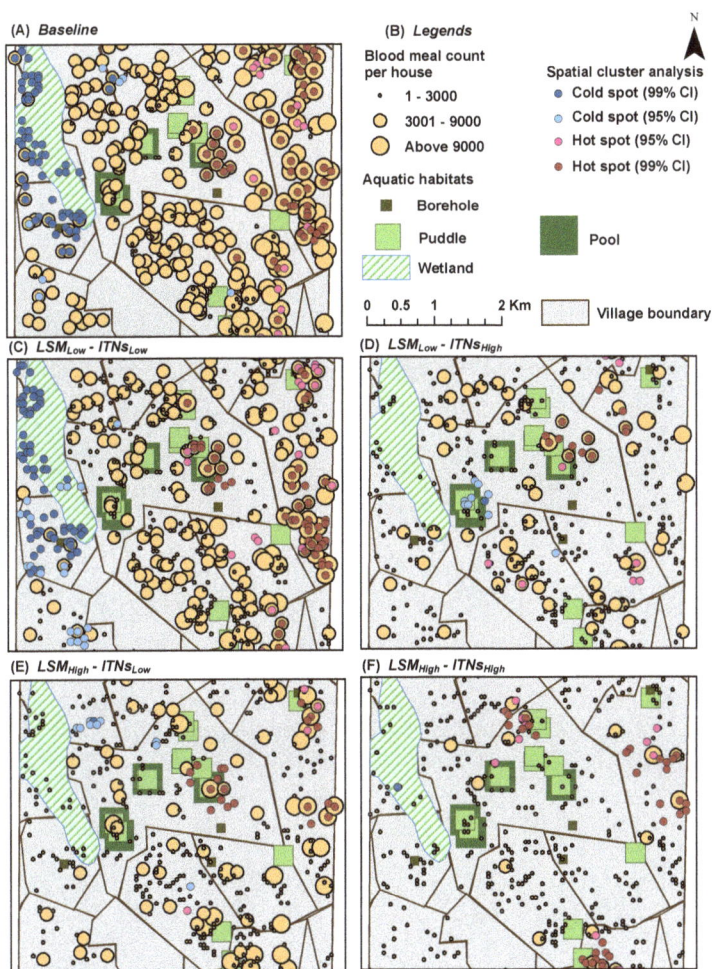

Figure 9. Maps for all scenarios for the *blood meal count per house* index. Each scenario represents the average results of 50 replicated simulations. Blood meal counts are categorized using the same quantitative scale, and are shown using graduated symbols which are proportional to the magnitudes. For clarity, houses and pit latrines are not shown. Hot spots and cold spots are spatially clustered using two confidence intervals (CIs) of 95% and 99%. (**A**) Baseline; (**B**) Legends; (**C–F**) The four intervention scenarios.

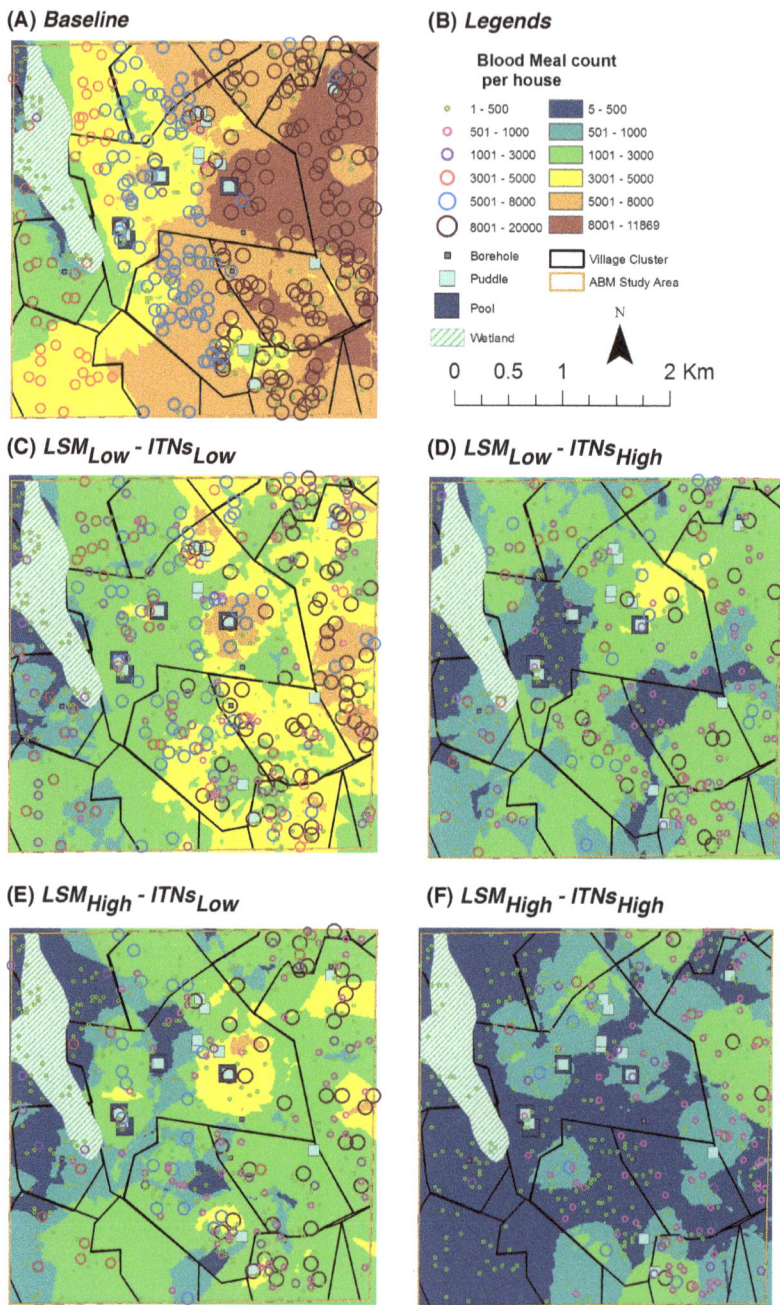

(A) *Baseline*

(B) *Legends*

Blood Meal count
per house

●	1 - 500		5 - 500
○	501 - 1000		501 - 1000
○	1001 - 3000		1001 - 3000
○	3001 - 5000		3001 - 5000
○	5001 - 8000		5001 - 8000
○	8001 - 20000		8001 - 11869

■ Borehole   ☐ Village Cluster
☐ Puddle   ☐ ABM Study Area
■ Pool
▨ Wetland

N

0   0.5   1   2 Km

(C) *LSM$_{Low}$ - ITNs$_{Low}$*

(D) *LSM$_{Low}$ - ITNs$_{High}$*

(E) *LSM$_{High}$ - ITNs$_{Low}$*

(F) *LSM$_{High}$ - ITNs$_{High}$*

Figure 10. Kriged maps for all scenarios for the *blood meal count per house* index. (**A**) Baseline; (**B**) Legends; (**C–F**) The four intervention scenarios.

The *blood meal count per house* index results show similar trends as observed for the *oviposition count per aquatic habitat* index results. Similar trends are also noticed from the kriged maps presented in Figure 10. The baseline map possesses the highest frequencies of higher values, hot spots, and cold spots (Figure 9A), all frequencies are reduced (with the introduction of a few cold spots in the lower-left area) when both interventions have *low* coverage levels (Figure 9C), ITNs (*high* coverage level) are more effective than LSM with further reduction in frequencies of higher values (Figure 9D,E), and very few higher values, hot spots, and cold spots remain when both interventions have *high* coverage levels (Figure 9F). Interestingly, the $LSM_{Low} - ITNs_{High}$ map shows some cold spots in an area where a few aquatic habitats with higher carrying capacities exist, and the $LSM_{High} - ITNs_{Low}$ map possesses very few cold spots. Similar trends are also observed in the corresponding kriged maps (Figure 10).

In general, statistically significant higher values are detected over the north east and the south east portions of the maps, as these portions contain more number of houses (hence more blood meal counts per house). This is also evident from the kriged maps. The central portions depict mostly random distribution of values which are not detected as hot spots.

4. Discussion

This study has presented a landscape epidemiology modeling framework to integrate the simulation results from a spatial ABM of malaria-transmitting mosquitoes with a GIS and then to apply spatial statistics techniques on the model outputs. Some of the key features, characteristics, and limitations of the framework are highlighted below.

4.1. Stochasticity and Initial Conditions

The ABM involves substantial stochasticity in the forms of probability-based distributions and equations. The mosquito agents' decisions and actions are often simulated using random draws from certain distributions. These sources of randomness are used to represent the diversity of model characteristics. To rule out any stochasticity effects introduced by these probabilistic events, 50 replicated simulation runs are performed for each simulation in each of the five scenarios, and their aggregate measures are reported in the form of averages.

To verify whether 50 replicated runs are enough for each simulation, we ran as many as 120 replicates of each simulation (using the versions of the ABM available at that point in time) in the earlier phases of model development (to be specific, during the verification, validation, and replication phases). After analyzing the results, it became apparent that roughly 30 replicates were enough to rule out most issues regarding stochasticity, initial seed bias, bifurcation, and other chaos factors. We also verified that the average could be treated as a deterministic measure for the mosquito abundance outputs of the ABM. In addition, the replication study also helped in model-to-model comparison and cross-model validation of the different versions (developed by individual authors) of the ABMs. Some of these results were presented in [14,15,68,69].

The initial uniform random assignment of female agents to arbitrary aquatic habitats does not affect the current emerging outcomes of the ABM. This was previously ensured as part of the verification and validation (V&V) studies of the ABMs by considering longer running times and with multiple initial random seeds to check for robustness [13,14,68,69]. In fact, this holds true for both cases of *with* and *without* the landscape approach, *i.e.*, when the simulations are run in spatial and non-spatial modes, respectively (these results are not included in the current paper).

4.2. Emergence

In general, an important characteristic of an ABM is its capability to capture emergent phenomena resulting from the interactions of the individual agents from the bottom up (after the simulation reaches equilibrium or steady state). To this regard, our ABM exhibits the emerging spatial distribution of mosquito agents once the simulations reach equilibrium on or after day 50. The emergence is primarily governed by two factors: (1) the assigned carrying capacities of the aquatic habitats; and (2) the spatial heterogeneity of the landscapes, which translates to the distributions and densities of houses and aquatic habitats. In the simulations, the 50-days warm-up period ensures that the model has reached steady state, and should not be treated as an absolute value. Each generation of the mosquitoes requires ≈15 days to become mature, and it takes ≈2–3 generations for the initial model to reach equilibrium. Thus, a 50-days warm-up period would have been sufficient in most cases. Note that the interventions (LSM and ITNs) are applied after day 100, and continued up to the end of the simulation. This longer period (100-days) also guards against oscillatory spikes in the abundance, which may occur due to several factors such as generation-to-generation oscillation tendency, density-dependence and skip-oviposition effects, short hiatus in egg-laying, *etc.* [15,16].

4.3. Complexity

In many complex systems, cause and effect relationships are usually not proportional to each other; as a result, manipulation attempts are often resisted, which may lead to an unexpected systemic shift or phase transition (the so-called *tipping points* or *critical points*) [70]. In the spatial ABM, such tipping points may occur with certain combinations of the intervention parameters. For this study, the coverage levels of 0.2 and 0.8 were used for both interventions. They should be treated as representative sample points which resemble two points closer to the opposite ends of the 0.0–1.0 coverage continuum (hence, representing coverage levels on the two extremes of low and high, respectively). Earlier, we tested the ABM by running simulations with varying levels of coverages including 0.2, 0.4, 0.6, 0.8, *etc.* (along with varying levels of repellence and mortality/insecticidal effect for the ITNs) [15]. Within these ranges and parameter settings, the simulations approached several tipping points with specific combinations of the three parameters. For example, in a landscape with high density of houses, 90% reductions in mosquito abundance were achieved with LSM coverage of 0.6, ITNs coverage of 0.87, and ITNs mortality of 0.5 [15].

4.4. Data Resolution (Granularity)

The choice of spatial, temporal, and spectral resolutions determines the degree of precision, realism, and general applicability of the models [7]. Even with the recent rapid advances in computing power, these factors cannot always be maximized simultaneously. Although the resolution of the co-ordinates recorded in a modern GIS may now be of the order of only a few metres, the modeled resolution must be carefully decided so that it reflects the specific study, its objectives, and the objects being mapped: it should be sufficiently high to allow meaningful inferences to be made from the results, but not too high to include irrelevant details. For this study, the spatial resolution (granularity) of the landscapes is chosen as 50 m × 50 m. This selection is based on several factors, some of which include the spatial GIS data availability, the number of maximum cells which can be practically processed by the ABM (within bounded run-time), the limited flight ability and perceptual ranges of mosquitoes, *etc.* The selected granularity may seem to be low (with a cell-size of 50 m × 50 m), particularly given the other assumptions on the distances that a mosquito agent can fly. However, in the future, with the availability of higher resolution spatial data and an advanced version of the current ABM capable of processing multiple spatial nodes in parallel (e.g., by using the message passing interface (MPI) technique), we plan to simulate landscapes with higher spatial resolution.

Due to the lack of detailed spatial data for aquatic habitats and demographic data for human populations and houses, arbitrary carrying capacities and occupants are assigned to the habitats and houses, respectively. However, the current study does ensure that the relative magnitudes of aquatic capacities follow the biological reality of the environment being modeled; for example, a pool cell possesses higher CC than a wetland cell, as described in Table 2. The flexible architecture of the modeling framework also provides an easy plug-in mechanism of such data from relevant future studies into the models.

4.5. Spatial Analysis

In hot spot analysis, the higher frequency of cold spots for the oviposition counts and blood meal counts along the wetland may seem counter-intuitive (see Figures 7 and 9). However, this anomaly can be explained by considering two primary factors: (1) the distributions of and the relative distances between the two types of resources (houses and aquatic habitats) along the wetland; and (2) the tiny carrying capacities assigned to each wetland cell (10 per cell, see Table 2). Both these indices (oviposition and blood meal counts) will have higher values depending on the successful completion of the cycles of alternate feeding and laying of eggs by adult female mosquito agents (the gonotrophic cycle). However, along the wetland (more noticeably along the western edge of the wetland where a larger density of cold spots are present), despite the presence of a few nearby houses, the lack of any nearby higher-capacity water bodies and the collective lower capacity (of the wetland cells and a few pit latrine cells, see Figure 4B) prevent the female mosquitoes to complete their gonotrophic cycles. As a result, higher frequencies of cold spots are generated along the wetland for both indices. Also, in most cases, cold spots are absent along the south east portion of the wetland since it is closer to both types of resources (in this case, large pools and houses). Eventually, this translates to the degree of ease with which adult female mosquitoes may find resources, and can also be quantitatively measured by considering the *average travel time (ATT)* required by a female mosquito to complete each gonotrophic cycle (ATT is inversely proportional to resource-densities; for more details, see, e.g., [13]).

As the spatial distribution results show, there is a strong correlation between the *a priori* distribution of houses and aquatic habitats and the emerging distribution of hot and cold spots. Thus, in general, the hot spots of our output indices occur near the clusters of houses and aquatic habitats. Since there are 395 pit latrines distributed almost all over the study area (in fact, covering almost all the house clusters), in effect, there are indeed some aquatic habitats near almost every house-habitat cluster. Recall that the flight heuristics do not distinguish among the types of aquatic habitats (*i.e.*, mosquito agents select the habitats randomly), and the agents do not engage in a directional flight during the simulations until and unless the aquatic habitats are found in the neighboring cells (see Section 2.5).

The strong spatial correlation, although not quantitatively measured in this study, is evident at some portions of the study area where there are some houses with no aquatic habitat in the vicinity (*i.e.*, without enough pit latrines nearby). For example, as shown in Figure 4B, both the eastern portion of the south-west quadrant and most of the eastern edge of the wetlands portray two house clusters with very few or no aquatic habitats (including pit latrines) in the vicinity. As a consequence, these areas contain almost no hot spots, as depicted in the hot spot analysis results (see Figures 5, 7, and 9), which hold true for both cases of *with* and *without* the mosquito control interventions.

For the entire study area, kriging analysis produces predicted values for unmeasured spatial locations, which are derived from the surrounding weighted measured values. Most of the spatial trends observed by the hot spot analysis are also visible in the kriging analysis results.

4.6. Habitat-Based Interventions

In this study, habitats and houses were selected using random sampling for the vector control interventions. However, given the power of ABMs, other sophisticated, habitat-based strategies for interventions are also equally applicable. For example, latrines and boreholes for LSM, or a firewall of ITNs at the village boundary can be excellent choices to target first or in a limited-resource setting.

Some of the habitat-based strategies were investigated by using targeted and non-targeted LSM in a previous work [15]. The targeted interventions removed the aquatic habitats within 100, 200, and 300 m of surrounding houses, while the corresponding non-targeted interventions randomly removed the same numbers of habitats. In general, with LSM applied in isolation, the results agreed with the findings of previous research that LSM coverage of 300 m surrounding all houses can lead to significant reductions in abundance, and, while targeting aquatic habitats to apply LSM, distance to the nearest houses can be an important measure. Similar research questions are also being investigated with spatial repellents (e.g., mosquito coils). However, given the constraints, we did not include the results of habitat-targeted interventions in this paper.

4.7. Miscellaneous Issues

In the ABM, the human population is modeled as static (*i.e.*, humans do not move in space), all humans are assumed to be identical, and human mortality is not implemented. This may be one of the reasons for the unusually high blood meal counts per house (in the range of thousands). In the future, with the inclusion of explicit parasite population (as agents) and the availability of detailed demographic data of human populations and houses in Asembo, we plan to parameterize and calibrate the model to reflect a more realistic scenario for the specific region.

The *oviposition count per aquatic habitat* output index is designed to reflect the aquatic habitat heterogeneity in the landscapes. In this regard, alternative choices are available (e.g., *eggs count per aquatic habitat*). However, the former is a better representative of habitat heterogeneity, because it intrinsically considers the degree of ease with which mosquitoes can find the aquatic habitats (distance-based foraging), rather than merely focusing on the size or carrying capacity of a habitat.

Although the modeling framework described in this paper utilizes an ABM of malaria-transmitting mosquitoes, the approach is generally applicable to a wider range of other infectious vector-borne diseases (VBD) including dengue, yellow fever, *etc.*, provided that the disease epidemiology has already been modeled using some standard mechanisms (e.g., mathematical, agent-based, *etc.*). In addition to the three output indices used in this study, other widely used disease epidemiology variables such as incidence, prevalence or mortality can also be mapped and spatially analyzed using the current framework.

In general, robustness of a modeling framework depends on several factors, including the choices for model parameters. For the current model, these may include the flight ability and perceptual ranges of mosquitoes, the carrying capacity of aquatic habitats, the detailed demographic data for human populations and houses, *etc.* In the future, once the models are fully calibrated, we envisage the modeling framework to become *more robust*.

5. Conclusions

In this paper, a landscape epidemiology modeling framework is presented which integrates the outputs of simulation runs from an established spatial malaria ABM with a GIS. For a study area in Kenya, five landscape scenarios are constructed with varying coverage levels of two mosquito-control interventions. For each scenario, maps are presented to show the average distributions of three output indices obtained from the results of 750 simulation runs. Hot spot analysis detects statistically significant hot spots and cold spots, and kriging analysis produces predicted values for unmeasured spatial locations for the entire study area. The integration of epidemiological simulation-based results with the GIS-based spatial analyses techniques within a single modeling framework can be a valuable tool for simulation modelers, epidemiologists, disease control managers, and public health officials by assisting these stakeholders in refining research questions and surveillance needs, and in guiding control efforts and field studies. The integrated modeling framework combines expert knowledge bases from entomological, epidemiological, simulation-based, and geo-spatial domains. Although it utilizes an ABM of malaria-transmitting mosquitoes, the approach is generally applicable to a wider range of other infectious vector-borne diseases.

Acknowledgments: This project was supported in part by the Bill and Melinda Gates Foundation Malaria Transmission Consortium (MTC) grant No. 45114. We would also like to profoundly thank the two anonymous reviewers for their constructive comments and suggestions.

Author Contributions: S.M.N.A., G.R.M., and F.H.C. contributed to the design and implementation of the ABM. S.M.N.A., R.R.A. and D.A.P. integrated the ABM with a GIS. S.M.N.A. and R.R.A. performed the hot spot analysis. M.S.R. and S.N. performed the kriging analysis. S.M.N.A. drafted the manuscript. G.R.M. and F.H.C. supervised the study. All authors read and approved the final manuscript. M.S.R. is currently on a sabbatical leave from BUET.

Acknowledgments: Abbreviations

Conflicts of Interest: The authors declare no conflict of interest.

ABM: Agent-Based Model
ABMS: Agent-Based Modeling and Simulation
ATT: Average Travel Time
CC: Carrying Capacity
CI: Confidence Interval
EAL: Event Action List (diagram)
GIS: Geographic Information System
GMMs: Genetically Modified Mosquitoes
GPS: Global Positioning System
IBM: Individual-Based Model
IRS: Indoor Residual Spraying
ITNs: Insecticide-Treated Nets
LLINs: Long-Lasting Impregnated Nets
LSM: Larval Source Management
MPI: Message Passing Interface
OOP: Object-Oriented Programming
SIT: Sterile Insect Technique
TBV: Transmission Blocking Vaccines
VBD: Vector-Borne Disease
V&V: Verification and Validation
WHO: World Health Organization

References

1. Elliott, P.; Wakefield, J.C.; Best, N.G.; Briggs, D.J. *Spatial Epidemiology: Methods and Applications*; Oxford University Press: Oxford, UK, 2000.
2. Bithell, J.F. A classification of disease mapping methods. *Stat. Med.* **2000**, *19*, 2203–2215.
3. Pfeiffer, D.U.; Robinson, T.P.; Stevenson, M.; Stevens, K.B.; Rogers, D.J.; Clements, A.C.A. *Spatial Analysis in Epidemiology*; Oxford University Press: Oxford, UK, 2008.
4. Kitron, U. Landscape ecology and epidemiology of vector-borne diseases: Tools for spatial analysis. *J. Med. Entomol.* **1998**, *35*, 435–445.
5. Reisen, W.K. Landscape epidemiology of vector-borne diseases. *Annu. Rev. Entomol.* **2010**, *55*, 461–483.
6. Emmanuel, N.N.; Loha, N.; Okolo, M.O.; Ikenna, O.K. Landscape epidemiology: An emerging perspective in the mapping and modelling of disease and disease risk factors. *Asian Pac. J. Trop. Dis.* **2011**, *1*, 247–250.

7. Kitron, U. Risk maps: Transmission and burden of vector-borne diseases. *Parasitol. Today* **2000**, *16*, 324–325.

8. Meentemeyer, R.K.; Cunniffe, N.J.; Cook, A.R.; Filipe, J.A.; Hunter, R.D.; Rizzo, D.M.; Gilligan, C.A. Epidemiological modeling of invasion in heterogeneous landscapes: Spread of sudden oak death in California (1990–2030). *Ecosphere* **2011**, *2*, doi:10.1890/ES10-00192.1.

9. Jacquez, G.M. Spatial analysis in epidemiology: Nascent science or a failure of GIS? *J. Geogr. Syst.* **2000**, *2*, 91–97.

10. Arifin, S.M.N.; Arifin, R.R.; de Alwis Pitts, D.; Madey, G.R. Integrating an agent-based model of malaria mosquitoes with a geographic information system. In Proceedings of the 25th European Modeling and Simulation Symposium (EMSS), Athens, Greece, 25–27 September 2013.

11. World Health Organization (WHO). Malaria. 2014. Available online: http://www.who.int/mediacentre/factsheets/fs094/en/ (accessed on 1 May 2015).

12. Zhou, Y.; Arifin, S.M.N.; Gentile, J.; Kurtz, S.J.; Davis, G.J.; Wendelberger, B.A. An agent-based model of the *Anopheles gambiae* mosquito life cycle. In Proceedings of the 2010 Summer Computer Simulation Conference, Ottawa, ON, Canada, 11–15 July 2010; pp. 201–208.

13. Arifin, S.M.N.; Davis, G.J.; Zhou, Y. A spatial agent-based model of malaria: Model verification and effects of spatial heterogeneity. *Int. J. Agent Technol. Syst.* **2011**, *3*, 17–34.

14. Arifin, S.M.N.; Davis, G.J.; Zhou, Y. Modeling space in an agent-based model of malaria: Comparison between non-spatial and spatial models. In Proceedings of the 2011 Workshop on Agent-Directed Simulation, Boston, MA, USA, 3–7 April 2011; pp. 92–99.

15. Arifin, S.M.N.; Madey, G.R.; Collins, F.H. Examining the impact of larval source management and insecticide-treated nets using a spatial agent-based model of *Anopheles gambiae* and a landscape generator tool. *Malar. J.* **2013**, *12*, doi:10.1186/1475-2875-12-290.

16. Arifin, S.M.N.; Zhou, Y.; Davis, G.J.; Gentile, J.E.; Madey, G.R.; Collins, F.H. An agent-based model of the population dynamics of *Anopheles gambiae*. *Malar. J.* **2014**, *13*, doi:10.1186/1475-2875-13-424.

17. Peng, R.D. Reproducible research in computational science. *Science* **2011**, *334*, 1226–1227.

18. Jasny, B.R.; Chin, G.; Chong, L.; Vignieri, S. Data replication & reproducibility. Again, and again, and again . . . . *Science* **2011**, *334*, doi:10.1126/science.334.6060.1225.

19. ArcGIS Desktop: Release 9.3. Available online: http://www.esri.com/ (accessed on May 2015).

20. Ross, R. *The Prevention of Malaria*; Dutton: New York, NY, USA, 1910.

21. Gimnig, J.E.; Ombok, M.; Otieno, S.; Kaufman, M.G.; Vulule, J.M.; Walker, E.D. Density-dependent development of *Anopheles gambiae* (Diptera: Culicidae) larvae in artificial habitats. *J. Med. Entomol.* **2002**, *39*, 162–172.

22. Koenraadt, C.J.M.; Takken, W. Cannibalism and predation among larvae of the *Anopheles gambiae* complex. *Med. Vet. Entomol.* **2003**, *17*, 61–66.

23. Sumba, L.; Okoth, K.; Deng, A.; Githure, J.; Knols, B.; Beier, J.; Hassanali, A. Daily oviposition patterns of the African malaria mosquito *Anopheles gambiae* Giles (Diptera: Culicidae) on different types of aqueous substrates. *J. Circadian Rhythms* **2004**, *2*, Art. 6.

24. Munga, S.; Minakawa, N.; Zhou, G.; Barrack, O.; Githeko, A.; Yan, G. Effects of larval competitors and predators on oviposition site selection of *Anopheles gambiae* sensu stricto. *J. Med. Entomol.* **2006**, *43*, 221–224.

25. Churcher, T.; Dawes, E.; Sinden, R.; Christophides, G.; Koella, J.; Basanez, M.G. Population biology of malaria within the mosquito: Density-dependent processes and potential implications for transmission-blocking interventions. *Malar. J.* **2010**, *9*, doi:10.1186/1475-2875-9-311.

26. Sumba, L.A.; Ogbunugafor, C.B.; Deng, A.L.; Hassanali, A. Regulation of oviposition in *Anopheles gambiae* ss: Role of inter- and intra-specific signals. *J. Chem. Ecol.* **2008**, *34*, 1430–1436.

27. Jannat, K.N.E.; Roitberg, B.D. Effects of larval density and feeding rates on larval life history traits in *Anopheles gambiae* s.s. (Diptera: Culicidae). *J. Vector Ecol.* **2013**, *38*, 120–126.

28. Himeidan, Y.; Temu, E.; Rayah, E.E.; Munga, S.; Kweka, E. Chemical cues for malaria vectors oviposition site selection: Challenges and opportunities. *J. Insects* **2013**, *2013*, doi:10.1155/2013/685182.

29. Java. Available online: http://www.java.com/en/ (accessed on 1 May 2015).

30. The Eclipse Foundation. Available online: http://www.eclipse.org/ (accessed on 1 May 2015).

31. Gu, W.; Novak, R.J. Agent-based modelling of mosquito foraging behaviour for malaria control. *Trans. R. Soc. Trop. Med. Hyg.* **2009**, *103*, 1105–1112.

32. Phillips-Howard, P.A.; Nahlen, B.L.; Alaii, J.A.; ter Kuile, F.O.; Gimnig, J.E.; Terlouw, D.J.; Kachur, S.P.; Hightower, A.W.; Lal, A.A.; Schoute, E.; *et al.* The efficacy of permethrin-treated bed nets on child mortality and morbidity in western Kenya I. Development of infrastructure and description of study site. *Am. J. Trop. Med. Hyg.* **2003**, *68*, 3–9.

33. Nahlen, B.L.; Clark, J.P.; Alnwick, D. Insecticide-treated bed nets. *Am. J. Trop. Med. Hyg.* **2003**, *68*, 1–2.

34. McElroy, P.D.; ter Kuile, F.O.; Hightower, A.W.; Hawley, W.A.; Phillips-Howard, P.A.; Oloo, A.J.; Lal, A.A.; Nahlen, B.L. All-cause mortality among young children in western Kenya. VI: The Asembo Bay Cohort Project. *Am. J. Trop. Med. Hyg.* **2001**, *64*, 18–27.

35. Kazura, J.W., Ed. The western Kenya insecticide-treated bed net trial. *Am. J. Trop. Med. Hyg.* **2003**, *68*, 1–173.

36. ESRI 2011. ArcGIS Desktop: Release 10. Available online: http://www.esri.com/ (accessed on 1 May 2015).

37. Fillinger, U.; Lindsay, S. Larval source management for malaria control in Africa: Myths and reality. *Malar. J.* **2011**, *10*, doi:10.1186/1475-2875-10-353.

38. Killeen, G.F.; Smith, T.A. Exploring the contributions of bed nets, cattle, insecticides and excitorepellency to malaria control: A deterministic model of mosquito host-seeking behaviour and mortality. *Trans. R. Soc. Trop. Med. Hyg.* **2007**, *101*, 867–880.

39. Yakob, L.; Yan, G. Modeling the effects of integrating larval habitat source reduction and insecticide treated nets for malaria control. *PLoS ONE* **2009**, *4*, doi:10.1371/journal.pone.0006921.

40. Hancock, P.A. Combining fungal biopesticides and insecticide-treated bednets to enhance malaria control. *PLoS Comput. Biol.* **2009**, *5*, doi:10.1371/journal.pcbi.1000525.

41. Chitnis, N.; Schapira, A.; Smith, T.; Steketee, R. Comparing the effectiveness of malaria vector-control interventions through a mathematical model. *Am. J. Trop. Med. Hyg.* **2010**, *83*, 230–240.

42. Griffin, J.T.; Hollingsworth, T.D.; Okell, L.C.; Churcher, T.S.; White, M.; Hinsley, W.; Bousema, T.; Drakeley, C.J.; Ferguson, N.M.; Basáñez, M.G.; *et al.* Reducing *Plasmodium falciparum* malaria transmission in Africa: A model-based evaluation of intervention strategies. *PLoS Med.* **2010**, *7*, doi:10.1371/journal.pmed.1000324.

43. Eckhoff, P. A malaria transmission-directed model of mosquito life cycle and ecology. *Malar. J.* **2011**, *10*, doi:10.1186/1475-2875-10-303.

44. White, M.; Griffin, J.; Churcher, T.; Ferguson, N.; Basanez, M.G.; Ghani, A. Modelling the impact of vector control interventions on *Anopheles gambiae* population dynamics. *Parasites Vectors* **2011**, *4*, doi:10.1186/1756-3305-4-153.

45. Okumu, F.; Chipwaza, B.; Madumla, E.; Mbeyela, E.; Lingamba, G.; Moore, J.; Ntamatungro, A.; Kavishe, D.; Moore, S. Implications of bio-efficacy and persistence of insecticides when indoor residual spraying and long-lasting insecticide nets are combined for malaria prevention. *Malar. J.* **2012**, *11*, doi:10.1186/1475-2875-11-378.

46. Greenwood, B.M.; Fidock, D.A.; Kyle, D.E.; Kappe, S.H.; Alonso, P.L.; Collins, F.H.; Duffy, P.E. Malaria: Progress, perils, and prospects for eradication. *J. Clin. Investig.* **2008**, *118*, 1266–1276.

47. Phuc, H.K.; Andreasen, M.H.; Burton, R.S.; Vass, C.; Epton, M.J.; Pape, G.; Fu, G.; Condon, K.C.; Scaife, S.; Donnelly, C.A.; *et al.* Late-acting dominant lethal genetic systems and mosquito control. *BMC Biol.* **2007**, *5*, doi:10.1186/1741-7007-5-11.

48. Klassen, W. Introduction: Development of the sterile insect technique for African malaria vectors. *Malar. J.* **2009**, *8*, doi:10.1186/1475-2875-8-S2-I1.

49. Hancock, P.; Thomas, M.; Godfray, H. An age-structured model to evaluate the potential of novel malaria-control interventions: A case study of fungal biopesticide sprays. *Proc. R. Soc. B Biol. Sci.* **2009**, *276*, 71–80.

50. Marshall, J.M.; Taylor, C.E. Malaria control with transgenic mosquitoes. *PLoS Med.* **2009**, *6*, e1000020.

51. Carter, R. Transmission blocking malaria vaccines. *Vaccine* **2001**, *19*, 2309–2314.

52. Hawley, W.A.; ter Kuile, F.O.; Steketee, R.S.; Nahlen, B.L.; Terlouw, D.J.; Gimnig, J.E.; Shi, Y.P.; Vulule, J.M.; Alaii, J.A.; Hightower, A.W.; *et al.* Implications of the western Kenya permethrin-treated bed net study for policy, program implementation, and future research. *Am. J. Trop. Med. Hyg.* **2003**, *68*, 168–173.

53. ArcGIS: Hot Spot Analysis, 2014. Available online: http://resources.arcgis.com/en/help/main/10.2/index.html (accessed on 1 May 2015).

54. ESRI 2015. GIS Dictionary. Available online: http://support.esri.com/ (accessed on 1 May 2015).

55. Diggle, P.; Moyeed, R.; Rowlingson, B.; Thompson, M. Childhood malaria in the Gambia: A case-study in model-based geostatistics. *Appl. Stat.* **2002**, *51*, 493–506.

56. Best, N.; Richardson, S.; Thomson, A. A comparison of Bayesian spatial models for disease mapping. *Stat. Methods Med. Res.* **2005**, *14*, 35–59.

57. Carrat, F.; Valleron, A.J. Epidemiologic mapping using the "Kriging" method: Application to an influenza-like epidemic in France. *Am. J. Epidemiol.* **1992**, *135*, 1293–1300.

58. Berke, O. Exploratory disease mapping: Kriging the spatial risk function from regional count data. *Int. J. Health Geogr.* **2004**, *3*, doi:10.1186/1476-072X-3-18.

59. Lai, P.C.; So, F.M.; Chan, K.W. *Spatial Epidemiological Approaches in Disease Mapping and Analysis*; CRC Press: Boca Raton, FL, USA, 2008.

60. De Carvalho Alves, M.; Pozza, E.A. Indicator Kriging modeling epidemiology of common bean anthracnose. *Appl. Geomat.* **2010**, *2*, 65–72.

61. Alexeeff, S.E.; Schwartz, J.; Kloog, I.; Chudnovsky, A.; Koutrakis, P.; Coull, B.A. Consequences of Kriging and land use regression for PM2.5 predictions in epidemiologic analyses: Insights into spatial variability using high-resolution satellite data. *J. Expo. Sci. Environ. Epidemiol.* **2014**, *25*, 138–144.

62. Malaria Atlas Project. Available online: http://www.map.ox.ac.uk/ (accessed on 1 May 2015).

63. Gething, P.W.; Patil, A.P.; Hay, S.I. Quantifying aggregated uncertainty in *Plasmodium falciparum* malaria prevalence and populations at risk via efficient space-time geostatistical joint simulation. *PLoS Comput. Biol.* **2010**, *6*, e1000724.

64. Howes, R.E.; Piel, F.B.; Patil, A.P.; Nyangiri, O.A.; Gething, P.W.; Dewi, M.; Hogg, M.M.; Battle, K.E.; Padilla, C.D.; Baird, J.K.; *et al.* G6PD deficiency prevalence and estimates of affected populations in malaria endemic countries: A geostatistical model-based map. *PLoS Med.* **2012**, *9*, e1001339.

65. Piel, F.B.; Patil, A.P.; Howes, R.E.; Nyangiri, O.A.; Gething, P.W.; Dewi, M.; Temperley, W.H.; Williams, T.N.; Weatherall, D.J.; Hay, S.I.; *et al.* Global epidemiology of sickle haemoglobin in neonates: A contemporary geostatistical model-based map and population estimates. *Lancet* **2013**, *381*, 142–151.

66. Gundogdu, K.S.; Guney, I. Spatial analyses of groundwater levels using universal Kriging. *J. Earth Syst. Sci.* **2007**, *116*, 49–55.

67. Batistella Pasini, M.P.; Dal'Col Lúcio, A.; Cargnelutti Filho, A. Semivariogram models for estimating fig fly population density throughout the year. *Pesqui. Agropecu. Bras.* **2014**, *49*, 493–505.

68. Arifin, S.M.N.; Davis, G.J.; Zhou, Y. Verification & validation by docking: A case study of agent-based models of *Anopheles gambiae*. In Proceedings of the Summer Computer Simulation Conference (SCSC), Ottawa, ON, Canada, 11–14 July 2010.

69. Arifin, S.M.N.; Davis, G.J.; Kurtz, S.J.; Gentile, J.E.; Zhou, Y. Divide and conquer: A four-fold docking experience of agent-based models. In Proceedings of the Winter Simulation Conference (WSC), Baltimore, MD, USA, 5–8 December 2010.

70. Helbing, D. *Social Self-Organization*; Springer: Berlin, Germany, 2012.

Agent-Based Models as "Interested Amateurs"

Peter George Johnson

Abstract: This paper proposes the use of agent-based models (ABMs) as "interested amateurs" in policy making, and uses the example of the SWAP model of soil and water conservation adoption to demonstrate the potential of this approach. Daniel Dennett suggests experts often talk past or misunderstand each other, seek to avoid offending each other or appearing ill-informed and generally err on the side of under-explaining a topic. Dennett suggests that these issues can be overcome by including "interested amateurs" in discussions between experts. In the context of land use policy debates, and policy making more generally, this paper suggests that ABMs have particular characteristics that make them excellent potential "interested amateurs" in discussions between our experts: policy stakeholders. This is demonstrated using the SWAP (Soil and Water Conservation Adoption) model, which was used with policy stakeholders in Ethiopia. The model was successful in focussing discussion, inviting criticism, dealing with sensitive topics and drawing out understanding between stakeholders. However, policy stakeholders were still hesitant about using such a tool. This paper reflects on these findings and attempts to plot a way forward for the use of ABMs as "interested amateurs" and, in the process, make clear the differences in approach to other participatory modelling efforts.

Reprinted from *Land*. Cite as: Johnson, P.G. Agent-Based Models as "Interested Amateurs". *Land* **2016**, *4*, 281–299.

1. Introduction

Policy making is a complex process [1–3] involving many actors. This is especially true of land use policy in which many different stakeholders interact. In any policy domain, individual actors often have little control over the process [4]. Most, if not all, of these actors are experts in their policy area; their (and their organisation's) combination of experience in policy making and the domain area mean that they often have detailed knowledge and strong opinions on what policies may and should be pursued and which actors should be included in the process. Despite this, we know policy making and policies themselves are not always successful. Why might this be?

There are, of course, many reasons for this, but one may be that these "policy experts" are not working together as effectively as may be possible or necessary. New ideas may be consistently ignored or out-of-date assumptions may go unscrutinised. Why might experts be unable to interact successfully? Dennett [5] suggests that when experts on a subject debate or discuss that subject, they assume the expertise of others and do not discuss basic concepts. The result is that they often "talk past" each other and fail to identify differences in assumptions and key understandings of the topic or system under discussion. This can also be the result of experts not wanting to offend one another or appearing ill-informed by asking for explanations of basic positions and assumptions. In either case, the experts end up erring on the side of under-explaining or discussing the topic at hand.

Dennett's proposed solution to this general problem is to use lay audiences, or curious non-experts (here called "interested amateurs"), to force the discussion to be focussed on assumptions and to err on the side of over-explaining issues under discussion. For Dennett, an academic philosopher, this means bringing undergraduate students into discussions and debates and asking them to query anything they find unclear.

This paper suggests that it is agent-based models (ABMs), via their overall design, agent rules, assumptions and results, that can play the role of "interested amateur" in policy making and, thus, potentially aid the interaction of policy experts. Moreover, it is suggested that they have an unusual combination of characteristics, such as specificity, intuitive appeal and representation of causation, that makes them excellent candidates for this role. Their specificity encourages detail in discussions, whilst their intuitive appeal keeps ideas tractable and the bigger picture within reach. As models are not people expressing opinions, but artefacts without emotions, it is suggested that participants in discussions are more likely to make strong critiques of a model than an expert or a person playing the role of an "interested amateur". It is this critique that brings otherwise hidden beliefs and assumptions into the open. These assertions will be explored using the example of the SWAP (Soil and Water Conservation Adoption) model of soil and water conservation (SWC) adoption amongst small-scale farmers in developing countries. The SWAP model was used with SWC policy stakeholders in Ethiopia (a policy area with well-documented interaction problems [6,7]). A workshop with policy stakeholders was held and a qualitative analysis used to understand if and how an ABM could act as an "interested amateur".

There is already a considerable literature on the use of models, and specifically ABMs, in participatory policy making contexts. This increasingly diverse field [8] is excellently overviewed by Voinov and Bousquet [9] and Matthews *et al.* [10]. Most relevant and notable within this literature is the companion modelling (or ComMod) approach developed at the French Agricultural Research Centre for International Development (see [11,12] for overviews of the approach). The approach's "charter" [13] outlines the principles upon which it is based. The approach places the utmost importance on interaction between modellers and stakeholders from the beginning of a project, with many iterations. The focus is placed on learning between researchers and stakeholders, and between stakeholders themselves, and using the process to come to decisions and/or build decision-making capacity. There are numerous examples of the application of ComMod, including water management in Bhutan [14], natural resource management [15] and forest management [16] in the Philippines and fishery management in Thailand [17].

ComMod has been very popular and successful in a range of contexts. However, there are some situations in which it may not be the best approach to take. Models created using the ComMod methodology are co-constructed by a group of stakeholders who, as a result, all have ownership of a model. This means that the model is an "insider"; it is part of their work and likely reflects their view of the world. In this sense, a ComMod model cannot play the role of an "interested amateur", as it is not an "outsider". This means that stakeholders are less likely to make strong criticisms of the model or include elements in its design that they do not see as relevant, but that others, outside the ComMod process, may view as important. Furthermore, any model created by stakeholders is likely to reflect and reinforce their current thinking. A diverse group of stakeholders can share and influence each other's thinking, but it is unlikely ideas from outside these bounds will be included in the model. This lack of both "outsiderness" and an inclusion of critique and thinking from outside the stakeholder group are not typically considered as weaknesses of the ComMod approach, and in many circumstances with specific aims, they are not. It is not the intention here to suggest that they are problematic in all situations, but to suggest that it is worthwhile considering what value a model, that is an "outsider" and that contains thinking from outside the current policy practice, may have in participatory contexts. This underpins the aim of this paper to explore the potential of ABMs to be used in participatory contexts in a different way: as "interested amateurs".

The rest of this paper is structured as follows. In Section 2, the SWAP model and the context it was used in are presented alongside findings from a workshop with stakeholders in Ethiopia. This serves as a demonstration of the use of an ABM as an "interested amateur". In Section 3, a more general discussion is put forward on when and how we might use ABMs as "interested amateurs". Finally, Section 4 concludes.

2. The SWAP Model

This section presents both the SWAP model itself and the approach and findings of a stakeholder workshop used to explore its use as an "interested amateur".

2.1. Model Description

A description of the model is given here which is sufficient for the purpose of this article; however, in the interest of space, this is not comprehensive. A complete description, including an ODD (Overview, Design Concepts, Details) protocol [18,19], is given by Johnson [4], and the model can be downloaded with the full code at http://modelingcommons.org/browse/one_model/4117. The model was developed in the open source environment NetLogo [20]. Using NetLogo enabled the model to be built in an environment with a well-established community and in a relatively naturalistic programming language. The SWAP model is a relatively simple ABM. There are two types of agents: farmer agents and extension agents. Farmer agents make a decision between using two generic farming methods: non-SWC methods or SWC methods. This decision and the design of the agent rules more broadly are based on a framework developed and tested in the literature by DeGraaff *et al.* [21]. This framework breaks up the decision into multiple steps and attaches different factors to these. It also allows for the intensity of adoption, rather than a simple dichotomous choice. First, farmers must accept the need for SWC, then they must decide on how much of their farm to adopt SWC, and finally, once adopted, they must continue to decide to maintain adoption (see Figure 1). The agents' basic decision is intended to be as close an implementation of the DeGraaff framework as possible.

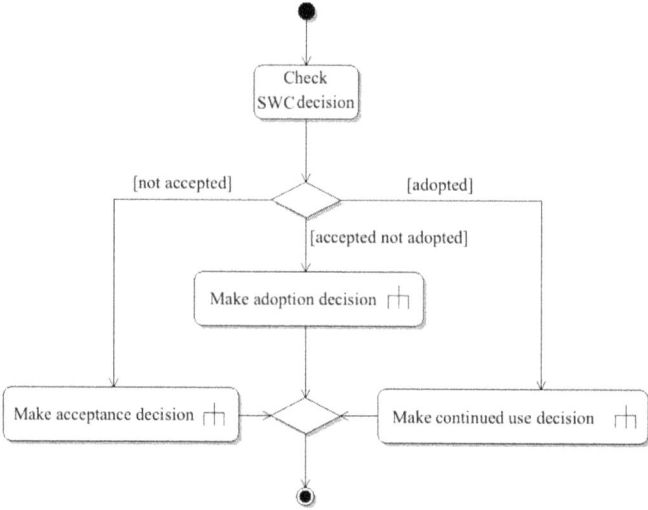

Figure 1. Farmer agents' basic decisions.

The acceptance decision is the most complex, based on the DeGraaff *et al.* framework [21], and requires the following eight steps given in pseudo-code:

```
(1) Run symptoms recognised?
    if ( farm soil quality is low )
    and ( decision maker works on farm )
    and ( farmer knows the land well )
```

        then [recognise symptoms]

(2) Run effects recognised?
    if (farmer not too old)
    and (farmer knows the land well)
    and (farmer is well educated)
    and (farmer has extension contact)
    and (farmer has low cultural inertia)
    then [recognise effects]

(3) Run degradation taken seriously?
    if (farmer has extension contact)
    and (farmer owns the land)
    then [take degradation seriously]

(4) Run aware of SWC methods?
    if (farmer has knowledge of methods)
    and (farmer has extension contact)
    then [be aware of methods]

(5) Run able to undertake SWC?
    if (farmer can hire labour)
    and (farmer not too old)
    and (farmer has extension contact)
    and (farmer can access credit)
    and (farmer owns the land)
    then [able to undertake SWC]

(6) Run willing to undertake SWC?
    if (discount rate is low)
    and (farmer has low cultural inertia)
    and (farmer sympathetic to gov/NGOs)
    and (farmer has a family successor)
    and (farmer is not too old)
    and (decision maker works on farm)
    then [willing to undertake SWC]

(7) Run ready to undertake SWC?
    if (not too risk averse)
    and (farmer has enough savings)
    and (farmer has enough income)
    then [ready to undertake SWC]

(8) Run accept SWC

```
set acceptance score to
[ accepted but not adopted ]
```

The various parameters required to implement this pseudo-code are either dichotomous (e.g., the decision maker works on the farm, the farmer has an extension contact, the farmer owns their land, the farmer can access credit or labour, the farmer has a successor), assigned a score between zero and one hundred (e.g., cultural inertia, knowledge of methods, sympathy to non-governmental organizations or government) or given an appropriate value (e.g., age is in years, years of education is in years, the discount rate, which denotes the rate at which farmers discount future costs and benefits against current costs and benefits, is given as a number between zero and one). These are then set using real-world data where available.

Each of the eight steps must be met for the farmer to proceed to the next stage of the decision. However, at each time step, there is a 10% chance that the agent will jump straight through the acceptance decision without meeting its criteria and move onto the adoption decision; this represents an element of chance in the decisions. This figure was reached after the calibration and sensitivity analysis.

Once they have accepted the need for SWC, farmers must decide on the intensity of their adoption, *i.e.*, how much of their farm they wish to adopt. The amount of land they adopt conservation on is determined by their level of savings (intended to represent an abstract form of capital, savings must meet a minimum threshold), their contact with extension workers (a contact is required for any adoption) and their risk aversion score (less risk-averse agents will adopt at a higher level).

Finally, if they have already adopted SWC measures, farmers must decide whether to increase or decrease adoption or indeed stop using SWC. If their "income" is higher than their "consumption requirement", they will increase adoption by 20%. If their "income" is lower than their "consumption requirement", they will reduce their adoption by 20%. If adoption falls at a very low level, they will simply stop using SWC. The presence of adoption will increase soil quality, which, in turn, will increase an agents "income" parameter. "Consumption requirement" in the model is a constant (per individual) multiplied by the number of people in a farmer's household, and "income" is a function of the soil quality and farmer knowledge.

Farmers also interact in the model, either: in farmer peer-groups, such as church or community groups, through influential individuals, such as community leaders or government chosen "model" farmers, or through extension agents (see Figure 2).

In farmer groups, farmers become more similar, all influencing each other equally as peers. Via influential individuals, those with higher "influence scores" make those near them more like themselves. Extension agents (representing development agents, as described in the next section) move around the model space increasing the chance of farmers adopting SWC when they are nearby, as extension contact is key to several of the decision stages. These interaction types are not described in the DeGraaff *et al.* framework, but are derived from a wider reading of the SWC literature. They are included, as it was felt it was important to represent the social interaction of farmers, as well as their individual decision making.

Figure 2. Farmer interactions.

During each time step, the agents take it, in turn, to make decisions in a randomised order. In one time step, an agent can start a decision and carry it out, but only one stage at a time; they cannot cycle through all of the decision stages at once (*i.e.*, an agent can decide that they recognise the existence of degradation, but cannot then also suddenly be aware of methods to combat it; or an agent can decide that they do accept the need for SWC, but then cannot also decide how much to adopt). This separation of the decision process over time reflects the idea that farmers do not go from not being aware of or considering SWC to suddenly adopting. The time step is intended to represent a period of around one to three months. This is a reasonable period for which to assume agents would make these decisions in the real world (*i.e.*, a farmer does not consider whether to change practices every day or week).

The spatial environment represents a non-specific area of land made up of many "patches". Each patch of land represents a field, with the collection of patches closest to each farmer agent being their farm. The environment is modelled in this way so that farmers can decide on the intensity of adoption on their farm, rather than making a simple dichotomous choice. Each patch has a parameter reflecting its soil quality, with a score out of one hundred. This is used to allow the feedback between decisions and soil quality and *vice versa*.

Though not the focus here, and not presented at the workshop, the model has been calibrated and validated against three case studies using real-world data and a pattern-oriented modelling approach [22–24]. A presentation of this process can be found in Johnson [4].

2.2. The Workshop

The workshop was held on 20 June 2013, at the International Livestock Research Institute (ILRI) Info-Centre in Addis Ababa, Ethiopia. In Ethiopia, as in much of Sub-Saharan Africa, land and soil degradation are increasingly problematic environmental, social and economic problems [25,26]. In the face of stagnating agricultural productivity, farmers have tended to expand production onto inappropriate and steep land, resulting in soil degradation and erosion [7]. Ethiopia's population now exceeds 80 million, with 75%–85% of the population making a livelihood in an agriculture industry characterised by low input-low output rainfed systems focussed on subsistence [7]. This has resulted in a strong perception that soil erosion poses a serious threat to Ethiopia's future despite widespread awareness amongst policy makers both in the country and externally. Policies in Ethiopia intended to increase farmers' adoption of SWC measures are understood to have been unsuccessful owing to:

"misguided policy, authoritarian and top-down approaches guided by targets and coercion to mobilise labour, blanket approaches across vastly different agro-ecological and socio-economic contexts, or inappropriate technologies" [6] (p. 5).

This reflects Ethiopia's political past under the Derg and, more recently, slow progress in moving towards more participatory policy making [7]. Ludi *et al.* [6] also highlight the difficulty of the work of "development agents" that are intended to provide a bridge between government and farmers, stating that they are:

"caught between farmers and government, with the difficult task of reconciling top-down plans and quotas with local concerns and needs. They transmit information down to farmers but struggle to pass ideas and reflections back from farmers to higher levels" (p. 19).

This second quote provides an excellent summary of the motivation for the use of the SWAP model as an "interested amateur" in this example. It is suggested that the model, via its overall design, agent rules, assumptions and results, can help to address this struggle in passing ideas and information up the policy hierarchy, between different "experts" on the system at hand. In a hierarchical and often sensitive policy landscape, the model can ease tensions by being the artefact that takes the criticism and critique of stakeholders, but still allows for focused, detailed and tractable discussion on various levels of the system. The workshop aimed, using a qualitative approach, to explore how the model performed in this role and how participants viewed the potential for the model to be used as an "interested amateur".

2.2.1. Participants

Potential participants were identified based on their positions in the regional Bureaus of Agriculture (responsible for agricultural policy implementation, coordination and evaluation) and the non-governmental organisations (NGOs) working with them. This "mid-level" position was ideal for the workshop aims, as the participants had experience working with stakeholders both at the local and national levels and, so, were well positioned to comment and reflect on the potential for poor interaction amongst stakeholders "up" and "down" the policy process. Table 1 outlines the participants' positions and expertise.

Table 1. List of workshop participants. SWC, soil and water conservation.

No.	Organisation	Expertise/Position
1	ORDA	Project Design and Action Research Officer
2	BoA Amhara Region	Soil and Water Conservation Specialist
3	BoA Amhara Region	Livestock Expert in Watershed Study Case Team
4	BoA Amhara Region	Agronomist in Integrated Watershed Planning team
5	BoA Amhara Region	Livestock and Forage Development Advisor
6	GIZ-SLM Amhara Region	SWC Engineering Specialist
7	GIZ-SLM Oromia Region	Senior Cluster Advisor
8	BoA Oromia Region	Watershed Development Planning Expert
9	BoA Oromia Region	Agricultural Engineer for SWC

NB: ORDA = Organisation for Rehabilitation and Development in Amhara; NGO. BoA = Bureau of Agriculture. GIZ-SLM = Deutsche Gesellschaft fur Internationale Zusammenarbeit (Sustainable Land Management Project); a non-Ethiopian Government Programme.

The Amhara and Oromia regions were well represented, as they are the two most populated regions. The main omission was participants from Tigray, the Ethiopian region with arguably the most political influence and with a long history of soil degradation. This may because the distance from Tigray to Addis Ababa deterred potential participants. Though there was a majority of Bureau of Agriculture participants, there were also enough non-Ethiopian government programme and NGO participants, such that their voices would not be drowned out or ignored. It was the general characteristic of a mix of participants, rather than specific groups or types of participant, that was important for the aims of the workshop. A mixed group meant it was unlikely the participants would all have very similar views. Had the group been more homogeneous, it is unlikely that the approach would have had a fair chance of drawing out misunderstandings and differences of opinion, as they would be much less likely to exist.

It is possible that the findings from the workshop are biased by the characteristics of the group of participants that took part. The fact that they were willing to take part and travel quite far in some cases suggests that they were already interested in visiting ILRI, in the researchers' work and/or tools, like the SWAP model. Generally, it is fair to assume that they are more engaged with researchers and interested than a typical mid-level policy stakeholder. The final participant list was also not comprehensive in the sense that it covered all regions or types of organisation working on SWC. This means that it is difficult to attempt to generalise the findings beyond government and large NGO actors or to other regions. Despite these potential drawbacks, the findings of the workshop can still be used to demonstrate the potential of the "interested amateur" approach, make attempts at understanding how policy stakeholders view tools, such as the SWAP model, and how they might fit into their work.

All of the participants spoke English to a functional level, and most spoke well. There were very few occasions during which translation into Amharic was required. However, the participants did on occasion switch to talking in Amharic with each other. This was obviously more convenient and natural for them, but meant that the non-Amharic speaking organisers could not understand what they were saying. There did not appear any obvious reason for this change in language in terms of the content of the discussion (e.g., a sensitive or complex topic); rather, it appeared that the participants did this when they wanted to say something quickly or with more clarity, though it is impossible to be 100% certain. When this persisted for more than a few sentences, humour was used to attempt to return to English, though this was rarely necessary.

2.2.2. Workshop Structure

The workshop was split into four substantive sessions, in addition to an introduction and wrap-up. The sessions were in the format of an initial short (approximately ten minutes) presentation, an extended discussion in break-out groups of four to five participants and a final whole-group "report-back" on discussions. Participants were asked to make notes on their discussions using flip-charts. These were used to refer back to after the workshop and as prompts during the whole-group feedback sections. Though timing slots were detailed in the workshop materials given to participants, they were left intentionally flexible, and where possible, time was extended or shortened to accommodate the natural flow of discussion. Indeed, on the day, the timings were not stuck to closely.

Of the four sessions, two were generic in nature, relating to experiences of the policy process, and two were directly related to the SWAP model. The first of these introduced participants to the model and built a discussion on SWC using the model.

The aim of this session was two-fold: first, to get a basic sense of the views of the participants on the model and, second, to demonstrate the use of the model as an "interested amateur". This was done by asking the participants to critique the model and explore their views, both when they agreed or differed. The framework and underlying assumptions were used as the main focus here, rather than the results or live "running" of the model. The second of the two sessions introduced the envisaged use of the SWAP model and built discussion on the participants' view of this. The aim was to understand whether the participants agreed that the SWAP model could be used as an "interested amateur", explore any other potential uses and understand what barriers there may be to its use.

2.2.3. Presenting the Model

If we are to suggest that ABMs have particular characteristics that make them good "interested amateurs", it is important to carefully consider how a model is presented to stakeholders. For this workshop, it was decided to present the SWAP model in two ways: first, to give an overview of the purpose, assumptions and results of the model in a short presentation, including videos of the model running "live". Figures 3 and 4 give a sense of what was presented in this section, showing a screenshot of the model interface and some of the results of the model when different interaction type scenarios are compared. Secondly, the framework of individual farmer decisions and interaction using handouts with diagrams (using Unified Modelling Language), pseudo-code and text (not dissimilar to those used in this paper) was presented. The most detailed attention was given to exploring the individual farmer decision making and interaction rules, rather than exploring model results or different analyses of outputs.

The participants were then given the task of critiquing the model in small groups. This meant that the participants received a focussed introduction to the model in a presentation and, then, a self-led critical exploration of the model using the handout materials. Giving the presentation first meant that the participants were able to get a sense of the overall purpose of the model, its components and results. Beyond this, they were also able to get a sense of what information on the model was being handed out and to what level of detail they could consider the model, but without having to actually go through all of the information themselves.

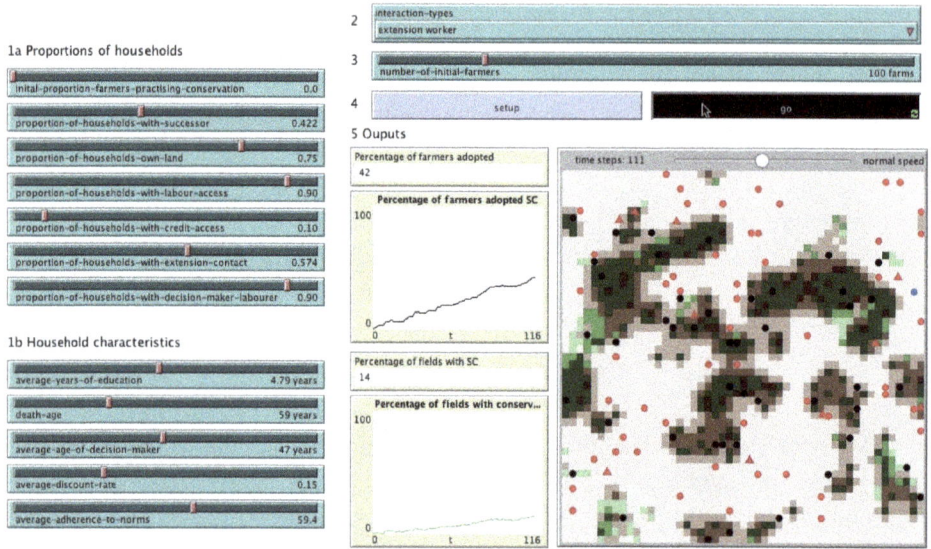

Figure 3. Screen shot of the model being run "live" for the participants.

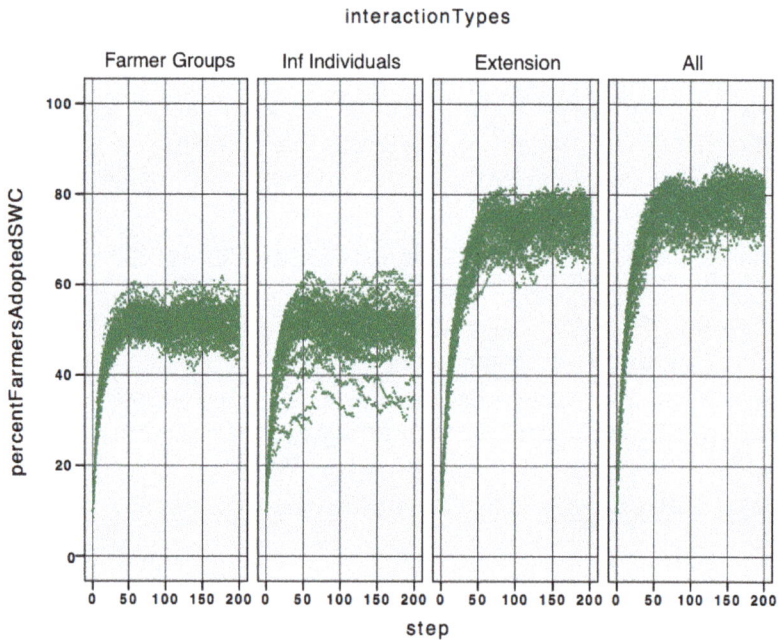

Figure 4. An example of model results shown to the participants: this graphs shows the percent of farmers adopting SWC under different interaction type scenarios for multiple runs of the model.

It was this premise of a quick overview, followed by a self-led task with depth available when required, that inspired the approach taken. It was during the break-out group discussions that the detail of the model really came to the fore. As the participants asked questions and made comments, the handouts were used to give the finer-level granular detail. Much use was made of the handout materials, which suggests that the participants did engage with the detail of the model.

2.3. Workshop Findings

This section first reports on the atmosphere at the workshop, using this to give a sense of how participants engaged with the model. Next, it addresses three specific questions key to the use of an ABM as an "interested amateur".

2.3.1. Atmosphere

In the first session, the participants engaged with the model in a lively way; discussion started quickly, with minimal prompting. The vibrant discussion continued throughout the session, with only minor prompting, and indeed continued beyond the allotted time. The session overran by approximately thirty minutes. The buoyant and sustained discussion was an excellent sign of the participants' engagement with the model and its detail. They appeared interested in the model, and fears of difficulty with facilitation were quickly dispelled.

The discussions were good natured and friendly. Humour was used to deal with the organiser's position as a clear outsider. This made the discussion open, if a little informal. Arguably, the lack of formality was as positive as the setting (on a Western NGO campus), and the political and cultural sensitivity of some of the issues under discussion (e.g., land tenure, ethnicity) meant the discussion may have become difficult and constrained. Furthermore, the informality maximised the chance that the participants would be less guarded about their opinions, and the model could begin to become the "interested amateur" as envisaged. The organisers presented themselves clearly as a non-expert on SWC, hoping to get help from the participants. In this sense, continued inspiration was drawn from the concept of the "interested amateur": not only was the model playing this part, but so were they.

It is important to note that the participants seemed to like ABM in general. They did not show any apprehension or distaste for the methodology, which was likely unfamiliar to all. It was not required to go into a detailed discussion or defence of ABM in general. The participants' positive reaction to ABM supports the idea that ABM has an intuitive appeal. The participants also appeared to gain a strong grasp of what an ABM was and what it can do; though they asked many questions about the approach and model, they did not ask any (or make comments) that showed misunderstanding of the methodology. Again, this could easily not have been the case and was encouraging from the start.

The second session, aimed at gathering direct opinions on how the SWAP model could be used, had been planned in a similar way to the previous sessions, with the group breaking into smaller discussion groups before coming back together. However, as time had run over in the morning, the session was streamlined into one larger group discussion. This meant that the session seemed more formal, with the participants all facing the front as notes were taken on a flip-chart. This format appeared to inhibit the discussion; the participants were less engaged than when in smaller groups. The subject matter may have played a role in this, too; the topic was more hypothetical and removed from the participants' current work and experience. The topic was more explicitly selfish in terms of the organisers getting information from the participants without much potential benefit for the participants. This is likely to have also reduced the participants' engagement in discussions. Having the session after lunch also gave the session a sense of lethargy that was not present in the morning. Perhaps of most note was that this session, though focused on discussion about the model's use, did not make use of the model itself, as in the other session. This could provide the perfect example of how, without the model to aid discussion, the same group of participants were less engaged and discussion was less buoyant. Despite this change in atmosphere, the discussions did bring out some interesting points and were certainly of use.

2.3.2. Can An ABM Be Used as an "Interested Amateur" in the Context of SWC Policy?

The participants recognised the vast majority of the factors in the farmers' behaviour framework and recognised the forms of interaction under which the model assumes that farmers act. The participants agreed that all of the factors identified in the framework were relevant, but to varying degrees. They felt some were less important than others, because, as a generic set of factors, some were less applicable to their specific region or Ethiopia as a whole. The participants were critical of some parts of the model, particularly the factors that they felt were inappropriate or less important, such as the use of the word "tribe" (one of the socio-economic factors identified in the DeGraaff framework) and the lack of a detailed biophysical representation.

There were some areas of discussion on which the participants did not come to a consensus. These included the prevalence of off-farm employment and/or activity and the prevalence of rented or short-term use of land. These differences became clear due to the explicit causation detailed in the model assumptions; they were challenged by some participants, but not others. It was the resulting debate on the direction of causation and the current status of these parameters (*i.e.*, how many farmers rent or own land) that brought out the differences in beliefs. There were also many contradictions in the discussion. For example, the same participant expressing one opinion early in the first session and, then, a mutually exclusive opinion later. This occurred because the presence of the model led participants to discuss a range of topics and to return and shift between topics in a way that they did not choose. Had the participants been in more control of the direction of discussion, without the model to lead them, it is possible that they could have easily avoided exposing these inconsistencies.

These differences and inconsistencies in opinion were clearly highlighted by the presence of the model in the participants' discussions. Whilst it is entirely possible that they may have reached these issues without the model, it is certain that the framework of agent behaviours, the granular detail it provides and participants' willingness to criticise the model led the participants directly to the main issues of contention. Having the model as the focus of discussion gave the participants an easy target at which to make their criticisms and assertions, in the full sight of others. In this sense, the model served as an excellent "interested amateur".

2.3.3. Can An ABM's Level of Detail Focus Discussion, whilst Still Keeping Concepts and Ideas Tractable?

The participants were quick to use the step-by-step and line-by-line nature of the agent rules as a guide for their discussion. This meant that they went through each step and its associated factors in a systematic manner. This certainly gave the discussion a level of detail that was valuable. At times, the discussion became very focussed on specific issues, and the participants made a lot of notes on each element of the model. The participants also went off on tangents on occasion. However, they appeared to never lose sight of the basic question of why farmers adopt SWC, returning to it without the need of prompting. Very little effort was required to keep the discussion on track, or on topic, as the model served as a natural chairperson. The main problem with the level of detail was that it meant that the session overran. This was mainly due to the depth to which the participants went through the model rules and assumptions. This highlighted the models' ability to force participants into a detailed discussion. Despite the overrun in time, all of the planned topics of discussion were covered. This was in part due to the model lending a clear structure to the discussions, allowing the participants to identify the next area of discussion easily. Once it was clear that time was overrunning, the participants appeared to check for the upcoming areas of discussion indicated by the model and insist that they wished to cover them also. In this sense, the model was successful in keeping the concepts and discussion tractable, if not concise.

2.3.4. Did Stakeholders See Value in the SWAP Model as An "Interested Amateur"?

The participants did recognise the value of the model as an "interested amateur", and agreed that it had shown differences in opinion amongst them and inconsistencies in their beliefs. Despite this, they were quick to suggest that the real value of using the model would be to those nearer the "bottom" of the policy process and working closely with farmers on a more regular basis. This appeared to be based on two beliefs. First, as the the model represents farmers' decision making, the participants appeared to see an intuitive appeal in using the model with farmers. Secondly, they seemed less keen on the idea that those "further up" the policy process needed to understand, or discuss, farmer behaviour in such detail; appearing to believe this was beneath them in some sense.

Of all the topics covered during the workshop, the only one for which almost completely negative views were expressed was the question of whether participants could use the SWAP model themselves. Beyond suggesting the model would be most useful to those nearer the "bottom" of the policy process, they were quick to suggest that it was not in their remit to "innovate" in the methods they use and that they would need to be instructed by their superiors to use such a tool as the SWAP model. It is not clear whether this is a genuine bureaucratic/administrative barrier to their use of such tools or whether this is a polite excuse, which avoids the need to be more critical of the potential to use the model as an "interested amateur".

3. Discussion

The SWAP model has shown us one example of how ABMs might be used as "interested amateurs" and begun to identify the barriers that may stand in the way of their use. This section will now attempt to outline more generally when and how this approach may be appropriate and consider some of the main challenges.

3.1. When to Use "Interested Amateurs"

There are two key issues that should help identify when using the "interested amateur" approach will be appropriate: firstly, when interaction, and the quality of interaction, between different policy stakeholders has been identified as problematic. This is a commonly-cited problem, in many policy domains, both in developing and developed countries. In land use policy, with a relatively high number of policy stakeholders, this is a particularly relevant issue. The approach has clear benefits in bringing together stakeholders and focussing discussions. However, this is true of other participatory approaches, namely companion modelling. Thus, secondly, what differentiates the "interested amateur" approach is that it allows the use of the model as an outsider, which can be an object for critique. The model becomes a "guess" at the behaviour of a system, which is easy to attack, both because it is an outsider (an amateur), but also because it is clearly not perfect or overly complex. Other participatory approaches may not allow for this type of attack or critique, as the simulation has been co-constructed, so that participants may be more hesitant to criticise it, because it is constructed by themselves and other stakeholders and is also less easily dismissed as an outsider. A model developed outside the immediate policy process is also more likely to contain thinking that is not being included in that process and so provoke criticism or new discussion. It is also this outsider status that allows a model, which can be perceived as a sophisticated technical object, to be an "amateur". The main challenge in this case is that the benefits of stakeholders having ownership of the model are lost. This was reflected in the experience with the SWAP model; it was critiqued, but participants were hesitant about using it themselves (*i.e.*, longer term engagement was non-existent). This decision between using an approach that allows being an outsider and encouraging critique and that which allows ownership and encourages future use will be the second key starting point for any researcher or practitioner considering when to use the "interested amateur" approach.

3.2. How to Use "Interested Amateurs"

At this point, having suggested that researchers and practitioners may wish to use the "interested amateur" approach, it is helpful to make a few suggestions of how to go about doing this. Firstly, it is likely a sensible strategy to base the agent behaviour on a theory users may be familiar with or a middle-range theory with a strong intuitive appeal; for example, a theory that has been developed in the literature for the topic at hand or a theory that has been developed for the central type of decision the agents in a model are making. For the SWAP model, this meant using the DeGraaff *et al.* framework of farmers' decision to adopt SWC measures [21]. Alternatively, a middle-range type framework, such as the Consumat approach [27,28], could have been used, because it closely relates to the decision process that the agents in the ABM are going through and has an intuitive appeal (*i.e.*, it makes conceptual sense to beginners). Using theories like this will give the model an immediate and intuitive appeal, making the model not appear as a "black box". This will suit its use as an 'interested amateur" and make it easier to communicate to stakeholders. The alternative of using more probabilistic or rational utility maximisation type behaviour rules will be less useful in discussions, as they will appear further removed from reality and make discussions more technical.

A less common tactic in designing an ABM, but one that will improve the use of an "interested amateur" type model, is to include (and use in the decision rules) many parameters; indeed, more than one might typically hope to include in a relatively simple model. This goes against the KISS (keep it simple stupid) principle [29], which advocates keeping a model simple, with as few parameters as possible. This approach is intended to make a model more tractable when seeking to understand results, emergent phenomena or running experiments. However, in the context of using a model as an "interested amateur", it is likely to focus discussions on those few parameters included, at the expense of others. If it is our goal to draw out false assumptions and misunderstanding, this lack of breadth in the discussion will hamper the likelihood of success. Thus, it is suggested that those using an ABM as an "interested amateur" seek to include more parameters, so that the model has more detail on which to focus participants' discussions and, thus, enhances the granularity that makes an ABM such an excellent "interested amateur". This suggestion also reflects the focus on the design and assumptions of agent rules rather than model outputs, as seen in the SWAP example.

Finally, a key area of consideration should be how to communicate the model to stakeholders. In the example of the SWAP model, we saw how a presentation was combined with a self-led task and handouts. This allowed for a clear overview, with detail accessible when discussion and critique required. However, if the benefits of using an ABM are to be gained fully, we must constantly reconsider how our models may be communicated, to capture their intuitive appeal, but also the level of detail in factors and their interaction. They must also be presented in a way that makes them amenable to critique. It is not our job as model developers to imbue our models with a sense of overt or undeniable credibility. Indeed, the more unimpressive a model appears, without actually being so poor that stakeholders dismiss it, the more likely it is to invite the critique that can be so valuable in its role as an "interested amateur". Giving users the chance to "play" with the model may also be a fruitful choice in some cases. Finally, one element that was not explored with the SWAP model, but that may be worthwhile, is considering how to communicate emergent phenomena and/or the results of the model more comprehensively and show how the micro-level assumptions of the model link to its macro-level results. Using various simple scenarios and comparing results is one potential avenue. Again, any approach used should be aimed at using the detail and intuitive appeal of the model, whilst keeping the model amenable to criticism.

By basing model development on existing academic literature, using many model parameters and considering carefully how the model can be positioned and presented to users, we will be able to maximise the chance that the model is accepted as a credible outsider and, thus, invite critique, but also contain thinking from outside the immediate policy process in which it is being used.

4. Conclusions

This paper has suggested that policy makers are experts in their policy area and that experts often have problems interacting effectively, owing to various pressures, which lead them to under-explain issues under discussion. It is suggested this is particularly true of land use policy in which there are many different stakeholders with a range of experiences and goals. The concept of the "interested amateur" has been used as inspiration for how ABMs might be used to help counter this problem. Dennett [5] suggests that "interested amateurs" can be included in experts' discussions to encourage the over-explaining of issues, with resulting benefits to the quality and effectiveness of discussions. This paper's main argument is that ABMs have the potential to play the role of "interested amateurs" in policy making processes. This is because of their unusual combination of characteristics; offering a high level of detail, intuitive appeal and explicit representation of causality. Furthermore, as models, they are more amenable to criticism resulting in debate than human facilitators.

This novel approach to the use of ABMs has been demonstrated with the example of the SWAP model of SWC. The use of the SWAP model at a workshop with SWC policy stakeholders showed how an ABM can be successful in generating and focussing discussion, inviting critique and allowing for the recognition of points of contention. However, the example also highlights the barriers to the use of a model over which policy stakeholders have no ownership.

Finally, some suggestions for when and how researchers and practitioners might wish to use an ABM as an "interested amateur" have been put forward. These highlight a key challenge for future research: to resolve the tension between the ownership and amenability to critique of a model. Both have benefits, but they appear mutually exclusive. Participatory modelling approaches that bridge this gap would be of great potential benefit to policy making processes.

Acknowledgments: This work was supported by the Economic and Social Research Council (Grant No. ES/J500148/1). Additional support was received from the International Livestock Research Institute and the International Water Management Institute.

Conflicts of Interest: The authors declare no conflict of interest. The funding sponsors had no role in the design of the study; in the collection, analyses or interpretation of data; in the writing of the manuscript; nor in the decision to publish the results.

References

1. Jäger, J. Current thinking on using scientific findings in environmental policy making. *Environ. Model. Assess.* **1998**, *3*, 143–153.

2. Matthies, M.; Giupponi, C.; Ostendorf, B. Environmental decision support systems: Current issues, methods and tools. *Environ. Model. Softw.* **2007**, *22*, 123–127.

3. Poch, M.; Comas, J.; Rodríguez-Roda, I.; Sànchez-Marrè, M.; Cortés, U. Designing and building real environmental decision support systems. *Environ. Model. Softw.* **2004**, *19*, 857–873.

4. Johnson, P.G. The SWAP Model: Policy and Theory Applications for Agent-Based Modelling of Soil and Water Conservation Adoption. Ph.D. Thesis, University of Surrey, Surrey, UK, 2014.

5. Dennett, D.C. *Intuition Pumps and Other Tools for Thinking*; W. W. Norton & Company: New York, NY, USA, 2013.

6. Ludi, E.; Belay, A.; Duncan, A.; Snyder, K.; Tucker, J.; Cullen, B.; Belissa, M.; Oljira, T.; Teferi, A.; Nigussie, Z.; *et al. Rhetoric versus Realities: A Diagnosis of Rainwater Management Development Processes in the Blue Nile Basin of Ethiopia*; CPWF Research for Development (R4D) Series 5: CGIAR Challenge Program on Water and Food (CPWF): Colombo, Sri Lanka, 2013.

7. Merrey, D.J.; Gebreselassie, T. *Promoting Improved Rainwater and Land Management in the Blue Nile (Abay) Basin of Ethiopia*; NBDC Technical Report 1; International Livestock Research Institute (ILRI): Nairobi, Kenya, 2011.

8. Smajgl, A. Challenging beliefs through multi-level participatory modelling in Indonesia. *Environ. Model. Softw.* **2010**, *25*, 1470–1476.

9. Voinov, A.; Bousquet, F. Modelling with stakeholders. *Environ. Model. Softw.* **2010**, *25*, 1268–1281.

10. Matthews, R.B.; Gilbert, N.G.; Roach, A.; Polhill, J.G.; Gotts, N.M. Agent-based land-use models: A review of applications. *Landsc. Ecol.* **2007**, *22*, 1447–1459.

11. Bousquet, E.F., Trébuil, G., Hardy, B., Eds. *Companion Modeling and Multi-Agent Systems for Integrated Natural Resource Management in Asia*; International Rice Research Institute: Los Baños, Philippines, 2005.

12. Étienne, M., Ed. *Companion Modelling: A Participatory Approach to Support Sustainable Development*; Springer: Enschede, The Netherlands, 2014.

13. Barreteau, O.; Antona, M.; D'Aquino, P.; Aubert, S.; Boissau, S.; Bousquet, F.; Dare, W.; Etienne, M.; Le Page, C.; Mathevet, R.; *et al.* Our companion modelling approach. *J. Artif. Soc. Soc. Simul.* **2003**, *6*, 1.

14. Gurung, T.R.; Bousquet, F.; Trébuil, G. Companion modeling, conflict resolution, and institution building: Sharing irrigation water in the Lingmuteychu Watershed, Bhutan. *Ecol. Soc.* **2006**, *11*, 36.

15. Campo, P.; Bousquet, F.; Villanueva, T. Modelling with stakeholders within a development project. *Environ. Model. Softw.* **2010**, *25*, 1302–1321.

16. Campo, P.C.; Mendoza, G.A.; Guizol, P.; Villanueva, T.R.; Bousquet, F. Exploring management strategies for community-based forests using multi-agent systems: A case study in Palawan, Philippines. *J. Environ. Manag.* **2009**, *90*, 3607–3615.

17. Worrapimphong, K.; Gajaseni, N.; Le Page, C.; Bousquet, F. A companion modeling approach applied to fishery management. *Environ. Model. Softw.* **2010**, *25*, 1334–1344.

18. Grimm, V.; Berger, U.; Bastiansen, F.; Eliassen, S.; Ginot, V.; Giske, J.; Goss-Custard, J.; Grand, T.; Heinz, S.K.; Huse, G.; *et al.* A standard protocol for describing individual-based and agent-based models. *Ecol. Model.* **2006**, *198*, 115–126.

19. Grimm, V.; Berger, U.; De Angelis, D.L.; Polhill, J.G.; Giske, J.; Railsback, S.F. The ODD protocol: A review and first update. *Ecol. Model.* **2010**, *221*, 2760–2768.

20. Wilensky, U. *NetLogo*; Center for Connected Learning and Computer-Based Modeling, Northwestern University: Evanston, IL, USA, 1999. Available online: http://ccl.northwestern.edu/netlogo (accessed on 26 January 2015).

21. De Graaff, J.; Amsalu, A.; Bodnar, F.; Kessler, A.; Posthumus, H.; Tenge, A. Factors influencing adoption and continued use of long-term soil and water conservation measures in five developing countries. *Appl. Geogr.* **2008**, *28*, 271–280.

22. Wiegand, T.; Jeltsch, F.; Hanski, I.; Grimm, V. Using pattern-oriented modeling for revealing hidden information: A key for reconciling ecological theory and application. *Oikos* **2003**, *100*, 209–222.

23. Grimm, V.; Frank, K.; Jeltsch, F.; Brandl, R.; Uchmanski, J.; Wissel, C. Pattern-oriented modelling in population ecology. *Sci. Total Environ.* **1996**, *183*, 151–166.

24. Grimm, V.; Revilla, E.; Berger, U.; Jeltsch, F.; Mooij, W.M.; Railsback, S.F.; Thulke, H.H.; Weiner, J.; Wiegand, T.; De Angelis, D.L. Pattern-oriented modeling of agent-based complex systems: Lessons from ecology. *Science* **2005**, *310*, 987–91.

25. United Nations Environment Programme. *Global Environmental Outlook 4*; Progress Press Ltd.: Valleta, Malta, 2007.

26. United Nations Environment Programme. *Global Environmental Outlook 5*; Progress Press Ltd.: Valleta, Malta, 2011.

27. Jager, W.; Janssen, M.; Vlek, C. *Consumats in a Commons Dilemma: Testing the Behavioural Rules of Simulated Consumers*; COV Report No. 99-01; Center for Environmental and Traffic Psychology, University of Groningen: Groningen, The Netherlands, 1999.

28. Jager, W.; Janssen, M.A.; Vries, H.J.M.D.; Greef, J.D.; Vlek, C.A.J. Behaviour in commons dilemmas: Homo economicus and homo psychologicus in an ecological-economic model. *Ecol. Econ.* **2000**, *35*, 357–379.

29. Axelrod, R. *The Complexity of Cooperation: Agent-Based Models of Competition and Collaboration*; Princeton University Press: Princeton, NJ, USA, 1997.

MDPI AG

St. Alban-Anlage 66

4052 Basel, Switzerland

Tel. +41 61 683 77 34

Fax +41 61 302 89 18

http://www.mdpi.com

Land Editorial Office

E-mail: land@mdpi.com

http://www.mdpi.com/journal/land